Metallic or Metallic Oxide (Photo)catalysts for Environmental Applications

Metallic or Metallic Oxide (Photo)catalysts for Environmental Applications

Editors

Sophie Hermans
Julien Mahy

MDPI • Basel • Beijing • Wuhan • Barcelona • Belgrade • Manchester • Tokyo • Cluj • Tianjin

Editors
Sophie Hermans
Institute of Condensed Matter
and Nanosciences
Université catholique de
Louvain
Louvain-la-Neuve
Belgium

Julien Mahy
Department of Chemical
Engineering
University of Liège
Liège
Belgium

Editorial Office
MDPI
St. Alban-Anlage 66
4052 Basel, Switzerland

This is a reprint of articles from the Special Issue published online in the open access journal *Catalysts* (ISSN 2073-4344) (available at: www.mdpi.com/journal/catalysts/special_issues/metal_photo_environment).

For citation purposes, cite each article independently as indicated on the article page online and as indicated below:

LastName, A.A.; LastName, B.B.; LastName, C.C. Article Title. *Journal Name* **Year**, *Volume Number*, Page Range.

ISBN 978-3-0365-6389-3 (Hbk)
ISBN 978-3-0365-6388-6 (PDF)

© 2023 by the authors. Articles in this book are Open Access and distributed under the Creative Commons Attribution (CC BY) license, which allows users to download, copy and build upon published articles, as long as the author and publisher are properly credited, which ensures maximum dissemination and a wider impact of our publications.
The book as a whole is distributed by MDPI under the terms and conditions of the Creative Commons license CC BY-NC-ND.

Contents

About the Editors vii

Julien G. Mahy and Sophie Hermans
Metallic or Metallic Oxide (Photo)catalysts for Environmental Applications
Reprinted from: *Catalysts* **2022**, *12*, 345, doi:10.3390/catal12030345 1

Xiaolan Kang, Chrysanthi Berberidou, Augustinas Galeckas, Calliope Bazioti, Einar Sagstuen and Truls Norby et al.
Visible Light Driven Photocatalytic Decolorization and Disinfection of Water Employing Reduced TiO_2 Nanopowders
Reprinted from: *Catalysts* **2021**, *11*, 228, doi:10.3390/catal11020228 5

Endang Tri Wahyuni, Titi Rahmaniati, Aulia Rizky Hafidzah, Suherman Suherman and Adhitasari Suratman
Photocatalysis over N-Doped TiO_2 Driven by Visible Light for Pb(II) Removal from Aqueous Media
Reprinted from: *Catalysts* **2021**, *11*, 945, doi:10.3390/catal11080945 21

Sigrid Douven, Julien G. Mahy, Cédric Wolfs, Charles Reyserhove, Dirk Poelman and François Devred et al.
Efficient N, Fe Co-Doped TiO_2 Active under Cost-Effective Visible LED Light: From Powders to Films
Reprinted from: *Catalysts* **2020**, *10*, 547, doi:10.3390/catal10050547 37

Stanislav D. Svetlov, Dmitry A. Sladkovskiy, Kirill V. Semikin, Alexander V. Utemov, Rufat Sh. Abiev and Evgeny V. Rebrov
Synthesis of Thin Titania Coatings onto the Inner Surface of Quartz Tubes and Their Photoactivity in Decomposition of Methylene Blue and Rhodamine B
Reprinted from: *Catalysts* **2021**, *11*, 1538, doi:10.3390/catal11121538 59

Julien G. Mahy, Louise Lejeune, Tommy Haynes, Nathalie Body, Simon De Kreijger and Benjamin Elias et al.
Crystalline ZnO Photocatalysts Prepared at Ambient Temperature: Influence of Morphology on *p*-Nitrophenol Degradation in Water
Reprinted from: *Catalysts* **2021**, *11*, 1182, doi:10.3390/catal11101182 71

Vincent Rogé, Joffrey Didierjean, Jonathan Crêpellière, Didier Arl, Marc Michel and Ioana Fechete et al.
Tuneable Functionalization of Glass Fibre Membranes with ZnO/SnO_2 Heterostructures for Photocatalytic Water Treatment: Effect of SnO_2 Coverage Rate on the Photocatalytic Degradation of Organics
Reprinted from: *Catalysts* **2020**, *10*, 733, doi:10.3390/catal10070733 93

Anna Baranowska-Korczyc, Ewelina Mackiewicz, Katarzyna Ranoszek-Soliwoda, Jaroslaw Grobelny and Grzegorz Celichowski
Core/Shell Ag/SnO_2 Nanowires for Visible Light Photocatalysis
Reprinted from: *Catalysts* **2021**, *12*, 30, doi:10.3390/catal12010030 111

Patrycja Wilczewska, Aleksandra Bielicka-Giełdoń, Karol Szczodrowski, Anna Malankowska, Jacek Ryl and Karol Tabaka et al.
Morphology Regulation Mechanism and Enhancement of Photocatalytic Performance of BiOX (X = Cl, Br, I) via Mannitol-Assisted Synthesis
Reprinted from: *Catalysts* **2021**, *11*, 312, doi:10.3390/catal11030312 123

Fabien Drault, Youssef Snoussi, Joëlle Thuriot-Roukos, Ivaldo Itabaiana, Sébastien Paul and Robert Wojcieszak
Study of the Direct CO_2 Carboxylation Reaction on Supported Metal Nanoparticles
Reprinted from: *Catalysts* **2021**, *11*, 326, doi:10.3390/catal11030326 . . . . **143**

Tommy Haynes, Sharon Hubert, Samuel Carlier, Vincent Dubois and Sophie Hermans
Influence of Water-Miscible Organic Solvent on the Activity and Stability of Silica-Coated Ru Catalysts in the Selective Hydrolytic Hydrogenation of Cellobiose into Sorbitol
Reprinted from: *Catalysts* **2020**, *10*, 149, doi:10.3390/catal10020149 . . . . **153**

Hisham K. Al Rawas, Camila P. Ferraz, Joëlle Thuriot-Roukos, Svetlana Heyte, Sébastien Paul and Robert Wojcieszak
Influence of Pd and Pt Promotion in Gold Based Bimetallic Catalysts on Selectivity Modulation in Furfural Base-Free Oxidation
Reprinted from: *Catalysts* **2021**, *11*, 1226, doi:10.3390/catal11101226 . . . . **167**

About the Editors

Sophie Hermans

Sophie Hermans is an Inorganic Chemist who obtained her first degrees ('Licence en Chimie'and DEA) at UCLouvain in Belgium. She carried out her Ph.D. at the University of Cambridge (UK) under the supervision of Prof. Brian F. G. Johnson, working on mixed-metal clusters synthesis, then pursued postdoctoral studies in Cambridge (as JRF, Newnham College) with Prof. Sir John M. Thomas to immobilize the mixed-metal clusters in MCM-41 for heterogeneous catalytic applications. After moving back to UCLouvain with a FNRS 'Chargée de recherches'post, she started working on carbon-supported catalysts for sugar transformations and chemical functionalization of (nano)carbon surfaces. She obtained the FNRS 'Chercheur Qualifié'and Assistant Professor positions in 2005 and since then was promoted to 'Professeur Extraordinaire'and FNRS Research Director in 2020. Her research interests are still connected to inorganic molecular chemistry, carbon-based catalysts for biomass valorization, surface functionalization and nanostructured materials preparation.

Key Topics: Carbon, Supported catalysts, Functionalization, Nanomaterials, Biomass

ORCID : 0000-0003-4715-7964

Institutional website: https://uclouvain.be/en/research-institutes/imcn/most/research.html

Research Gate: https://www.researchgate.net/profile/Sophie_Hermans

URL Google Scholar: http://scholar.google.be/citations?user=_inQ2UQAAAAJ

Julien Mahy

Julien G. Mahy is a chemical engineer from the University of Liège (ULiège, Belgium). He made a Ph.D. thesis, under the supervision of Pr. Stéphanie Lambert and Pr. Benoît Heinrichs, focused on the development of a TiO2 aqueous sol-gel process in order to produce, at large scale, photocatalysts with hydrophilic properties and high activity, both under visible and UV/visible light, for water and air remediation. In 2018, after a short period working in industry in CRM group as project leader, he accepted a postdoctoral position in collaboration between the NCE (ULiège) and the "Institut für Energie- und Umwelttechnik e.V. (IUTA)", Duisburg (Germany) to work on the development of a quaternary treatment for water in order to degrade residual organic micropollutants using oxidation processes. From July 2019 to June 2021, he worked as a postdoctoral researcher at the Institute of Condensed Matter and Nanosciences (IMCN) at the Université catholique de Louvain under the supervision of Pr. Sophie Hermans. The work consisted in the development of low-cost sensors for air pollution detection by producing onion-like composite nanoparticles.

Since October 2021, he is FNRS postdoctoral researcher at the Université de Liège under the supervision of Pr. Stéphanie Lambert and Pr Nathalie Job and at the INRS, Québec, under the supervision of Pr. Patrick Drogui. His work consists in the development of inorganic materials and processes in environmental applications such as water and air decontamination process by photocatalysis, adsorption and electrochemistry.

Key Topics: TiO2, Photocatalysts, Water treatment, Sol-Gel process, Adsorption

ORCID : 0000-0003-2281-9626

Institutional website: https://www.chemeng.uliege.be/cms/c_3482790/fr/chemeng-repertoire?uid=u217056

Research Gate: https://www.researchgate.net/profile/Julien-Mahy?ev=hdr_xprf

URL Google Scholar: https://scholar.google.com/citations?hl=fr&user=ohH-zUgAAAAJ

Editorial

Metallic or Metallic Oxide (Photo)catalysts for Environmental Applications

Julien G. Mahy [1,2,*] and Sophie Hermans [1,*]

[1] Institute of Condensed Matter and Nanosciences (IMCN), Université Catholique de Louvain, Place Louis Pasteur 1, B-1348 Louvain-la-Neuve, Belgium
[2] Department of Chemical Engineering–Nanomaterials, Catalysis & Electrochemistry, University of Liège, B6a, B-4000 Liège, Belgium
* Correspondence: julien.mahy@uliege.be (J.G.M.); sophie.hermans@uclouvain.be (S.H.)

During the last century, industrialization intensified in a growing number of countries around the world, and in various industries, particularly in the chemical, pharmaceutical, cosmetics, horticulture, food, and petroleum sectors. This intense industrialization has resulted in the emergence of a large variety of organic pollutants, such as dyes, aromatics, pesticides, solvents, EDCs (Endocrine Disrupting Chemicals), and PPCPs (Pharmaceuticals and Personal Care Products). Additionally, various litter is produced as wood or forest residues, waste from food crops (wheat straw, bagasse), horticulture (yard waste), or human waste from sewage plants. All this pollution and waste need to be treated and valorised in order to maintain a safe and clean environment. Numerous innovative catalytic and photocatalytic processes are being developed to eliminate these by-products of our industries. In this Special Issue, we are dealing with innovative (photo)catalytic processes for environmental applications based on metallic or metallic oxide materials. The papers concern either the (photo)catalytic transformation of various waste to produce high-value chemicals, or the (photo)catalytic degradation of pollutants. Advances highlighted in the issue are summarized in the following paragraphs focusing on these two main topics: photocatalysis for pollutant removal and catalytic valorisation of biomass.

1. Photocatalysis for pollutant removal

In this collection of articles, four different photocatalyst materials were explored for pollutant degradation: TiO_2, ZnO, SnO_2, and $BiOX$ (X = Cl, Br, I).

In Kang et al. [1], titania nanopowders were produced with many defects (such as vacancies, Ti^{3+}, N lattice heteroatoms) to increase their photocatalytic performance in the visible region for the degradation of malachite green (MG) and the disinfection of *Geobacillus stearothermophilus*. These defects were produced thanks to sensitization by (a) high temperature nitridation in NH_3 atmosphere and (b) reduction in H_2 atmosphere. It results in two types of materials: N-doped TiO_2, showing absorption in visible range, and C-doped TiO_2, with no visible absorption. As expected, the N-TiO_2 sample showed the best performance under visible light illumination. The MG solution was completely decolorized after 3 h of illumination, while the endospore suspension was inactivated by 70% after 5 h. It should be underlined that by contrast both commercial TiO_2 P25 and C-TiO_2 showed no activity towards the inactivation of the endospore suspension in the presence of visible light.

N-doping of TiO_2 was also performed by hydrothermal synthesis and using urea as the N source [2]. The research results confirmed that N doping had been successfully performed, which shifted TiO_2 absorption into the visible region, allowing it to be active under visible irradiation. This photocatalyst was used for Pb(II) removal thanks to photo-oxidation under visible light to form PbO_2. The highest photocatalytic oxidation of 15 mg/L Pb(II) in 25 mL of the solution could be reached by employing 15 mg TiO_2 doped with

10 wt.% of N, in 30 min and at pH 8. The stability of the removal activity was shown for three consecutive experiments. The highly effective process of Pb(II) photo-oxidation under benign conditions, producing less toxic and handleable PbO_2 with a recyclable photocatalyst, suggests an applicable method for Pb(II) remediation on an industrial scale.

A co-doping of titania with N and Fe has also given an increased pollutant removal rate under visible light. Indeed, Douven et al. [3] developed an ecofriendly synthesis of N,Fe-co-doped TiO_2 thanks to a sol-gel process in water. Crystalline materials were obtained without any calcination step and all samples displayed a higher visible absorption and specific surface area than P25. The photoactivity was assessed with the powder samples on p-nitrophenol showing 65% degradation after 24 h of visible illumination. Coatings of these materials were performed on steel substrates and their activity was assessed also on p-nitrophenol under UV-A and on Rhodamine-B under LED light. Degradation activity was maintained on both molecules. The possibility of producing photocatalytic films without any calcination step that are active under low-energy LED light constitutes, also, a great step forward for an industrial depollution development.

The development of methods to produce different shapes of photocatalysts is also of interest for industrial applications. In particular, the shaping of titania coating on the inner surface of quartz tubes was successfully performed by Svetlov et al. [4]. In this case, the introduction of two-phase (gas–liquid) flow with the gas flowing in the middle of the tube and a thin liquid film of synthesis sol flowing near the hot tube walls allows a TiO_2 coating deposition. The liquid flow rate in the annular flow regime allows to control the coating thickness between 3 and 10 micron and the coating porosity between 10% and 20%. By increasing the liquid flow rate, the coating porosity can be substantially reduced. The coatings obtained were active in the reactions of photocatalytic decomposition of methylene blue and rhodamine B under UV light with no observed catalyst deactivation.

ZnO is also a semiconductor material which presents good photocatalytic property for pollutant removal. In Mahy et al. [5], a green sol-gel process was used to produce ZnO with many different morphologies. Crystalline ZnO materials were obtained without any calcination. The most important parameter controlling the shape and size was found to be pH, thanks to a DoE plan and statistical analysis of the results. The photoactivity study on a model pollutant (p-nitrophenol) degradation shows that the resulting activity is mainly governed by the specific surface area of the material. A comparison with a commercial TiO_2 photocatalyst (Evonik P25) showed that the best ZnO nanoparticles obtained can reach similar photoactivity.

Composite materials with two semiconductors are also developed to enhance the photoactivity. In Rogé et al. [6], ZnO/SnO_2 heterostructures were directly synthesized in macroporous glass fibres membranes for water treatment. Hydrothermal ZnO nanorods have been functionalized with SnO_2 using an atomic layer deposition (ALD) process. The covering of ZnO by SnO_2 controlled the resulting photoactivity. Indeed, the highest degradation of methylene blue was obtained with a 40% coverage rate of SnO_2 over ZnO. This covering led also to a passivation of the surface and a better stability of the catalyst resulting in a more efficient photocatalyst in reuse.

Specific assemblies could also be produced to improve photocatalytic properties. Indeed, Baranowska–Korczyc et al. [7] produced core/shell Ag/SnO_2 nanowires for the degradation of organic compounds under visible light. In this study, Ag nanoparticles were coated with SnO_2 leading to visible absorption. Rhodamine B and malachite green were selected as model pollutants. Their degradation was investigated under 450 nm light and both pollutants were completely degraded after 90 and 40 min, respectively. The efficient photocatalytic process is attributed to two phenomena: surface plasmon resonance effects of AgNWs, which allowed light absorption in the visible range, and charge separations on the Ag core and SnO_2 shell interface of the nanowires which prevents recombination of photogenerated electron-hole pairs.

The last semiconductor photocatalyst explored in this Special Issue was Bismuth oxyhalides, BiOX (X = Cl, Br, I). These materials were synthesized via a mannitol-assisted solvothermal method [8]. The resulting materials had dominant (110) facets, first time reported with this type of synthesis using mannitol which acted simultaneously as a solvent, capping agent, and/or soft template. At the lowest mannitol concentration, it acted as a structure-directing agent, causing unification of nanoparticles, while at higher concentrations, it functioned as a solvent and soft template. The photoactivity of BiOX was evaluated on the oxidation of Rhodamine B and 5-fluorouracil, and on the reduction of Cr(VI). BiOCl and BiOBr photocatalysts presented the best photocatalytic activities leading to a total degradation or reduction in the model pollutants. This study demonstrated that BiOX prepared in mannitol solution could be useful for efficiently removing a wide range of micropollutants.

2. Biomass valorisation

In this second topic, different catalysts were developed for the transformation of biomass in high added-value molecules. Noble metals supported on various supports in particular were prepared, characterized, and tested.

In Drault et al. [9], various metals (Co, Ni, Zn, Ag, Cd, Cs, and Au) supported on silica were used as heterogeneous catalysts for the direct CO_2 carboxylation of furoic acid salts (FA, produced from furfural, derivative of inedible lignocellulosic biomass) to 2,5-furandicarboxylic acid (2,5-FDCA, a building block in the synthesis of green polymers). An experimental setup was firstly validated, and then several operation conditions were optimized, using heterogeneous catalysts instead of the semi-heterogeneous counterparts (molten salts). The preliminary results confirmed the possibility to decrease the reaction temperature to 230 °C, obtaining an acceptable conversion (74%) with the best catalyst, namely Ag/SiO_2.

Ru-based catalysts were also developed for cellulose valorisation. Haynes et al. [10] focused on the protection of active Ru nanoparticles supported on carbon black (CB) by various mesoporous silica coatings. The influence of key parameters, such as the protective layer pore size and the solvent nature of the catalytic reaction were investigated. The results showed that the hydrothermal stability was highly improved in ethanolic solution with low water content (silica loss: 99% in water and 32% in ethanolic solution); and that the silica layer pore sizes greatly influenced the selectivity of the reaction (shifting from 4% to 68% by increasing the pores sizes from 3.4 to 5 nm). The addition of an acidic co-catalyst (CB–SO_3H) led to sorbitol production through the hydrolytic hydrogenation of cellobiose (used as a model molecule of cellulose), demonstrating the high potential of the presented methodology to produce active catalysts in biomass transformations.

Finally, catalytic furfural valorisation was also investigated by Al Rawa et al. [11] with Au_x-Pt_y and Au_x-Pd_y bimetallic nanoparticles supported on TiO_2. In this work, furfural (FF) was converted into furoic acid (FA) and maleic acid (MA) by a base-free aerobic oxidation in water. By comparing the monometallic Au-, Pt-, and Pd-based catalysts to the bimetallic counterparts, the synergetic effect of alloying was evidenced. The monometallic catalysts were by far less active than the bimetallic catalysts in terms of FF conversion, and in the formation of FA, MA, and FAO intermediates. The highest selectivity (100%) to FA was obtained using a Au_3-Pd_1 catalyst, with 88% FA selectivity using 0.5% Au_3-Pt_1 with about 30% of FF conversion at 80 °C. Using Au-Pd-based catalysts, the maximum yield of MA (14%) and 5% of 2(5H)-furanone (FAO) were obtained by using a 2%Au_1-Pd_1/TiO_2 catalyst at 110 °C.

In conclusion, this Special Issue entitled "Metallic or metallic oxide (photo)catalysts for Environmental Applications" gives an overview of the latest advances in the development of metallic or metallic oxide (photo)catalytic materials, with environmental applications for the elimination of organic pollutants or the valorisation of biomass. These studies open the way for the development of new and green processes and present materials for a better environment.

Finally, we are grateful to all authors for their valuable contributions and to the editorial team of *Catalysts* for their kind support, especially to Zerlinda Tian for her constant help during this Special Issue submission and publication process.

Funding: This research received no external funding.

Institutional Review Board Statement: Not applicable.

Informed Consent Statement: Not applicable.

Data Availability Statement: Not applicable.

Acknowledgments: J.G.M. and S.H. are grateful to F.R.S.-F.N.R.S. for, respectively, his postdoctoral position and her Senior Research Associate position.

Conflicts of Interest: The authors declare no conflict of interest.

References

1. Kang, X.; Berberidou, C.; Galeckas, A.; Bazioti, C.; Sagstuen, E.; Norby, T.; Poulios, I.; Chatzitakis, A. Visible Light Driven Photocatalytic Decolorization and Disinfection of Water Employing Reduced TiO_2 Nanopowders. *Catalysts* **2021**, *11*, 228. [CrossRef]
2. Wahyuni, E.T.; Rahmaniati, T.; Hafidzah, A.R.; Suherman, S.; Suratman, A. Photocatalysis over N-Doped TiO_2 Driven by Visible Light for Pb(II) Removal from Aqueous Media. *Catalysts* **2021**, *11*, 945. [CrossRef]
3. Douven, S.; Mahy, J.G.; Wolfs, C.; Reyserhove, C.; Poelman, D.; Devred, F.; Gaigneaux, E.M.; Lambert, S.D. Efficient N, Fe Co-Doped TiO_2 Active under Cost-Effective Visible LED Light: From Powders to Films. *Catalysts* **2020**, *10*, 547. [CrossRef]
4. Svetlov, S.D.; Sladkovskiy, D.A.; Semikin, K.V.; Utemov, A.V.; Abiev, R.S.; Rebrov, E.V. Synthesis of Thin Titania Coatings onto the Inner Surface of Quartz Tubes and Their Photoactivity in Decomposition of Methylene Blue and Rhodamine B. *Catalysts* **2021**, *11*, 1538. [CrossRef]
5. Mahy, J.G.; Lejeune, L.; Haynes, T.; Body, N.; de Kreijger, S.; Elias, B.; Marcilli, R.H.M.; Fustin, C.A.; Hermans, S. Crystalline ZnO Photocatalysts Prepared at Ambient Temperature: Influence of Morphology on p-Nitrophenol Degradation in Water. *Catalysts* **2021**, *11*, 1182. [CrossRef]
6. Rogé, V.; Didierjean, J.; Crêpellière, J.; Arl, D.; Michel, M.; Fechete, I.; Dinia, A.; Lenoble, D. Tuneable Functionalization of Glass Fibre Membranes with ZNO/SNO_2 Heterostructures for Photocatalytic Water Treatment: Effect of SNO_2 Coverage Rate on the Photocatalytic Degradation of Organics. *Catalysts* **2020**, *10*, 733. [CrossRef]
7. Baranowska-Korczyc, A.; Mackiewicz, E.; Ranoszek-Soliwoda, K.; Grobelny, J.; Celichowski, G. Core/Shell Ag/Sno_2 Nanowires for Visible Light Photocatalysis. *Catalysts* **2022**, *12*, 30. [CrossRef]
8. Wilczewska, P.; Bielicka-Giełdoń, A.; Szczodrowski, K.; Malankowska, A.; Ryl, J.; Tabaka, K.; Siedlecka, E.M. Morphology Regulation Mechanism and Enhancement of Photocatalytic Performance of Biox (X = cl, Br, i) via Mannitol-Assisted Synthesis. *Catalysts* **2021**, *11*, 312. [CrossRef]
9. Drault, F.; Snoussi, Y.; Thuriot-Roukos, J.; Itabaiana, I.; Paul, S.; Wojcieszak, R. Study of the Direct CO_2 Carboxylation Reaction on Supported Metal Nanoparticles. *Catalysts* **2021**, *11*, 326. [CrossRef]
10. Haynes, T.; Hubert, S.; Carlier, S.; Dubois, V.; Hermans, S. Influence of Water-Miscible Organic Solvent on the Activity and Stability of Silica-coated Ru Catalysts in the Selective Hydrolytic Hydrogenation of Cellobiose into Sorbitol. *Catalysts* **2020**, *10*, 149. [CrossRef]
11. Al Rawas, H.K.; Ferraz, C.P.; Thuriot-Roukos, J.; Heyte, S.; Wojcieszak, R.; Paul, S. Influence of Pd and Pt Promotion in Gold Based Bimetallic Catalysts on Selectivity Modulation in Furfural Base-Free Oxidation. *Catalysts* **2021**, *11*, 1226. [CrossRef]

Article

Visible Light Driven Photocatalytic Decolorization and Disinfection of Water Employing Reduced TiO$_2$ Nanopowders

Xiaolan Kang [1,†], Chrysanthi Berberidou [2,*,†], Augustinas Galeckas [3], Calliope Bazioti [3], Einar Sagstuen [4], Truls Norby [1], Ioannis Poulios [2] and Athanasios Chatzitakis [1,*]

1. Centre for Materials Science and Nanotechnology, Department of Chemistry, University of Oslo, FERMiO, Gaustadalléen 21, NO-0349 Oslo, Norway; xiaolan.kang@kjemi.uio.no (X.K.); truls.norby@kjemi.uio.no (T.N.)
2. Department of Chemistry, Aristotle University of Thessaloniki, 54124 Thessaloniki, Greece; poulios@chem.auth.gr
3. Centre for Materials Science and Nanotechnology, Department of Physics, University of Oslo, P.O. Box 1048 Blindern, NO-0316 Oslo, Norway; augustinas.galeckas@fys.uio.no (A.G.); kalliopi.bazioti@smn.uio.no (C.B.)
4. Department of Physics, University of Oslo, P.O. Box 1048 Blindern, NO-0316 Oslo, Norway; einar.sagstuen@fys.uio.no
* Correspondence: cberber@chem.auth.gr (C.B.); a.e.chatzitakis@smn.uio.no (A.C.)
† Equally contributing authors.

Abstract: Defect-engineering of TiO$_2$ can have a major impact on its photocatalytic properties for the degradation of persisting and non-biodegradable pollutants. Herein, a series of intrinsic and extrinsic defects are induced by post annealing of crystalline TiO$_2$ under different reducing atmospheres. A detailed optoelectronic characterization sheds light on the key characteristics of the defect-engineered TiO$_2$ nanopowders that are linked to the photocatalytic performance of the prepared photocatalysts. The photodegradation of a model dye, malachite green, as well as the inactivation of bacterial endospores of the *Geobacillus stearothermophilus* species were studied in the presence of the developed catalysts under visible light illumination. Our results indicate that a combination of certain defects is necessary for the improvement of the photocatalytic process for water purification and disinfection under visible light.

Keywords: photocatalysis; defect-engineered TiO$_2$; solar light; decolorization and disinfection of water; advanced oxidation processes; metallic oxide nanoparticles

Citation: Kang, X.; Berberidou, C.; Galeckas, A.; Bazioti, C.; Sagstuen, E.; Norby, T.; Poulios, I.; Chatzitakis, A. Visible Light Driven Photocatalytic Decolorization and Disinfection of Water Employing Reduced TiO$_2$ Nanopowders. *Catalysts* **2021**, *11*, 228. https://doi.org/10.3390/catal11020228

Academic Editor: Sophie Hermans
Received: 27 December 2020
Accepted: 5 February 2021
Published: 9 February 2021

Publisher's Note: MDPI stays neutral with regard to jurisdictional claims in published maps and institutional affiliations.

Copyright: © 2021 by the authors. Licensee MDPI, Basel, Switzerland. This article is an open access article distributed under the terms and conditions of the Creative Commons Attribution (CC BY) license (https://creativecommons.org/licenses/by/4.0/).

1. Introduction

Semiconductor photocatalysis has attracted great attention in the last decades due to the potential and prospects it offers in a variety of fields including environmental remediation, energy production, and organic synthesis [1]. The exploitation of solar energy in the presence of nanostructured semiconductors and mild environmental conditions may serve as a reliable alternative to energy-intensive conventional technologies [2,3].

A well-established material for photocatalytic applications is TiO$_2$ (anatase) because of its high photocatalytic activity, resistance to photo-corrosion, biological inertness, and low cost. [4–6]. Several strategies have been applied to tackle the main disadvantage of TiO$_2$, which is its inability to be activated by visible light due to its high band gap energy. Strategies such as doping with transition-metal ions (Fe, W, Mo, V, and many others), dye sensitization (Ru-, Os-, Zn-complexes, etc.), as well as non-metal ion doping with nitrogen serving as probably the most popular and efficient dopant [7–9]. It has been found that substitutional doping of N introduces p states that, combined with the O 2p states of TiO$_2$, contribute to its bandgap narrowing. In addition, interstitial N doping couples with lattice O, creating antibonding states in the bandgap of the material [10,11]. These states are located close to the valence band (VB) of TiO$_2$, and ionized electrons coming from the VB can be accommodated. In turn, this increases the concentration of holes in the

VB. Additionally, electrons from these states can be photoexcited by longer wavelengths improving the performance of TiO_2 under visible light irradiation [12,13].

Recently, the colorful chemistry of TiO_2 has been revealed and especially important was the development of black TiO_2 by Chen and co-workers [14]. In the latter case, the formation of oxygen vacancies, Ti^{3+} states, as well as of core-shell crystalline-disorder structures, Ti-H and Ti-OH species, has extended the activation potential of TiO_2 even in the IR region of the light spectrum. Since the discovery of black TiO_2 in 2011, there have been numerous reports of the improved photocatalytic activity and electronic conductivity of TiO_2, leading to enhanced photocatalytic efficiencies both in the degradation of organic molecules, and the production of solar fuels [14–18].

In this work, TiO_2 nanopowders were synthesized via the sol-gel method and were then sensitized by (a) high temperature nitridation in NH_3 atmosphere and (b) reduction in H_2 atmosphere. The synthesized N-doped and black TiO_2 nanopowders were then tested for their efficiency on the decomposition of a model and non-biodegradable organic pollutant as well as on the inactivation of highly resistant biological targets, in the presence of visible light. Both N-doped and black TiO_2 are well-known material classes and they have been extensively studied in the literature. However, a direct comparison of their efficacy and efficiency in the visible light-driven water remediation coupled to their intrinsic and extrinsic defects (oxygen vacancies, Ti^{3+}, N lattice heteroatoms) is not common. Our work showcases how certain defects are associated with the photocatalytic performance of two important forms of TiO_2.

More specifically, an evaluation of photocatalytic efficiencies of the as prepared TiO_2 nanopowders under visible light was conducted based on the degradation of an organic, triarylmethane dye, malachite green (MG). Organic dyes are used in a wide range of applications including textile, leather, and paper production industries, laser technology, photoelectrochemical, and dye-sensitized solar cells [19–21]. It is estimated that approximately 7×10^5 tons of dyes are produced annually, while more than 10^5 tons per year are released in output flows, severely contaminating aquatic bodies. Organic dyes are in most cases soluble in water, non-biodegradable, and toxic to humans and ecosystems, rendering their efficient removal from waterbodies of the outmost importance [22,23]. MG, in particular, is annually produced in thousands of tons and consumed mainly in the textile and paper industry [24]. MG is also used as one of the most effective fungicides and ectoparasiticides in aquaculture and fisheries [25] and may persist in edible fish tissues for extended periods of time [26]. Furthermore, MG is used in medical applications as a disinfectant and a biological stain or counterstain for microscopic analysis of cell biology and tissue samples. In spite of its extended use, MG is a highly controversial material due to its toxic properties which are known to cause respiratory problems, carcinogenesis, mutagenesis and teratogenicity [27].

After investigation of the photocatalytic degradation of MG in the presence of the as prepared photocatalysts and visible irradiation, the study focused on the catalysts with the optimum photocatalytic properties and their potential to inactivate bacterial endospores of the *Geobacillus stearothermophilus* species. After prions [28] and along with certain protozoans, endospores demonstrate the highest resistance to harsh environmental conditions, enabling them to survive in dormant state for up to thousands of years and to germinate when conditions become favorable. Endospores, such as *G. stearothermophilus* exhibit a remarkably high degree of resistance to inactivation by most physical treatments and oxidizing agents [29,30]. Bacterial endospores can serve as excellent reference tools in inactivation studies, due to the remarkable stability of their suspensions [31].

2. Results

2.1. Characterizations

The morphology of the as-prepared TiO_2 samples was observed under ultra-high resolution cold-field emission SEM, as shown in Figure 1b–e. Photographs of the obtained powders are also presented in Figure 1a.

All the nanoparticles in the A-TiO$_2$ sample appear to aggregate and do not show regular morphology (Figure 1b). However, the post thermal treatments have a significant effect on the visual appearance and size of the obtained TiO$_2$ nanoparticles. The white color of the amorphous TiO$_2$ powder (A-TiO$_2$) did not change after heat treatment at 550 °C in air (C-TiO$_2$), but it turned yellow after nitridation (N-TiO$_2$) and grey after hydrogen reduction (H-TiO$_2$), as seen in Figure 1a. In general, bandgap engineering in order to alter the absorption properties of a semiconductor produces colored powders and especially the lattice N incorporation results in yellow-colored TiO$_2$. In the case of intrinsic doping of TiO$_2$, i.e., induction of intrinsic defects, the appearance of a grey color can be ascribed to Ti^{3+} and oxygen vacancies ($v_O^{\bullet\bullet}$) species that increase in concentration after treatment in reducing atmospheres. Furthermore, as evident from the comparison of nanocrystals in Figure 1b–e, the average particle size progressively increases in accord with the received amount of thermal energy (annealing T × treatment time).

Figure 1. Photographs of the investigated TiO$_2$ nanocrystals (**a**) and representative SEM images of A-TiO$_2$ (**b**), C-TiO$_2$ (**c**), H-TiO$_2$ (**d**), and N-TiO$_2$ (**e**), respectively. Scale bar: 300 nm.

To determine the crystal structure and possible phase changes after the hydrogenation and nitridation treatments, the X-ray diffraction (XRD) patterns of the TiO$_2$ samples annealed in the presence of different gases were collected and are presented in Figure 2. All the TiO$_2$ nanocrystals display almost the same XRD patterns, which can be indexed mainly to the pure anatase phase. More specifically, the peaks at 2θ = 25.2°, 37.7°, 48.0°, 53.8°, 55.2°, and 62.7° are indexed to the (101), (004), (200), (105), (211), and (118) crystal planes of anatase (JCPDS Card No.21-1272), respectively. The intense XRD diffraction peaks for the H- and N-TiO$_2$ at 2θ = 25.4° samples indicate their improved crystallinity, in accordance with the SEM observations above. The crystallite size of all the samples was calculated using the Scherrer equation based on the XRD spectra. The results show that the crystallite sizes are increasing following the trend: A-TiO$_2$ (5.4 nm) < C-TiO$_2$ (13.1 nm) < H-TiO$_2$ (21.6 nm) < N-TiO$_2$ (29.6 nm). The calculated values indicate that calcination under H$_2$ and NH$_3$ atmosphere promoted the growth of the TiO$_2$ naono- grains. Consequently, we can assume that the surface area of each powder sample decreases with the same trend. Compared to the C-TiO$_2$, the (101) diffraction peak (Figure 2 inset) of the rest three samples shows a slight shift towards a lower angle, which can be ascribed to lattice expansion [32]. It is reasonable to assume that this expansion is the result of the lattice disorder and the interstitial/substitutional N incorporation in the case of the H- and N-TiO$_2$ samples. Moreover, the XRD patterns of the H-TiO$_2$ and N-TiO$_2$ samples show a small amount of rutile phase. Indeed, the characteristic diffraction peaks at 27.5°, 36.1° and 41.2° are attributed to the (110), (101) and (111) facets of rutile, respectively (JCPDS Card No. 21-1276). The rutile phase in the C-TiO$_2$ sample is not obvious from the corresponding XRD pattern, but it may be hindered by the lower degree of crystallinity and obscured by the background signal.

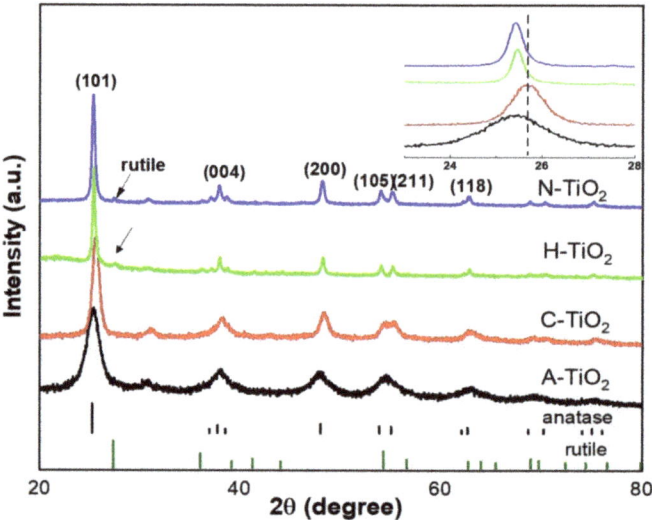

Figure 2. XRD patterns of the prepared TiO$_2$ nanocrystals.

(Scanning) Transmission Electron Microscopy ((S)TEM) was employed for a more detailed phase identification and structural analysis on the nanoscale. Figure 3 shows Selected-Area Electron Diffraction (SAED) patterns (a–c) and high-resolution Annular Bright Field (ABF) STEM images (d–f) from C-TiO$_2$, H-TiO$_2$ and N-TiO$_2$, respectively. The coexistence of both anatase and rutile phases is identified in all samples, with anatase being dominant. High-resolution imaging reveals the formation of nanoparticles of high crystal quality, terminated by well-defined facets. The maximum particle-size is ~15 nm for C-TiO$_2$, while a slight enlargement up to ~25 nm is observed when TiO$_2$ was annealed in H$_2$ and NH$_3$ atmospheres, corresponding very well with the sizes obtained by the XRD refinement. Additional ABF-STEM images indicative of the difference in the grain sizes can be seen in Figure S1 as well. Precise measurements of the d-spacings were performed on high-resolution images, as well as on the corresponding Fast Fourier Transform (FFT) diffractograms. A lattice expansion in the H-TiO$_2$ and N-TiO$_2$ nanoparticles is detected with respect to C-TiO$_2$. In particular, the average d-spacing of the anatase (101) lattice planes in C-TiO$_2$ is 0.347 ± 0.001 nm, while the values increase in the case of H-TiO$_2$ and N-TiO$_2$ to 0.355 ± 0.001 nm and 0.357 ± 0.001 nm, respectively. This lattice expansion justifies the left-shift of the XRD peaks presented in Figure 2. Furthermore, as shown at Figure 3d, the C-TiO$_2$ nanoparticles were terminated by highly crystalline sharp facets, while arrows at Figure 3e,f indicate the formation of a distorted surface layer of ~1 nm thickness in the case of H-TiO$_2$ and N-TiO$_2$. This phenomenon is well known and discussed for black TiO$_2$ and was also reported by Wang et al., where surface distortion in TiO$_2$ nanoparticles was also observed after hydrogen and nitrogen plasma treatments [17,33].

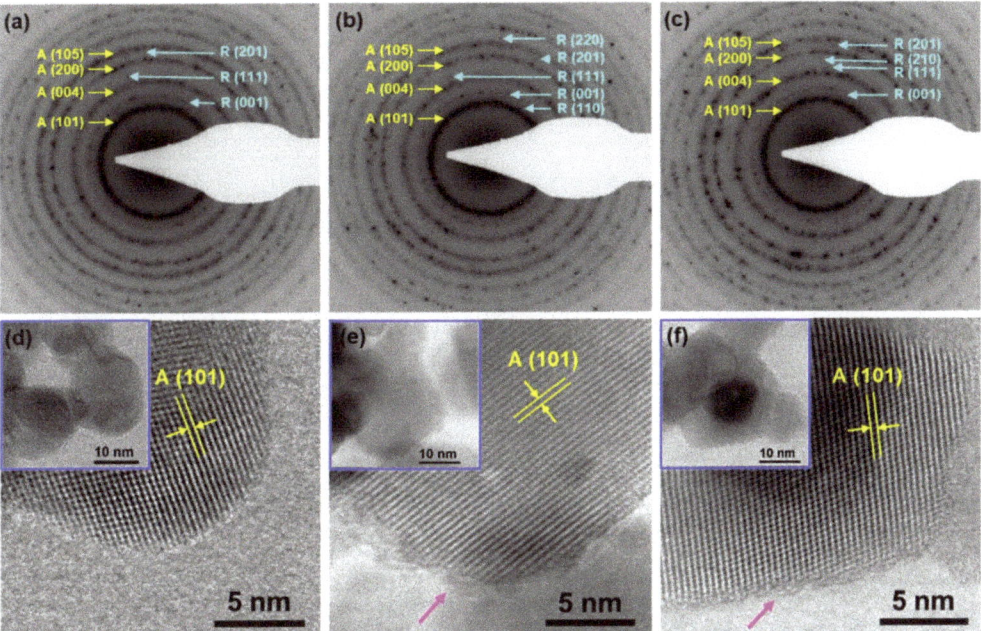

Figure 3. Selected-Area Electron Diffraction (SAED) patterns (**a**–**c**) and high-resolution ABF-STEM images (**d**–**f**) acquired from C-TiO$_2$, H-TiO$_2$ and N-TiO$_2$, respectively. Anatase (A) and rutile (R) phases coexist in all samples, with anatase being dominant. The insets in (**d**–**f**) show lower-magnification images, illustrating the overall morphology of the faceted nanoparticles. Measurements of the A(101) d-spacings reveal a lattice expansion in the case of H-TiO$_2$ and N-TiO$_2$ along with the formation of ~1 nm surface lattice distortion (annotated by pink arrows).

The absorption properties of the TiO$_2$ powders derived from the diffuse reflectance measurements at room temperature are summarized in Figure 4. The bandgap energy and the dominant type of optical transitions can be determined from the onset and linearity of the absorption edge by applying standard Kubelka–Munk [34] and Tauc [35] treatment to the DRS spectra in Figure 4a. The absorption spectra, represented by K-M function in Figure 4b, explain the coloration of TiO$_2$ powders apparent in Figure 1a. The dissimilar yet spectrally uniform absorption throughout the visible range leads to a white/grey appearance of the A-TiO$_2$, C-TiO$_2$ and H-TiO$_2$, whereas the yellow color of N-TiO$_2$ is a result of higher absorption in the blue-green region (400–500 nm). Note that unlike other samples, N-TiO$_2$ exhibits a prominent absorption band in the visible spectral region centered at

2.95 eV as indicated by the dashed curve in Figure 4b obtained after deconvolution of the fundamental edge. The investigated TiO$_2$ powders demonstrate a relatively minor variation of optical bandgaps as evidenced by the absorption edge positions indicated on the Tauc plot in Figure 4c. It is worth noting at this point that the direct application of the Tauc method for bandgap estimation is only appropriate for spectra with fundamental edge undistorted by manifestation of defects/dopants, localized states and phonon interaction collectively referred to as Urbach tail. In practice, most accurate estimates can be obtained through deconvolution of absorption spectra into major constituents, or by employing the so-called baseline method [36], which was used in the present work. The estimated bandgap values of the C-TiO$_2$ and H-TiO$_2$ powders appear similar and close to 3.3 eV, whereas A-TiO$_2$ and N-TiO$_2$ demonstrate slightly blue-shifted fundamental edges at ~3.33 eV. These values are consistent with the reported bandgaps of amorphous [37,38] and anatase TiO$_2$ [2], also suggesting that crystallization of anatase leads to a slight reduction of the bandgap in agreement with earlier studies [39].

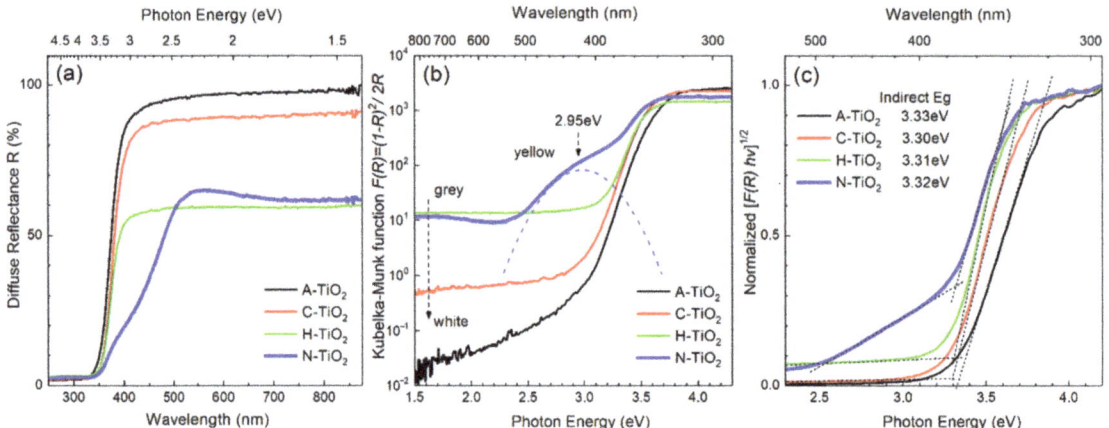

Figure 4. Optical absorption properties of the TiO$_2$ powders: diffuse reflectance (**a**), modified Kubelka-Munk function (**b**), Tauc plot assuming indirect (allowed) transitions (**c**).

In general, there are several factors that can affect the apparent absorption edge (optical bandgap) of the investigated TiO$_2$ powders, such as the level of crystallinity (lattice disorder/imperfections), strain-induced bandgap modulation (built-in strain in nanoparticles) and doping-induced bandgap modulation (bandgap-narrowing and Burstein-Moss effects). The bandgap states introduced by defects and dopants are a common cause of absorption extending into the visible range. Indeed, the red-shift of TiO$_2$ absorption edge is generally attributed to the localized gap states induced by Ti^{3+} and N species (interstitial and/or substitutional N ions, whose orbitals hybridize with orbitals from lattice oxygen) [40,41] as well as to oxygen vacancies [42]. These paramagnetic species have been evidenced in the case of H-TiO$_2$ and N-TiO$_2$ by our EPR measurement as shown and discussed later. In the case of N-TiO$_2$, the absorption band seen in the visible spectral region at 2.95 eV is unambiguously associated with oxygen vacancies (F-type color centers) according to [43].

The optical emission properties of the TiO$_2$ powders measured in vacuum at 300 K and 10 K are summarized in Figure 5. The high surface-to-volume ratio of the nanometer size particles that form TiO$_2$ powders leads to an enhanced role of surface-related effects as compared to bulk material or a film. The luminescence from surface-related sources of both intrinsic and extrinsic nature, such as sub-surface defects (e.g., $v_O^{\bullet\bullet}$) and adsorbed species on the surface (e.g., -OH), apparently dominate in the room-temperature spectra (PL300K) shown in Figure 5a, where several characteristic optical signatures are indicated by arrows.

For reference, their spectral positions are extrapolated as vertical dash-line markers and shared with low-temperature spectra (PL10K) presented in Figure 5b. At low temperatures, the non-radiative recombination pathways are suppressed, and surface-related features become overwhelmed by the increased emission originating from self-trapped exciton (STE) recombination [44–47] and luminescent defects in the bulk of nanoparticles, which collectively represent intrinsic properties of the anatase TiO_2. There is a consensus in the literature that the emission in the visible optical spectral range of anatase TiO_2 can be due to two distinct radiative recombination processes involving STEs and localized states within the bandgap related to surface defects [48]. Indeed, the broad PL band comprises of several emission components, as indicated by Gaussian deconvolution curves with shaded areas in Figure 5b. In addition to the dominant STE emission centered at around 2.37 eV, the other major contributions at 2.55 eV and 2.13 eV could be associated with N species (interstitial and substitutional) incorporated in the TiO_2 lattice; such an assumption was inferred from the calculations by Asahi and Morikawa in [49]. On the other hand, note that the deconvoluted components in Figure 5b are common to all TiO_2 samples, and thus their respective origins cannot be associated with specific outcomes of different growth/doping/post-processing of the powders. Instead, according to the literature reports, the emission peak at 3.16 eV can be attributed to phonon-assisted indirect band transitions, and the peak at 2.97 eV can be linked to STE localized on TiO_6 octahedra [50]. A series of peaks originate from $v_O^{\bullet\bullet}$–the peaks at 2.78 and 2.55 eV are associated with Ti^{3+} ions (Ec-0.5 eV), the peaks at 2.37 eV and 1.9 eV are linked to color centers F (oxygen vacancies occupied by two electrons, $v_O^{\bullet\bullet}$, Ec-0.8 eV) and F^+ (oxygen vacancy occupied by a single electron, $v_O^{\bullet}$), respectively [51–53]. Note that the peak at 2.37 eV could also be assigned to STEs as mentioned earlier. The emission around 2.13 eV is possibly due to transitions involving hydroxyl (OH^-) related deep acceptors [52].

As a final point, we note that a generally low PL efficiency is observed for all samples, which is consistent with the indirect bandgap of TiO_2 also inferring that photo-generated carriers recombine mostly non-radiatively via surface defect traps and mid-bandgap states. The concentration of such non-radiative centers determines quantum efficiency; hence the total PL yield can be considered as a measure of TiO_2 crystallinity. In this regard, the highest PL yields observed for the N-TiO_2 and H-TiO_2 powders imply the highest degree of crystal order among the investigated samples, which agrees very well with the XRD and TEM results discussed afore.

To further analyze the electronic defects in the different TiO_2 samples, EPR studies were performed and the results are presented in Figure 6. Several EPR signals have been identified, which are summarized in Table 1. It can be seen that all the studied samples have EPR signals arising from the presence of oxygen vacancies with varying intensities. As expected, the H-TiO_2 has the strongest EPR signal associated with oxygen vacancies, which is attributed to electrons localized at the vacancies induced by high temperature reduction in H_2 gas atmosphere. This is in accordance with previous studies on black TiO_2 nanotubes prepared in our labs [54]. Moreover, the presence of oxygen vacancies in C-TiO_2, with a much lower intensity though, is also reasonable and explains the n-type conductivity of TiO_2. In the case of N-TiO_2 several additional paramagnetic species have been detected, such Ti^{3+}, $O_2^{\bullet -}$, and $N_b^{\bullet}$. The latter, as argued by Livraghi et al., denotes the incorporation of N in the bulk of the TiO_2 crystal and it can be either placed interstitially or substitutionally [55]. The same authors have also identified the presence of superoxide radicals upon illumination of their N-doped TiO_2 samples [55], but these species have been previously detected in reduced TiO_2 surfaces [56,57]. Another interesting finding is the presence of Ti^{3+} only in the case of N-TiO_2, as indicated by the strong EPR signal at g = 1.9846. Such a signal arises when excess of electrons is introduced into the crystal and it is suggested that these additional electrons representing Ti^{3+} are delocalized over several Ti^{4+} lattice ions [58].

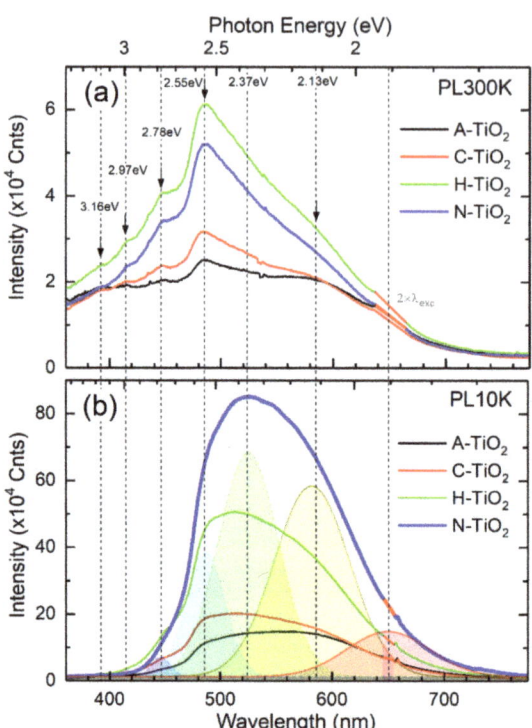

Figure 5. Optical emission properties of the TiO$_2$ powders: PL spectra measured in vacuum at 300 K (**a**) and 10 K (**b**). The key emission components are indicated by Gaussian deconvolution curves/shaded areas.

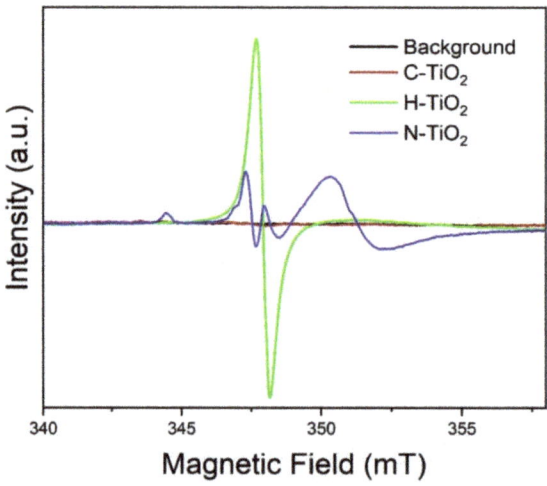

Figure 6. EPR spectra of paramagnetic species in C-TiO$_2$, H-TiO$_2$ and N-TiO$_2$. Spectra in this display have not been adjusted for slightly different microwave frequencies or for slightly different sample weights. The frequencies were 9755.2 MHz, 9755.0 MHz, 9753.3 MHz and 9754.3 MHz for background, C-TiO$_2$, H-TiO$_2$ and N-TiO$_2$, respectively.

Table 1. Spin Hamiltonian Parameters of the different paramagnetic species in C-TiO$_2$, H-TiO$_2$, and N-TiO$_2$ identified by EPR. The uncertainty is about ±0.0006.

Species * Sample	$v_O^{\bullet\bullet}$	Ti^{3+}	$O_2^{\bullet-}$ (g$_{zz}$)	$O_2^{\bullet-}$ (g$_{yy}$)	$N_b^{\bullet}$ (g$_1$)
C-TiO$_2$	2.0020	-	-	2.0088 **	2.0050 **
H-TiO$_2$	2.0033	-	-	-	-
N-TiO$_2$	2.0018	1.9846	2.0235	2.0087	2.0055

* The notation used herein is in accordance with Livraghi et al. [55]. ** It is underlined that the intensity of these signals is marginal but can still be differentiated from the background signal. A higher resolution graph is given in the Supplementary Information (SI) in Figure S1.

2.2. Evaluation of Photocatalytic Efficiencies

An evaluation of the photocatalytic efficacy and efficiency of the as prepared nanopowders was conducted in bench scale employing MG as a target molecule, at an initial concentration of 5 mg L^{-1} in the presence of 0.5 g L^{-1} of photocatalyst and visible light illumination. The performance of the prepared photocatalysts was compared against the state-of-the-art and commercially available TiO$_2$ P25 [59]. The concentration of the organic molecule present in the aqueous suspension (C), divided with the equilibrium concentration (C$_{eq}$), after 15 min of dark adsorption, is plotted as a function of the illumination time in Figure 7a. Degradation kinetics are described by a pseudo first-order equation based on the Langmuir–Hinshelwood model, according to which the initial degradation rates, r$_0$, were calculated and are given in Table 2. The prevalence of the N-TiO$_2$ followed by H-TiO$_2$ is revealed by the degradation curves of Figure 7. The initial degradation rate of N-TiO$_2$ is approximately three times higher than that of P25 or A-TiO$_2$. H-TiO$_2$ has a two times higher degradation rate in comparison to P25, while both have led to the complete decolorization of the MG solution within 300 min of illumination. On the other hand, visible light itself led to a mere 23% degradation of MG after 300 min.

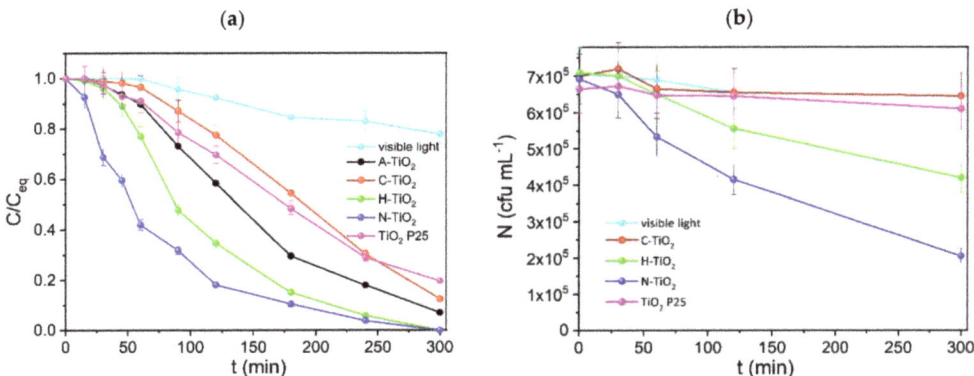

Figure 7. Photocatalytic decolorization of MG (**a**) and inactivation of *G. stearothermophilus* endospores (**b**) under visible light illumination. In all cases 0.5 g L^{-1} of the corresponding photocatalyst were used.

The photocatalytic inactivation of *G. stearothermophilus* was conducted in aqueous endospore suspensions in the presence of 0.5 g L^{-1} nanopowders and visible light illumination. The number of *G. stearothermophilus* endospores present in the suspension, N (cfu mL^{-1}) is plotted as a function of the illumination time (Figure 7b). As it can be seen, C-TiO$_2$ and P-25 were unable to inactivate *G. stearothermophilus* even after 5 h of visible light illumination. On the other hand, and similarly to the decolorization of MG, Figure 7b demonstrates the efficacy of the N-TiO$_2$ sample also in the oxidative attack on the endospore suspension. N-TiO$_2$ followed by H-TiO$_2$ led in both cases in enhanced photocatalytic efficiencies and after 300 min of illumination approximately 70 and 40% of the initial endospore content was inactivated. On the other hand, spore viability was not affected when experiments were performed in the dark (data are not shown), in the presence of all tested photocatalysts, demonstrating that spore death was induced by the photogenerated ROS.

Table 2. Initial degradation (r_0) rates during photocatalytic oxidation of 5 mg L^{-1} MG in the presence of 0.5 g L^{-1} photocatalyst and visible light illumination.

Photocatalyst	r_0 (mg L^{-1} min^{-1})
-	0.0002 ± 0.00008
A-TiO$_2$	0.007 ± 0.0009
C-TiO$_2$	0.002 ± 0.0003
H-TiO$_2$	0.012 ± 0.007
N-TiO$_2$	0.019 ± 0.002
TiO$_2$ P25	0.006 ± 0.001

3. Discussion

It is well known that the charge-carrier mobility and the band edge positions are key-factors which determine the photocatalytic reactivity of TiO$_2$. To achieve the solar degradation of organic pollutants, the photo-generated holes must be able to oxidize OH$^-$ to reactive •HO groups, while the reduction ability of conduction band (CB) electrons should be able to generate superoxide anion radicals (O$_2$•$^-$) and superoxide acid. For pure anatase, the bandgap is 3.2 eV while the conduction band minimum (CBM) and valence band maximum (VBM) are about −0.37 V and +2.83 V, respectively. The oxidation ability of the generated holes in the VB is so high that may induce a significant potential difference between the VB holes and most organic materials, which in turn undermines the transfer of holes to the target organic material. The superiority of the as prepared N-doped TiO$_2$ in the disinfection and decolorization of water is primarily attributed to its improved visible light absorption as identified by our DRS measurements. These are closely related to the different defects found by our well-aligned EPR and PL data. In addition, high-resolution ABF-STEM images on N-TiO$_2$ further showcase the formation of point defects, which are observed as nano-scale crystal lattice distortion (Figure S3). The presence of the N$_b$• species is known to introduce energy levels above the valence band of TiO$_2$ due to the hybridization of the O 2p with the N 2p orbitals [11,60]. Such hybridization leads to the uplifting of the density of states (DOS) in the VB region that eventually improves the mobility of charge carriers. Furthermore, the potential difference between holes and pollutants is also diminished due to the overlap between O 2p and N 2p states. Apart from the sub bandgap charge excitation in the N-doped material (i.e., visible light activation), an increased number of holes is expected that can contribute to its improved oxidizing power. It should be mentioned that X-ray Photoelectron Spectroscopy (XPS) measurements were also conducted (refer to SI and Figures S4 and S5) and the N 1 s peak at 399.91 eV implies that N is placed interstitially in the TiO$_2$ lattice [61]. Interstitial N incorporation correlates very well with the shift in the (001) peak in the XRD measurements but on the other hand, the same N 1 s peak is found for the rest of the samples, which is not uncommon even for undoped TiO$_2$ [10]. Ar$^+$ sputtering can probe further into the bulk of the material, but ion sputtering in TiO$_2$ is also known to preferentially remove oxygen and reduce the material (induction of Ti^{3+} states) [61]. For these reasons, we relied on the EPR and PL data for our conclusions.

In accordance with the increased holes concentration, the presence of Ti^{3+} introduces localized states below the CBM, which may finally render the formation of the conduction band tail. An excess of electrons and improved electron conductivity are achieved in the CB region, which can facilitate reducing reactions, e.g., of oxygen gas to superoxide radicals [60]. Such reactive oxygen species (ROS) can further contribute to the oxidation of organic pollutants and herein we provide clear evidence of such boosting phenomena.

This finding also suggests that both $v_O^{\bullet\bullet}$ and Ti^{3+} species are necessary for the improvement of the photocatalytic efficiency under visible light. This is the reason that H-TiO$_2$ is inferior compared to N-TiO$_2$. Our PL data at 10 K showed the improved crystallinity of the N-TiO$_2$, and the decreased amount of non-radiative recombination centers, implying the improved charge separation of the N-doped sample and its higher intrinsic photocatalytic activity. Moreover, we see that H-TiO$_2$ can completely decolorize the MG solution after 300 min, similarly as the N-TiO$_2$, but its efficiency in the endospore inactivation is 40%, while N-TiO$_2$ reached 70%. This observation further correlates with the notion that both defect species are necessary for improved visible light photocatalysis.

Finally, it should not be neglected that reduced oxides, operating under oxidizing conditions may not be stable and oxidize themselves. Recycling experiments involving the H-TiO$_2$ and N-TiO$_2$ samples have been performed for the decolorization of MG solutions. The results are given in Figure S6 and show the excellent stability of both samples after three consecutive photocatalytic cycles of 300 min each. The N-TiO$_2$ sample showed no efficiency degradation, while the H-TiO$_2$ sample retained 95% of each activity after 900 min of photocatalytic MG decolorization. The structural integrity of the recycled photocatalysts was also confirmed by XRD, where no changes were observed (Figure S7).

4. Materials and Methods

4.1. Materials

Malachite green carbinol hydrochloride (bis[4-(dimethylamino)phenyl]-phenylmethanol hydrochloride, $C_{23}H_{26}N_2O\bullet HCl$, CAS No: 123333-61-9, Mr: 382.93) was purchased from Sigma-Aldrich and was used as received.

TiO$_2$ P25 (Evonik GmbH, anatase/rutile = 3.6/1, BET: 50 m^2 g^{-1}, nonporous) was used for comparative purposes against the synthesized nanopowders. All other reagent-grade chemicals were purchased from Merck and were used without further purification. Doubly distilled water was used throughout the study.

B. stearothermophilus endospores were initially purchased impregnated on paper strips (ATCC 7953, Fluka, 10^6 cfu/strip). These endospores are highly resistant thermophiles and can grow at temperatures $\geq 60\,°C$. This particular strain is used as an indicator of proper autoclave performance and is specified by U.S. military specification MIL-S-36586 as GMP requirements of the U.S. Food and Drug Administration.

4.2. Synthesis of TiO$_2$ Nanopowders

TiO$_2$ was prepared according to the sol-gel method. The desired amounts of titanium isopropoxide (TTIP, $\geq$97%, Merck, KGaA, Darmstadt, Germany, 87560) and absolute ethanol (Fisher chemicals, E/0650DF/17, Fisher Scientific UK Ltd., Leicestershire, UK) were mixed in a beaker and placed in an ultrasonic bath for 15 min taking the necessary precautions to prevent addition of moisture in the TTIP/ethanol mixture. Then, an aqueous HCl 1 M solution was added dropwise (~1 mL min^{-1}) to the above solution under vigorous magnetic stirring. The molecular ratio of TTIP/ethanol/H$_2$O/acid was 1/25/8/0.007. The viscous content of the beaker was vigorously stirred for 16 h and then left to age for 16 h at room temperature yielding a gel, which was dried at 75 °C for 16 h. The resulting white product (A-TiO$_2$) was ground, washed 5 times with doubly distilled H$_2$O and dried at 75 °C for 16 h. This was followed by calcination at 550 °C with a heating rate of 2 °C min^{-1} for 60 min, cooling to room temperature, grinding, and storage at room temperature. This sample is denoted as C-TiO$_2$.

N-doped TiO$_2$ (N-TiO$_2$) was obtained by high temperature annealing in NH$_3$ gas atmosphere of the C-TiO$_2$ sample. Briefly, the as-prepared TiO$_2$ powder was put in a ProboStat™ cell (NORECS, Oslo, Norway) under NH$_3$ gas flow at atmospheric pressure and calcined at 550 °C for 4 h with a heating rate of 5 °C min^{-1}. Finally, hydrogenated TiO$_2$ (H-TiO$_2$) was obtained by annealing the A-TiO$_2$ sample under H$_2$ gas atmosphere at 550 °C with a heating rate of 5 °C min^{-1} for 2 h.

4.3. Characterizations

Micrographs were obtained with a Hitachi SU8230 ultra-high resolution cold-field emission scanning electron microscope (SEM) equipped with a secondary electron (SE) detector under an acceleration voltage of 3 kV. The TiO_2 phase in all the samples was confirmed using X-ray diffraction (XRD, Bruker D8 Discover, Cu Kα1, Bragg-Brentano) with a Cu Kα-filtered radiation source (λ = 1.5046 Å), step 0.01° (2θ) and a scan rate of 1° min^{-1}.

(S)TEM investigations were conducted on an FEI Titan G2 60–300 kV equipped with a CEOS DCOR probe-corrector, monochromator and Super-X EDX detectors. Observations were performed at 300 kV with a probe convergence angle of 24 mrad. The camera length was set to 77 mm and simultaneous STEM imaging was conducted with 3 detectors: HAADF (collection angles 98.7–200 mrad), ADF (collection angles 21.5–98.7 mrad), and ABF (collection angles 10.6–21.5 mrad). The resulting spatial resolution achieved was approximately 0.08 nm. For the (S)TEM study, the TiO_2 powders were dispersed in ethanol with ultrasonic support and drop-casted onto Cu-grids supported by holey-C films.

The optical absorption properties were investigated at room temperature by means of diffuse reflectance spectroscopy (DRS) using a ThermoScientific EVO-600 UV−Vis spectrophotometer. The emission properties were attained by measuring photoluminescence (PL) of the powder samples in vacuum at 10 K and 300 K using a 325 nm wavelength He-Cd cw-laser as an excitation source (power density < 10 W cm^{-2}). The emission was collected by a microscope and analyzed by an imaging spectrometer system (Horiba Jobin Yvon, iHR320 coupled to Andor iXon888 Ultra EMCCD) with a spectral resolution below 0.2 nm. Low-temperature measurements were realized using a closed-cycle He refrigerator (Janis, Inc. CCS450).

Electron paramagnetic resonance (EPR) spectra were recorded at room temperature using a BRUKER EleXsyS 560 spectrometer operating at X-band frequencies and an ER 4122 SHQE cavity. The sample was placed in a 4 mm OD Wilmad quartz tube located in the cavity center by a Teflon rod. Spectra were obtained at 5 mW microwave power (16 dB attenuation), receiver gain 76 dB and 0.4 mT modulation width at 100 kHz modulation frequency. The central magnetic field was fixed at 338 mT and the sweep width at 20 mT at 1 k points resolution. The field sweep time and the time constant were 84 s and 165 ms, respectively. Among the samples, the microwave frequency varied between 9753.3 and 9755.0 MHz. 10 spectra were added for each sample.

4.4. Preparation of G. stearothermophilus Endospore Suspension

The *G. stearothermophilus* endospore suspension was prepared as previously described [31]. After staining with the Schaeffer–Fulton method [62] and enumeration employing a reverse phase microscope, it was found that the suspension consisted of at least 98% endospores. To determine the initial concentration of the prepared endospore suspension, serial dilutions were performed in triplicates and the samples were plated and incubated on plates containing tryptic soy broth (TSB) as described in Section 4.5. It was found that the initial concentration of the purified endospore suspension was approximately 4×10^7 cfu mL^{-1}.

4.5. Enumeration of G. stearothermophilus Endospores

To evaluate the efficiency of photocatalytic inactivation of *G. stearothermophilus*, samples were collected in duplicates and spread on 9 cm agar plates containing TSB (casein peptone 17 g L^{-1}, soya peptone 3 g L^{-1}, K_2HPO_4 2.5 g L^{-1}, NaCl 5 g L^{-1}, glucose 2.5 g L^{-1}, bacteriological agar 15 g L^{-1}) under sterile conditions. The samples collected during the photocatalytic experiments were incubated for 5 min at 100 °C to induce sporulation of endospores to vegetative cells (thermal activation). 20 µL of each activated sample were added to 3 mL of melted TSB containing 0.7% (*w/v*) agarose (pure plate method) and were spread under sterile conditions onto the plates. The plates were then incubated for 24 h at 60 °C and BSE colonies were enumerated by direct counting.

4.6. Photocatalytic Experiments

The photocatalytic oxidation of MG was performed in a bench-scale photocatalytic reactor of 300 mL capacity, equipped with a central visible lamp (Osram Dulux S, 9 W/71, emission range: 400–550 nm, maximum emission: 450 nm) and inlet and outlet ports for bubbling CO_2 free air, under constant magnetic stirring, at a working volume of 250 mL. Potassium ferrioxalate served as a chemical actinometer [63] to determine the incident photon flow of the irradiation source, which was 57×10^{-6} Einstein L^{-1} min^{-1}. Photocatalytic experiments were conducted at initial pH of 5.0 ± 0.1 and constant temperature (25 ± 0.1 °C). Maximum adsorption of the dye onto the photocatalyst's surface, in the absence of light, was achieved within 30 min. The photocatalyst particles were removed from the MG solution by employing a 0.45 μm filter.

The photocatalytic inactivation of BSE endospores was performed in sterile polystyrene 6 well plates with lid (Greiner). Each well served as a cylindrical 12 mL photocatalytic reactor. The appropriate amount of the stock endospore suspension was diluted in sterile distilled H_2O and was added to a single well of the plate. The catalyst was then added at the desired concentration, resulting to a 10 mL working volume. The plate was illuminated by a system of 4 parallel lamps emitting visible irradiation (length: 45 cm, Osram L 15 W/840. Lumilux cool white, emission range: 400–700 nm, maximum emission: 550 nm), connected with a voltage regulator, placed 5 cm above the surface of the endospore suspension. The intensity of the incident irradiation at this distance was measured using a Photometer/Radiometer PMA 2140 (Solar Light Co., Glenside, PA, USA) equipped with a global range detector, which measures irradiance within the range of 400–1100 nm and found to be 2 ± 0.1 mW cm^{-2}. Photocatalytic inactivation took place at room temperature, under constant magnetic stirring. Samples of 20 μL were collected in duplicates at various time intervals in sterile Eppendorf tubes, diluted up to a final volume of 200 μL with sterile, distilled H_2O, and kept in the dark. After the end of the photocatalytic process the samples were spread on TSB plates and enumerated as described in Section 4.5. All experiments were repeated three times, under identical experimental conditions.

4.7. Analytical Methods

A UV-Visible spectrophotometer (UV-1700, PharmaSpec, Shimadzu, Randburg, South Africa) was employed to follow changes in the concentration of MG, based on the linear dependence between the initial concentration of the insecticide and its absorption at 617 nm. The standard deviation of the optical density values was within ±5%.

5. Conclusions

In this work, TiO_2 nanopowders were prepared via a sol-gel synthesis route followed by anatase formation after annealing at high temperature in air atmosphere (C-TiO_2). In order to obtain visible light activated photocatalysts, the sol-gel prepared powders were subsequently annealed in hydrogen gas (H-TiO_2) or ammonia stream (N-TiO_2). An extensive physico-chemical characterization revealed the presence of several lattices as well as electronic defects. Clearly, the N-TiO_2 sample showed an absorption shoulder stretching in the visible region and down to 2.5 eV, as showed by the DRS spectra. It also showed the highest degree of crystallinity and a suppressed number of non-radiative recombination centers, which are related to the identified defects via the EPR measurements ($v_O^{\bullet\bullet}$, Ti^{3+}, $O_2^{\bullet -}$ and $N_b^\bullet$). These defects are known to introduce energy states inside the bandgap of the material and facilitate its activation with light in the visible region, in accordance with the DRS and PL data. On the other hand, the hydrogenation of the C-TiO_2 sample led to an increased amount of oxygen vacancies, which are charge compensated by localized electrons in the form of Ti^{3+} species. The latter species were not seen by EPR but only a strong signal assigned to oxygen vacancies, thus leading to the conclusion that the excess electrons were not delocalized in the Ti atoms. The H-TiO_2 sample was still not visible light active and retained the bandgap of the C-TiO_2 of approx. 3.3 eV.

The efficiency of the synthesized photocatalytic nanopowders was assessed based on the decolorization and disinfection of water and two model systems were studied for this purpose: malachite green, a triarylmethane dye and *G. stearothermophilus* endospores, respectively. As expected, the N-TiO$_2$ sample showed the best performance in both cases under visible light illumination. The MG solution was completely decolorized after 3 h of illumination, while the endospore suspension was inactivated by 70% after 5 h. It should be underlined that both TiO$_2$ P25 and C-TiO$_2$ showed no activity towards the inactivation of the endospore suspension in the presence of visible light.

This work studies and exemplifies the colorful chemistry of defective TiO$_2$ and provides a direct comparison of the photocatalytic activity under visible light between white (C-TiO$_2$), grey (H-TiO$_2$) and yellow (N-TiO$_2$) TiO$_2$. The ability to tune and induce certain defects can have a major impact on the photocatalytic efficiency and intrinsic properties of the material. Photocatalysts activated by visible illumination are highly desirable, especially for practical applications regarding water purification and disinfection.

Supplementary Materials: The following are available online at https://www.mdpi.com/2073-4344/11/2/228/s1, Figure S1: Additional ABF-STEM images, Figure S2: A higher magnification of the EPR spectra of paramagnetic species in C-TiO$_2$, H-TiO$_2$ and N-TiO$_2$, Figure S3: Additional high resolution and filtered ABF-STEM images, Figure S4: Survey XPS spectra of A-TiO$_2$ (black), C-TiO$_2$ (blue), H-TiO$_2$ (green) and N-TiO$_2$ (red). Atomic percentages are given on the graph, Figure S5: XPS spectra of N 1 s of A-TiO$_2$ (black), C-TiO$_2$ (blue), H-TiO$_2$ (green) and N-TiO$_2$ (red), Figure S6: Recycling experiments and Figure S7: Post-operation XRD results.

Author Contributions: X.K.: Investigation (synthesis and characterization of the reduced TiO$_2$ samples), methodology, visualization, funding acquisition, writing original draft, review and editing. C.B. (Chrysanthi Berberidou): Conceptualization, investigation (synthesis of the original TiO$_2$ powders), methodology (photocatalytic experiments), writing original draft, review and editing. A.G.: Investigation, methodology, visualization, funding acquisition, writing original draft, review and editing. C.B. (Calliope Bazioti): Investigation, methodology, visualization, writing review and editing. E.S.: Investigation, methodology, writing review and editing. T.N.: Supervision, resources, writing review and editing. I.P.: Supervision, resources writing review and editing. A.C.: Conceptualization, investigation, methodology, funding acquisition, supervision, visualization, writing original draft, review and editing. All authors have read and agreed to the published version of the manuscript.

Funding: X.K. acknowledges the China Scholarship Council (File number 20180606014). A.G. acknowledges the FUNDAMeNT project (project number 251131) and A.C. acknowledges the PH2ON project (project number 288320), both sponsored by the Research Council of Norway (Norges Forskningsråd). The Research Council of Norway is also acknowledged for its support of the Norwegian Center for Transmission Electron Microscopy (NORTEM) (197405/F50).

Data Availability Statement: Data available in a publicly accessible repository.

Acknowledgments: The authors acknowledge Ingvild Julie Thue Jensen (SINTEF Industry) and Ylva Knausgård Hommedal for the XPS measurements.

Conflicts of Interest: The authors declare no conflict of interest.

References

1. Agbe, H.; Nyankson, E.; Raza, N.; Dodoo-Arhin, D.; Chauhan, A.; Osei, G.; Kumar, V.; Kim, K.H. Recent advances in photoinduced catalysis for water splitting and environmental applications. *J. Ind. Eng. Chem.* **2019**, *72*, 31–49. [CrossRef]
2. Chen, X.; Mao, S.S. Titanium dioxide nanomaterials: Synthesis, properties, modifications, and applications. *Chem. Rev.* **2007**, *107*, 2891–2959. [CrossRef] [PubMed]
3. 6-Nanostructured semiconductor composites for solar cells. In *Nanostructured Semiconductor Oxides for the Next Generation of Electronics and Functional Devices*; Zhuiykov, S., Ed.; Woodhead Publishing: Sawston, UK, 2014; pp. 267–320. [CrossRef]
4. Noman, M.T.; Ashraf, M.A.; Ali, A. Synthesis and applications of nano-TiO$_2$: A review. *Environ. Sci. Pollut. R* **2019**, *26*, 3262–3291. [CrossRef]
5. Xu, H.; Ouyang, S.X.; Liu, L.Q.; Reunchan, P.; Umezawa, N.; Ye, J.H. Recent advances in TiO$_2$-based photocatalysis. *J. Mater. Chem. A* **2014**, *2*, 12642–12661. [CrossRef]
6. Nakata, K.; Fujishima, A. TiO$_2$ photocatalysis: Design and applications. *J. Photoch. Photobio. C* **2012**, *13*, 169–189. [CrossRef]

7. Asahi, R.; Morikawa, T.; Ohwaki, T.; Aoki, K.; Taga, Y. Visible-Light Photocatalysis in Nitrogen-Doped Titanium Oxides. *Science* **2001**, *293*, 269–271. [CrossRef]
8. Kumara, N.T.R.N.; Lim, A.; Lim, C.M.; Petra, M.I.; Ekanayake, P. Recent progress and utilization of natural pigments in dye sensitized solar cells: A review. *Renew. Sustain. Energy Rev.* **2017**, *78*, 301–317. [CrossRef]
9. Di Paola, A.; Garcıa-López, E.; Ikeda, S.; Marcì, G.; Ohtani, B.; Palmisano, L. Photocatalytic degradation of organic compounds in aqueous systems by transition metal doped polycrystalline TiO_2. *Catal. Today* **2002**, *75*, 87–93. [CrossRef]
10. Asahi, R.; Morikawa, T.; Irie, H.; Ohwaki, T. Nitrogen-Doped Titanium Dioxide as Visible-Light-Sensitive Photocatalyst: Designs, Developments, and Prospects. *Chem. Rev.* **2014**, *114*, 9824–9852. [CrossRef] [PubMed]
11. Di Valentin, C.; Pacchioni, G.; Selloni, A.; Livraghi, S.; Giamello, E. Characterization of Paramagnetic Species in N-Doped TiO_2 Powders by EPR Spectroscopy and DFT Calculations. *J. Phys. Chem. B* **2005**, *109*, 11414–11419. [CrossRef]
12. Di Valentin, C.; Finazzi, E.; Pacchioni, G.; Selloni, A.; Livraghi, S.; Paganini, M.C.; Giamello, E. N-doped TiO_2: Theory and experiment. *Chem. Phys.* **2007**, *339*, 44–56. [CrossRef]
13. Spadavecchia, F.; Cappelletti, G.; Ardizzone, S.; Bianchi, C.L.; Cappelli, S.; Oliva, C.; Scardi, P.; Leoni, M.; Fermo, P. Solar photoactivity of nano-N-TiO_2 from tertiary amine: Role of defects and paramagnetic species. *Appl. Catal. B Environ.* **2010**, *96*, 314–322. [CrossRef]
14. Chen, X.; Liu, L.; Yu, P.Y.; Mao, S.S. Increasing Solar Absorption for Photocatalysis with Black Hydrogenated Titanium Dioxide Nanocrystals. *Science* **2011**, *331*, 746. [CrossRef]
15. Lin, T.; Yang, C.; Wang, Z.; Yin, H.; Lü, X.; Huang, F.; Lin, J.; Xie, X.; Jiang, M. Effective nonmetal incorporation in black titania with enhanced solar energy utilization. *Energy Environ. Sci.* **2014**, *7*, 967–972. [CrossRef]
16. Wang, Z.; Yang, C.; Lin, T.; Yin, H.; Chen, P.; Wan, D.; Xu, F.; Huang, F.; Lin, J.; Xie, X.; et al. Visible-light photocatalytic, solar thermal and photoelectrochemical properties of aluminium-reduced black titania. *Energy Environ. Sci.* **2013**, *6*, 3007–3014. [CrossRef]
17. Chatzitakis, A.; Sartori, S. Recent Advances in the Use of Black TiO_2 for Production of Hydrogen and Other Solar Fuels. *ChemPhysChem* **2019**, *20*, 1272–1281. [CrossRef]
18. Naldoni, A.; Allieta, M.; Santangelo, S.; Marelli, M.; Fabbri, F.; Cappelli, S.; Bianchi, C.L.; Psaro, R.; Dal Santo, V. Effect of Nature and Location of Defects on Bandgap Narrowing in Black TiO_2 Nanoparticles. *J. Am. Chem. Soc.* **2012**, *134*, 7600–7603. [CrossRef]
19. Grätzel, M. Dye-sensitized solar cells. *J. Photochem. Photobiol. C Photochem. Rev.* **2003**, *4*, 145–153. [CrossRef]
20. Sakthivel, S.; Neppolian, B.; Shankar, M.V.; Arabindoo, B.; Palanichamy, M.; Murugesan, V. Solar photocatalytic degradation of azo dye: Comparison of photocatalytic efficiency of ZnO and TiO_2. *Sol. Energy Mater. Sol. Cells* **2003**, *77*, 65–82. [CrossRef]
21. Wróbel, D.; Boguta, A.; Ion, R.M. Mixtures of synthetic organic dyes in a photoelectrochemical cell. *J. Photochem. Photobiol. A Chem.* **2001**, *138*, 7–22. [CrossRef]
22. Reza, K.M.; Kurny, A.S.W.; Gulshan, F. Parameters affecting the photocatalytic degradation of dyes using TiO_2: A review. *Appl. Water Sci.* **2017**, *7*, 1569–1578. [CrossRef]
23. Gusain, R.; Gupta, K.; Joshi, P.; Khatri, O.P. Adsorptive removal and photocatalytic degradation of organic pollutants using metal oxides and their composites: A comprehensive review. *Adv. Colloid. Interface Sci.* **2019**, *272*, 102009. [CrossRef]
24. Shedbalkar, U.; Jadhav, J.P. Detoxification of Malachite Green and Textile Industrial Effluent by Penicillium ochrochloron. *Biotechnol. Bioproc. E* **2011**, *16*, 196–204. [CrossRef]
25. Lieke, T.; Meinelt, T.; Hoseinifar, S.H.; Pan, B.; Straus, D.L.; Steinberg, C.E. Sustainable aquaculture requires environmental-friendly treatment strategies for fish diseases. *Rev. Aquacultue* **2020**, *12*, 943–965. [CrossRef]
26. Yaoping, H.; Zhijin, G.; Junfei, L. Fluorescence detection of malachite green in fish tissue using red emissive Se,N,Cl-doped carbon dots. *Food Chem.* **2021**, *335*, 127677.
27. Srivastava, S.; Sinha, R.; Roy, D. Toxicological effects of malachite green. *Aquat. Toxicol.* **2004**, *66*, 319–329. [CrossRef] [PubMed]
28. Berberidou, C.; Xanthopoulos, K.; Paspaltsis, I.; Lourbopoulos, A.; Polyzoidou, E.; Sklaviadis, T.; Poulios, I. Homogenous photocatalytic decontamination of prion infected stainless steel and titanium surfaces. *Prion* **2013**, *7*, 488–495. [CrossRef]
29. Nicholson, W.L.; Munakata, N.; Horneck, G.; Melosh, H.J.; Setlow, P. Resistance of Bacillus endospores to extreme terrestrial and extraterrestrial environments. *Microbiol. Mol. Biol. R* **2000**, *64*, 548–572. [CrossRef]
30. Schottroff, F.; Pyatkovskyy, T.; Reineke, K.; Setlow, P.; Sastry, S.K.; Jaeger, H. Mechanisms of enhanced bacterial endospore inactivation during sterilization by ohmic heating. *Bioelectrochemistry* **2019**, *130*, 107338. [CrossRef]
31. Berberidou, C.; Paspaltsis, I.; Pavlidou, E.; Sklaviadis, T.; Poulios, I. Heterogenous photocatalytic inactivation of B. stearothermophilus endospores in aqueous suspensions under artificial and solar irradiation. *Appl Catal. B Environ.* **2012**, *125*, 375–382. [CrossRef]
32. Xu, Y.; Zhang, C.; Zhang, L.; Zhang, X.; Yao, H.; Shi, J. Pd-catalyzed instant hydrogenation of TiO_2 with enhanced photocatalytic performance. *Energy Environ. Sci.* **2016**, *9*, 2410–2417. [CrossRef]
33. Wang, H.; Xiong, J.; Cheng, X.; Chen, G.; Kups, T.; Wang, D.; Schaaf, P. Hydrogen–nitrogen plasma assisted synthesis of titanium dioxide with enhanced performance as anode for sodium ion batteries. *Sci. Rep.* **2020**, *10*, 11817. [CrossRef] [PubMed]
34. Kubelka, P.; Munk, F. Ein Beitrag Zur Optik Der Farbanstriche. *Zhurnal Tekhnicheskoi Fiz.* **1931**, *12*, 593–601.
35. Tauc, J.; Grigorovici, R.; Vancu, A. Optical Properties and Electronic Structure of Amorphous Germanium. *Physica Status Solidi* **1966**, *15*, 627–637. [CrossRef]

36. Makuła, P.; Pacia, M.; Macyk, W. How To Correctly Determine the Band Gap Energy of Modified Semiconductor Photocatalysts Based on UV–Vis Spectra. *J. Phys. Chem. Lett.* **2018**, *9*, 6814–6817. [CrossRef]
37. Zhang, M.; Lin, G.; Dong, C.; Wen, L. Amorphous TiO_2 films with high refractive index deposited by pulsed bias arc ion plating. *Surf. Coat. Technol.* **2007**, *201*, 7252–7258. [CrossRef]
38. Bendavid, A.; Martin, P.J.; Takikawa, H. Deposition and modification of titanium dioxide thin films by filtered arc deposition. *Thin Solid Film.* **2000**, *360*, 241–249. [CrossRef]
39. Ohtani, B.; Kakimoto, M.; Miyadzu, H.; Nishimoto, S.; Kagiya, T. Effect of surface-adsorbed 2-propanol on the photocatalytic reduction of silver and/or nitrate ions in acidic titania suspension. *J. Phys. Chem.* **1988**, *92*, 5773–5777. [CrossRef]
40. Kalathil, S.; Khan, M.M.; Ansari, S.A.; Lee, J.; Cho, M.H. Band gap narrowing of titanium dioxide (TiO_2) nanocrystals by electrochemically active biofilms and their visible light activity. *Nanoscale* **2013**, *5*, 6323–6326. [CrossRef]
41. Liu, X.; Gao, S.; Xu, H.; Lou, Z.; Wang, W.; Huang, B.; Dai, Y. Green synthetic approach for Ti^{3+} self-doped TiO_{2-x} nanoparticles with efficient visible light photocatalytic activity. *Nanoscale* **2013**, *5*, 1870–1875. [CrossRef]
42. Pan, X.; Yang, M.-Q.; Fu, X.; Zhang, N.; Xu, Y.-J. Defective TiO_2 with oxygen vacancies: Synthesis, properties and photocatalytic applications. *Nanoscale* **2013**, *5*, 3601–3614. [CrossRef] [PubMed]
43. Kuznetsov, V.N.; Serpone, N. On the Origin of the Spectral Bands in the Visible Absorption Spectra of Visible-Light-Active TiO_2 Specimens Analysis and Assignments. *J. Phys. Chem. C* **2009**, *113*, 15110–15123. [CrossRef]
44. Watanabe, M.; Sasaki, S.; Hayashi, T. Time-resolved study of photoluminescence in anatase TiO_2. *J. Lumin.* **2000**, *87–89*, 1234–1236. [CrossRef]
45. Tang, H.; Berger, H.; Schmid, P.E.; Lévy, F.; Burri, G. Photoluminescence in TiO_2 anatase single crystals. *Solid State Commun.* **1993**, *87*, 847–850. [CrossRef]
46. Sildos, I.; Suisalu, A.; Aarik, J.; Sekiya, T.; Kurita, S. Self-trapped exciton emission in crystalline anatase. *J. Lumin.* **2000**, *87–89*, 290–292. [CrossRef]
47. Najafov, H.; Tokita, S.; Ohshio, S.; Kato, A.; Saitoh, H. Green and Ultraviolet Emissions From Anatase TiO_2 Films Fabricated by Chemical Vapor Deposition. *Jpn. J. Appl. Phys.* **2005**, *44*, 245–253. [CrossRef]
48. Zhang, W.F.; Zhang, M.S.; Yin, Z.; Chen, Q. Photoluminescence in anatase titanium dioxide nanocrystals. *Appl. Phys. B* **2000**, *70*, 261–265. [CrossRef]
49. Asahi, R.; Morikawa, T. Nitrogen complex species and its chemical nature in TiO_2 for visible-light sensitized photocatalysis. *Chem. Phys.* **2007**, *339*, 57–63. [CrossRef]
50. Saraf, L.V.; Patil, S.I.; Ogale, S.B.; Sainkar, S.R.; Kshirsager, S.T. Synthesis of Nanophase TiO_2 by Ion Beam Sputtering and Cold Condensation Technique. *Int. J. Mod. Phys. B* **1998**, *12*, 2635–2647. [CrossRef]
51. Serpone, N. Is the Band Gap of Pristine TiO_2 Narrowed by Anion- and Cation-Doping of Titanium Dioxide in Second-Generation Photocatalysts? *J. Phys. Chem. B* **2006**, *110*, 24287–24293. [CrossRef]
52. Mathew, S.; Prasad, A.K.; Benoy, T.; Rakesh, P.P.; Hari, M.; Libish, T.M.; Radhakrishnan, P.; Nampoori, V.P.N.; Vallabhan, C.P.G. UV-visible photoluminescence of TiO_2 nanoparticles prepared by hydrothermal method. *J. Fluoresc* **2012**, *22*, 1563–1569. [CrossRef]
53. Lei, Y.; Zhang, L.D.; Meng, G.W.; Li, G.H.; Zhang, X.Y.; Liang, C.H.; Chen, W.; Wang, S.X. Preparation and photoluminescence of highly ordered TiO_2 nanowire arrays. *Appl. Phys. Lett.* **2001**, *78*, 1125–1127. [CrossRef]
54. Liu, X.; Carvalho, P.; Getz, M.N.; Norby, T.; Chatzitakis, A. Black Anatase TiO_2 Nanotubes with Tunable Orientation for High Performance Supercapacitors. *J. Phys. Chem. C* **2019**, *123*, 21931–21940. [CrossRef]
55. Livraghi, S.; Paganini, M.C.; Giamello, E.; Selloni, A.; Di Valentin, C.; Pacchioni, G. Origin of Photoactivity of Nitrogen-Doped Titanium Dioxide under Visible Light. *J. Am. Chem. Soc.* **2006**, *128*, 15666–15671. [CrossRef]
56. Shiotani, M.; Moro, G.; Freed, J.H. ESR studies of 0–2 adsorbed on Ti supported surfaces: Analysis of motional dynamics. *J. Chem. Phys.* **1981**, *74*, 2616–2640. [CrossRef]
57. Che, M.; Tench, A.J. Characterization and Reactivity of Molecular Oxygen Species on Oxide Surfaces. In *Advances in Catalysis*; Eley, D.D., Pines, H., Weisz, P.B., Eds.; Academic Press: Cambridge, MA, USA, 1983; Volume 32, pp. 1–148.
58. Livraghi, S.; Chiesa, M.; Paganini, M.C.; Giamello, E. On the Nature of Reduced States in Titanium Dioxide As Monitored by Electron Paramagnetic Resonance. I: The Anatase Case. *J. Phys. Chem. C* **2011**, *115*, 25413–25421. [CrossRef]
59. Kang, X.; Song, X.Z.; Han, Y.; Cao, J.; Tan, Z. Defect-engineered TiO_2 Hollow Spiny Nanocubes for Phenol Degradation under Visible Light Irradiation. *Sci. Rep.* **2018**, *8*, 5904. [CrossRef] [PubMed]
60. Dong, F.; Zhao, W.; Wu, Z.; Guo, S. Band structure and visible light photocatalytic activity of multi-type nitrogen doped TiO_2 nanoparticles prepared by thermal decomposition. *J. Hazard. Mater.* **2009**, *162*, 763–770. [CrossRef] [PubMed]
61. Chatzitakis, A.; Grandcolas, M.; Xu, K.; Mei, S.; Yang, J.; Jensen, I.J.T.; Simon, C.; Norby, T. Assessing the photoelectrochemical properties of C, N, F codoped TiO_2 nanotubes of different lengths. *Catal. Today* **2017**, *287*, 161–168. [CrossRef]
62. Schaeffer, A.B.; Fulton, M.D. A simplified method of staining endospores. *Science* **1933**, *77*, 194. [CrossRef]
63. Braun, A.M.; Maurette, M.T.; Oliveros, E. *Photochemical Technology*; Wiley: New York, NY, USA, 1991.

Article

Photocatalysis over N-Doped TiO$_2$ Driven by Visible Light for Pb(II) Removal from Aqueous Media

Endang Tri Wahyuni *, Titi Rahmaniati, Aulia Rizky Hafidzah, Suherman Suherman and Adhitasari Suratman

Chemistry Department, Faculty of Mathematic and Natural Sciences, Gadjah Mada University, Sekip Utara P.O. Box Bls 21, Yogyakarta 55281, Indonesia; titi.rahmaniati@mail.ugm.ac.id (T.R.); a.rizky.hafidzah@mail.ugm.ac.id (A.R.H.); suherman.mipa@ugm.ac.id (S.S.); adhitasari@ugm.ac.id (A.S.)
* Correspondence: endang_triw@ugm.ac.id; Tel.: +62-274-545188

Abstract: The photocatalysis process over N-doped TiO$_2$ under visible light is examined for Pb(II) removal. The doping TiO$_2$ with N element was conducted by simple hydrothermal technique and using urea as the N source. The doped photocatalysts were characterized by DRUVS, XRD, FTIR and SEM-EDX instruments. Photocatalysis of Pb(II) through a batch experiment was performed for evaluation of the doped TiO$_2$ activity under visible light, with applying various fractions of N-doped, photocatalyst mass, irradiation time, and solution pH. The research results attributed that N doping has been successfully performed, which shifted TiO$_2$ absorption into visible region, allowing it to be active under visible irradiation. The photocatalytic removal of Pb(II) proceeded through photo-oxidation to form PbO$_2$. Doping N into TiO$_2$ noticeably enhanced the photo-catalytic oxidation of Pb(II) under visible light irradiation. The highest photocatalytic oxidation of 15 mg/L Pb(II) in 25 mL of the solution could be reached by employing TiO$_2$ doped with 10%w of N content 15 mg, 30 min of time and at pH 8. The doped-photocatalyst that was three times repeatedly used demonstrated significant activity. The most effective process of Pb(II) photo-oxidation under beneficial condition, producing less toxic and handleable PbO$_2$ and good repeatable photocatalyst, suggest a feasible method for Pb(II) remediation on an industrial scale.

Keywords: doping N; TiO$_2$; Pb(II); photocatalytic-oxidation; visible light

Citation: Wahyuni, E.T.; Rahmaniati, T.; Hafidzah, A.R.; Suherman, S.; Suratman, A. Photocatalysis over N-Doped TiO$_2$ Driven by Visible Light for Pb(II) Removal from Aqueous Media. *Catalysts* **2021**, *11*, 945. https://doi.org/10.3390/catal11080945

Academic Editor: Sophie Hermans

Received: 26 May 2021
Accepted: 27 July 2021
Published: 5 August 2021

Publisher's Note: MDPI stays neutral with regard to jurisdictional claims in published maps and institutional affiliations.

Copyright: © 2021 by the authors. Licensee MDPI, Basel, Switzerland. This article is an open access article distributed under the terms and conditions of the Creative Commons Attribution (CC BY) license (https://creativecommons.org/licenses/by/4.0/).

1. Introduction

Lead (Pb), along with Cd, Hg, Cr(VI) and As, is categorized as the most toxic heavy metal [1]. Lead in the form of Pb(II) ion can be present intensively in wastewater of many industries, such as storage batteries, mining, metal plating, painting, smelting, ammunition, oil refining and the ceramic glass [2–7]. Pb(II) ion is non-biodegradable and tends to accumulate in living tissues, causing various diseases and disorders, such as anemia encephalopathy (brain disfunction), cognitive impairment, kidney and liver damage, and toxicity to the reproductive system [2–6]. The permissible limit of Pb(II) in drinking water is 0.005 mg/L according to the current US Environmental Protection Agency (USEPA) standard, while WHO determines the limit is 0.01 mg/L [3]. In fact, the actual concentration of lead in wastewater is as high as several hundred milligram per liter. Therefore, the removal of lead from wastewater before the heavy metal ions contacted with unpolluted natural water bodies is important and urgent.

In recent years, adsorption has become one of the techniques that is frequently applied for Pb(II) removal, because it is a very effective technique in terms of initial cost, simplicity of design, ease of operation and insensitivity to toxic substances [2–6]. Several adsorbents that have been devoted for removal of Pb(II) include palm tree leaves [2], mesoporous activated carbon [3], modified natural zeolite [4], magnetic natural zeolite [5] and sulfonated polystyrene [6]. However, when the adsorbent is saturated with concentrated Pb(II) ions, it becomes solid waste with higher toxicity, which creates new environmental problems.

Currently, remediation of Pb(II) containing water by the photo-oxidation process through photo-Fenton method [7], under TiO_2 photocatalyst [8], and photo-assisted electrochemical [9] have also been developed. Furthermore, oxidation of Pb(II) by manganese salt has been studied [10]. Oxidation seems to be the most interesting method since Pb(II) oxidation resulted in the non-toxic precipitate PbO_2 and is more easily handled [7–10].

With respect to the photocatalysis process, TiO_2 as a photocatalyst has recently received considerable attention for the removal of persistent organic pollutants (POPs), including phenols [11], tetracycline [12], dyes [13] and linier alkyl benzene sulphonate [14] due to its cost-effective technology, non-toxicity, quick oxidation rate and chemical stability [11–26]. The practical application of TiO_2 is, however, significantly restricted by its low visible absorption due to its wide bandgap energy (E_g) $\approx$ 3.20 eV [15–26] and immediate charge (electron and hole) recombination [19–22]. With such wide a band gap, TiO_2 can only be excited by photons with wavelengths shorter than 385 nm emerging in UV region [15–26]. In fact, UV light only occupies a small portion (about 5%) of the sunlight spectrum [15,19–21,23,24], which limits TiO_2 application under low-cost sunlight or visible irradiation. Moreover, the fast charge recombination leads to a less effective photocatalysis process [19–22].

Therefore, an effort has made to overcome these deficiencies by doping TiO_2 structure with either metal elements [23–26] or non-metal elements [15–22]. Non-metal dopant has been reported to give be effective in narrowing the gap or decreasing the band gap energy (E_g) of TiO_2, compared to the metal dopant. This is because the size of non-metal elements are smaller than the metal dopants [15–22], allowing them to be inserted into the lattice of TiO_2 crystal facilely. Among the non-metal dopants that have been examined, N is the most interesting due to the very effective shift of the band gap energy into the lower value [16–21]. It has also been reported that N-doped TiO_2 materials exhibit strong absorption of visible light irradiation and significant enhancement of photocatalytic activity [16–24].

The doping N into TiO_2 has been performed by sol-gel [16,18,20,21], hydrothermal [21], coprecipitation [19], and plasma-assisted electrolysis [17] methods. Furthermore. according to the related research reports that have been reviewed by Gomez et al. [21], hydrothermal is believed to be the best method, since a large amount of N can be doped, which cannot occur by using other methods. In addition, in this method, the longer time for photocatalyst ageing is not required and the direct using TiO_2 powder is possible, suggesting a simple and low-cost process [23]. The hydrothermal method has been employed for doping transition metals [23], but to the best of our knowledge, it has not been explored for N doping.

Many studies of using N-doped TiO_2 under visible light for degradation of dyes [17] and para-nitrophenol [18], as well oxidation of NO_2 gas [19] and H_2 gas production [16], have been used. However a research of using N-doped TiO_2 for catalysis of the Pb(II) photo-oxidation and removal under visible irradiation is not traceable.

Under the circumstances, preparation of N-dopedTiO_2 by hydrothermal method is addressed and photocatalytic activity of N-doped TiO_2 is systematically evaluated for photo-oxidation of Pb(II) in the aqueous media driven by visible light. Moreover, the efficiency of the photocatalysis process strongly depends on the operating parameters, including the content of the dopant [16,17,19,20,24], reaction time [12,14,15,23,24], the photocatalyst mass [12,14,24] and the solution pH [8,12,14,24]. Therefore, it is necessary to find the optimal process parameters through laboratory treatability tests. This study is hoped to contribute to the development of doped TiO_2 photocatalysts and toxic metal remediation technology.

2. Results and Discussion

2.1. Characterization of TiO_2–N Photocatalyst

2.1.1. By Diffuse Reflectance UV/Vis. (DRUV/Vis.) Method

The DRUV/visible spectra of the photocatalyst samples were displayed as Figure 1. From the spectra, the wavelengths of the absorption edge could be determined that were displayed in Table 1. As expected, doping TiO_2 with N atom can shift the absorption into longer wavelength emerging visible region, due to the narrowing their gaps. The narrowing resulted from the insertion of N atom into the lattice of TiO_2 crystal. The evident narrowing gap was represented by decreasing band gap energy (Eg) values, as exhibited in Table 1. It is also notable that an increasing amount of introduced N caused Eg to decline more effectively. Some studies have also reported similar findings [15,19,20,24,27].

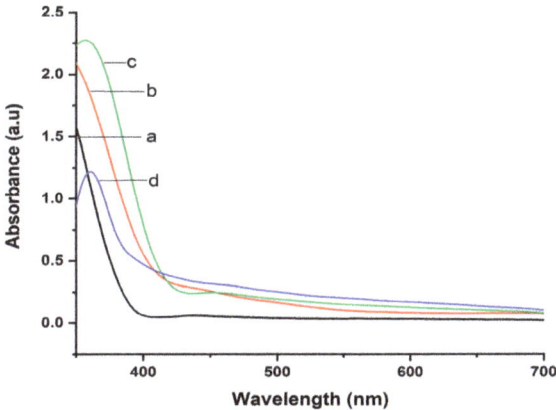

Figure 1. The DRUV Spectra of (**a**) TiO_2, (**b**) TiO_2–N (5) (**c**) TiO_2–N (10) and (**d**) TiO_2–N (15).

Table 1. The effect of N doped into TiO_2 on band gap energy.

Photocatalyst	Wavelength (nm)	Band Gap Energy (eV)
TiO_2	387.5	3.20
TiO_2–N(5)	405.2	3.06
TiO_2–N(10)	411.9	3.01
TiO_2–N(15)	418.9	2.96

The diminution of Eg implied that N atoms have successfully been doped in the TiO_2 crystal [19–21]. A doping generally can take place through substitutional and/or interstitial mechanisms and can be distinguished based on the lowering Eg values [27]. According to Ansari et al. [27], the decreasing Eg into 3.06 from 3.20 eV is due to the presence of substitutional mechanism, that is, replacing oxygen of TiO_2 by nitrogen dopant; meanwhile, lowering Eg toward less than 2.50 eV assigns to the interstitial mode, referring to the addition of nitrogen atoms into the TiO_2 crystal. However, some others [19–21] have been proposed that reduce Eg into 3.0–2.9 eV and can be a form of interstitial doping. Reducing Eg up to 3.06–2.96 eV as observed in this present study reveals that N doping involves a combination of substitutional and interstitial mechanisms [27].

2.1.2. By X-ray Diffraction (XRD) Method

The XRD patterns of TiO_2 and TiO_2–N photocatalysts are displayed in Figure 2. Several 2θ values of 25.25°, 37.52°, 48.02°, 53.58°, 54.88°, 62.61°, 68.78°, 70.33° and 75.07° are observed, which match with Miller index as (101), (004), (200), (105), (211), (204), (116), (204) and (215) of the lattice planes for anatase TiO_2 as listed in JCPDS card no. 21-1272 [16,17,20].

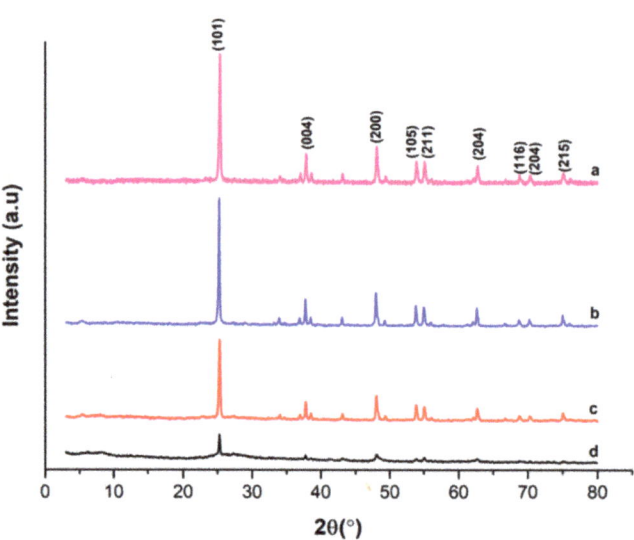

Figure 2. The XRD patterns of (**a**) TiO$_2$, (**b**) TiO$_2$–N (5), (**c**) TiO$_2$–N (10) and (**d**) TiO$_2$–N (15).

In the XRD patterns of all N-doped photocatalysts, additional phase, except anatase, is not observed, indicating that the N dopant did not agglomerate to reach the macro level [20]. This finding is in accordance with the result reported previously [16–20].

It is also observable that doping N caused a decrease in the XRD intensities, and the intensities gradually declined with the enhancement of N amount. The decrease of the intensities was partially due to TiO$_2$ crystallinity damage, due to the incorporation of N dopant in the TiO$_2$ lattice [16,21]. The increase of the partially damaged generated more amorphous phase, as demonstrated clearly by the XRD pattern of TiO$_2$–N (15) having highest N content. A similar finding was also delivered by Li et al. [16] and Mahy et al. [18] and was reviewed by Gomez et al. [21]. The slight crystallinity damage to TiO$_2$–N (5) was possible as a result of an interstitial doping [27]. Meanwhile, the more significant damaged as found in TiO$_2$–N (10) and TiO$_2$–N (15) may be caused by substitution of oxygen in the TiO$_2$ lattice by nitrogen dopant atom, or by a simultaneous interstitial and substitutional mechanisms [27]. This XRD data should agree with the trend of Eg decreasing.

2.1.3. By Fourier Transform Infra Red (FTIR) Method

In Figure 3, it is apparent that some characteristic bands of urea at the wavenumbers of 3448 cm^{-1} are associated with O-H stretching, 3346 cm^{-1} is related to N-H deformation, 1661 cm^{-1} is assigned to C=O bond, 1604 cm^{-1} originated from N-H (NH$_2$) and 1465 cm^{-1} is due to C-N vibration [28]. Furthermore, the spectra of all TiO$_2$–N samples are similar to those of un-doped TiO$_2$, where several characteristic peaks of TiO$_2$ are observed at around 3400, 1630 and 700–500 cm^{-1} of the wavenumbers. Many studies also reported the same IR spectra [13,15,16,18,21]. The peaks appearing at 3400 were attributed to the Ti–OH bond, a band at ~1630 cm^{-1} to the OH bending vibration of chemisorbed and/or physisorbed water molecule on the surface of TiO$_2$, and the strong band in the range of 700–500 cm^{-1} to stretching vibrations of Ti–O–Ti bond [13,15,16,18,21].

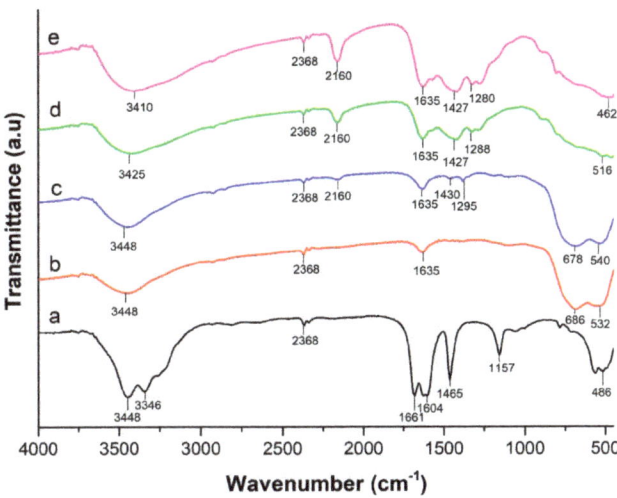

Figure 3. FTIR spectra of (**a**) urea, (**b**) TiO$_2$, (**c**) TiO$_2$–N (5), (**d**) TiO$_2$–N (10) and (**e**) TiO$_2$–N (15).

In comparison with the undoped TiO$_2$, the spectra of all TiO$_2$–N samples display additional peaks at around 2160, 1427–1430 and 1288–1295 cm^{-1}. The peaks look shaper for TiO$_2$–N with higher N content. The peaks at 1426–1430 and 1270–1290 cm^{-1} were attributed to the vibrations of the N-Ti bond [16,21]. The appearance of the Ni-Ti bond in the samples suggests that N atoms have been incorporated into the TiO$_2$ lattice. The peak at 2155–2160 cm^{-1} can be assigned for carbide (C$_2^{2-}$ or –C≡C–) species that possibly corresponded to the residual of the incompletely decomposed urea [16], during the hydrothermal at 150 °C and calcination at 400 °C. According to Li et al. [16], the complete removal of the carbide species and urea decomposition into N element can be obtained by calcination at the temperature of 450 °C. Lastly, there is also a weak peak at 2368 cm^{-1} appearing in the spectra of all samples that was neither from TiO$_2$ nor from urea. Hence, the peak possibly originated from the laboratory impurity contaminating KBr, which was used for pelleting the samples.

2.1.4. By Scanning Electron Microscope (SEM)

In order to investigate the surface morphology of the N-doped TiO$_2$ photocatalyst, their SEM images were taken, which are displayed in Figure 4. The particles of TiO$_2$ are seen clearly as crystalline forms. The less crystalline phase is demonstrated by the images of all N-doped TiO$_2$, and the amorphous form is observed for TiO$_2$–N (15) with highest amount of N dopant. The similar images due to the higher dopant content have also been reported by Khan et al. [19]. The less crystallin phase may infer the incorporation of N into the TiO$_2$ structure [16,21]. This phenomena matched well with the XRD data.

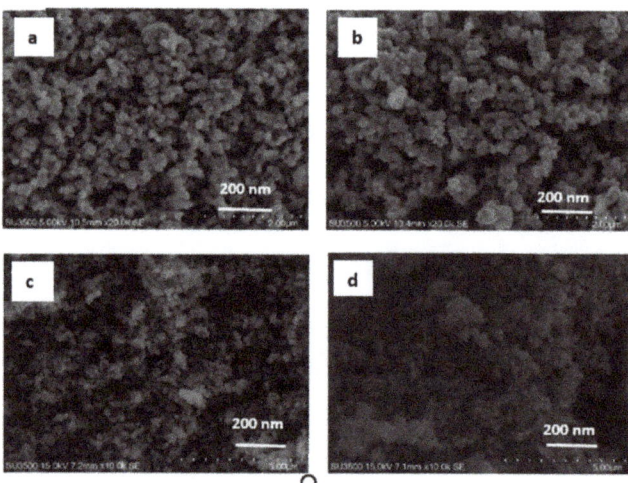

Figure 4. The SEM images of (**a**) TiO$_2$, (**b**) TiO$_2$–N (5), (**c**) TiO$_2$–N (10) and (**d**) TiO$_2$–N (15).

2.2. Photocatalytic Activity TiO$_2$–N under Visible Light in the Removal of Pb(II)

2.2.1. Influence of N Doping

The activity of the doped photocatalyst was evaluated by applying it to the photocatalysis of Pb(II) under visible light process, as well as under dark and UV irradiated conditions for comparison. The results of the Pb(II) removal are illustrated in Figure 5.

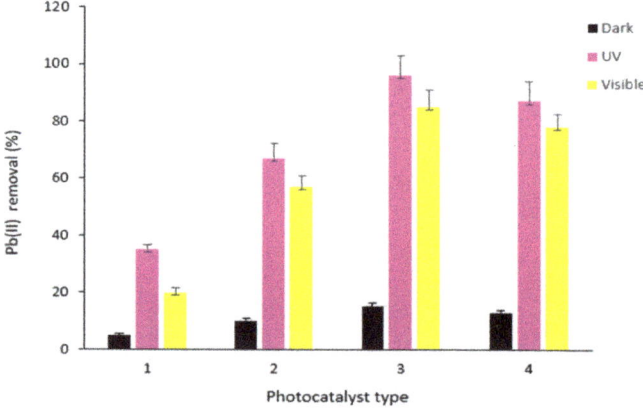

Figure 5. The effectiveness of the Pb(II) photo-oxidation over (**1**) TiO$_2$, (**2**) TiO$_2$–N (5), (**3**) TiO$_2$–N (10) and (**4**) TiO$_2$–N (15); under dark condition; and UV and visible light irradiation (photocatalyst weigh = 15 mg, volume of Pb(II) solution = 25 mL, Pb(II) concentration = 15 mg/L, reaction time = 30 min and solution pH = 7).

It can be seen in the figure that N-doped TiO$_2$ posed higher activity in the Pb(II) photocatalytic removal both under visible and UV lights compared to the undoped TiO$_2$ activity. The same trend has also been acquired by many authors [16–20,23,24]. The figure also shows that the Pb(II) ions could be removed under dark conditions, due to the adsorption of Pb^{2+} on the TiO$_2$ surface. It is inferred that the photocatalytic removal is initiated and/or accompanied by the adsorption step [18]. The enhancement of the visible photocatalytic-oxidation was promoted by lowering band gap energy (Eg), that allowed TiO$_2$ to be activated by visible light generating a lot of OH radicals for oxidation [16–20,23,24]. The reactions of the formation of OH radicals and Pb(II) photo-oxidation were presented as Equation (1), Equation (2) [11–14] and Equation (3) [7–9]. The effective photocatalytic-oxidation of Pb(II) under visible light provides a potential and promising method to be applied on a larger scale for industrial wastewater treatment.

$$TiO_2\text{-}N + hv \rightarrow TiO_2\text{-}N + h^+ + e^- \tag{1}$$

$$H_2O + h^+ \rightarrow OH + H^+ \tag{2}$$

$$Pb^{2+} + 2 \cdot OH \rightarrow PbO_2 + 2H^+ \tag{3}$$

$$TiO_2\text{-}N + h^+ + e^- \rightarrow TiO_2\text{-}N + heat \tag{4}$$

Doping N could also considerably improve the photocatalyst activity under UV irradiation, which can be promoted by slower recombination of e^- and h^+ (electron and hole) pair, that is, generated during light exposure, as presented in Equation (4). It should be noted that the recombination proceeds naturally and can cause photo-oxidation or reduction reactions. Therefore, inhibition of the recombination needs to be afforded. The recombination can be retarded by doping, since the N dopant can act as electron-hole separation center [17,19,21] by capturing the electrons, which can prevent the recombination. Hence, doping N atom essentially narrows the band gap of TiO_2 for the photo-excitation or red shift and simultaneously delays the recombination rate of photogenerated electron–hole pair. The synergic role gives rise to the high effective photo-oxidation.

Moreover, the photocatalytic oxidation of Pb(II) under UV light is seen to be more effective than that of under visible light. The Eg of TiO_2 is 3.2 eV that is equal to UV light, enabling TiO_2 to be activated by UV irradiation, and hence more OH radicals could be provided. In contrast, the energy of visible light t is lower than the Eg of TiO_2, consequently TiO_2 was less active in generation of OH radicals [19], which resulted in the lower photo-oxidation.

Figure 5 also illustrates that higher amount of N-doped raised the photo-oxidation efficiency and reached maximum at 10% of N. The improvement was promoted by increasing their activity under visible light due to the lower Eg. Additionally, more N dopant content may improve the retardation of the recombination, further enhancing the Pb(II) oxidation. On the contrary, at the N content beyond its optimum value, the Pb(II) oxidation seems to be detrimental. The excess of N dopant (15% N) considerably decreased the crystallinity of TiO_2, which extended the amorphous phase, as described above. The greater amorphous portion in TiO_2 structure could adsorb more water molecules that prevent the formation of OH radicals [19]. It is also possible that the residue of urea in TiO_2–N (15) covered the active sites on the TiO_2 surface, thereby inhibiting the OH radical formation. These conditions explain the decrease in the photo-oxidation. Same finding was also found by some studies [16–20]. Based on their Eg values, 15% of N-doped photocatalysts posed lowest Eg, suggesting the highest visible light absorption, and this showed the highest photo-oxidation effectiveness. In fact, the highest photo-oxidation result was shown by TiO_2–N with 10% of N content. It is therefore obvious that both Eg value and the content of N dopant played role in the photocatalysis process. In this case, the N content exhibited higher role than the Eg value did.

2.2.2. Influence of Irradiation Time

As seen in Figure 6, prolonged irradiation time up to 30 min could considerably improve the photo-oxidation, but upon further expansion of the irradiation time longer than 30 min, a plateau of the photo-oxidation is observed. The similar trend data was also noted previously [12,14,15,23,25].

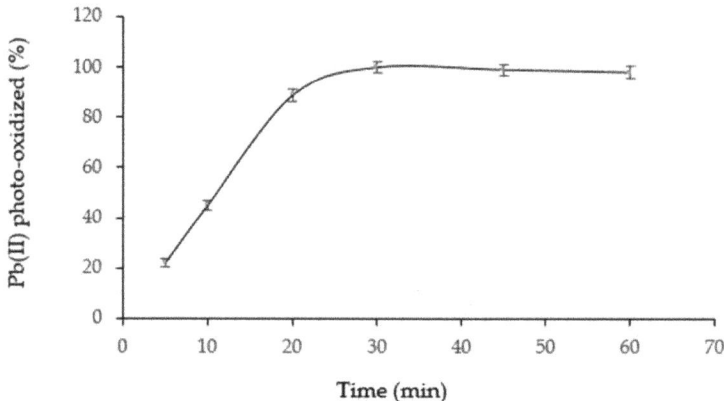

Figure 6. The influence of the irradiation time on the Pb(II) photo-oxidation (photocatalyst weight = 15 mg, volume of Pb(II) solution = 25 mL, Pb(II) concentration = 15 mg/L, and solution pH = 7).

With the extended time, the contact between the light and TiO_2 proceeded to be more effective, producing more OH radicals. In addition, with the prolong time, a greater collision frequency occurred between the hydroxyl radicals and ion Pb(II) in the solution. This conductive condition, therefore, promoted more effective photooxidation of Pb(II). The longer process than 30 min resulted in a greater amount of the PbO_2 deposited on the surface of TiO_2–N. Covering the TiO_2 surface by PbO_2 can limit the visible light tht reaches the active surface of TiO_2. As a consequence, the production of the OH radicals could not be enhanced [23], so that the OH radicals available remained same, leading to constant photo-oxidation. Moreover, the short optimum time for the most effective Pb(II) removal is beneficial in terms of application on an industrial scale.

2.2.3. Influence of Photocatalyst Weight on the Photo-Oxidation of Pb(II)

The photo-oxidation of Pb(II) sharply increases with the elevation weight of the photocatalyst, as demonstrated by Figure 7. The increase of the photocatalyst weight provided more OH radicals that could enhance photodegradation.

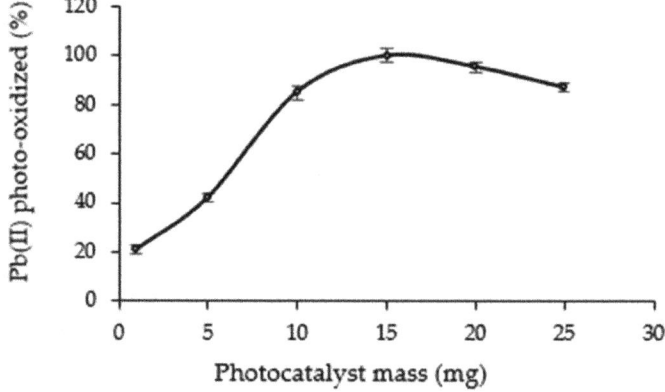

Figure 7. The influence of the photocatalyst mass on the Pb(II) photo-oxidation (volume of Pb(II) solution = 25 mL, Pb(II) concentration = 15 mg/L, reaction time = 30 min and solution pH = 7).

This trend data is in perfect accordance with the data resulted by some studies [8,12,14,23]. The larger weight that exceeded the optimum level caused a detrimental in photodegradation. The excessive photocatalyst elevated the turbidity of the mixture that inhibited the light penetration [8,14,23]. The other possible reason is that the large photocatalyst dose allowed for agglomeration that depleted the active photocatalyst surface; thus, less photo-oxidation efficiency was obtained [12]. The mass producing maximum photo-oxidation is found as 15 mg for 25 mL of the Pb(II) solution, which is equal to 0.6 g/L of the photocatalyst dose. Such a dose is believed to be cost-effective on a larger scale.

2.2.4. Influence of Solution pH

The initial pH is one of the most effective parameters in photocatalytic processes, which influences on the adsorption of the substrate on the photocatalyst surface and hence the degradation. The success of the adsorption is dictated by the suitability of the substrate and the photocatalyst surface charges [12]. It is notable in Figure 8, raising solution pH up to 8 was found to noticeably enhance the photo-oxidation. The further elevation of the pH decreased the photo-oxidation.

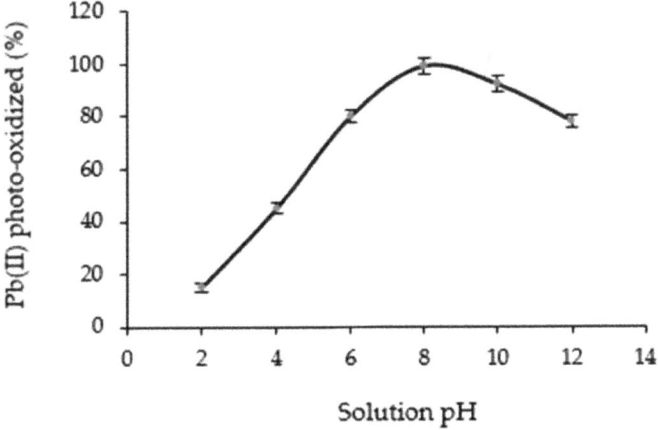

Figure 8. The influence of solution pH on the Pb(II) photo-oxidation (photocatalyst weigh = 15 mg, volume of Pb(II) solution = 25 mL, Pb(II) concentration = 15 mg/L and reaction time = 30 min).

In the solution with low pH, the surface of TiO_2 was protonated to form positive charge (TiO_2H^+), inhibiting the OH radicals formation [8,12,14,23]. On the other side, the cationic of Pb^{2+} was formed predominantly in the solution [7,8], which repulsed the adsorption of Pb^{2+} on the surface of the protonated N-doped TiO_2. These simultaneous conditions led to less effective photo-oxidation. When the pH was gradually increased up to 8, the protonation of TiO_2 gradually depleted, which generated a greater amount of OH radicals. In such a pH range, a large number of Pb^{2+} were available [7,8], enabling it to interact with TiO_2 surface effectively. This condition was very conducive to reaching high photo-oxidation. At pH higher than 8 (base condition), the surface of TiO_2 was found as anionic TiO^- species that were inhibited to release OH radicals [8,12,14,23]. In the base solution, $Pb(OH)_2$ precipitate was formed [7,8], which constrained the light penetration, consequently limiting the photocatalyst to produce the OH radicals. The lower number of OH radicals available and the presence of the precipitate explained the significant decline of the photo-oxidation. From the data, it is found that the most effective process took place at neutral pH and has the potential to be applied on an industrial scale.

2.2.5. Detection of PbO$_2$ Produced from the Photo-Oxidation

In order to detect the result of Pb(II) oxidation, the SEM-EDX analysis of TiO$_2$–N (10) before and after being used for Pb(II) photo-oxidation was executed, and the SEM images and the EDX spectra were displayed as Figures 9 and 10, respectively. At the SEM image of TiO$_2$–N (10) after being used for photo-oxidation, the small particles (arrow signs) over the surface of TiO$_2$ are observable, which were not seen in the image of TiO$_2$–N (10) before it was used. The particles were very likely of PbO$_2$ resulting from the photo-oxidation of Pb(II).

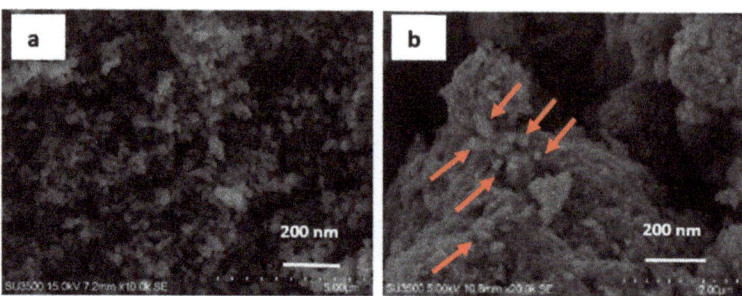

Figure 9. The SEM images of (**a**) TiO$_2$–N (10) before and (**b**) TiO$_2$–N (10) after being used for Pb(II) photo-oxidation.

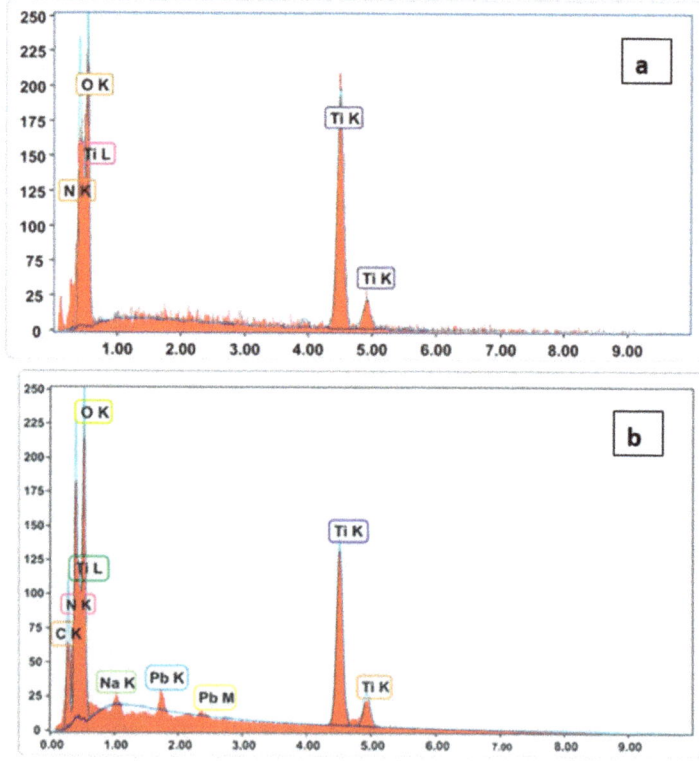

Figure 10. The EDX spectra of (**a**) TiO$_2$–N (10) before and (**b**) TiO$_2$–N (10) after used for Pb(II) photo-oxidation.

The presence of PbO$_2$ was strongly supported by EDX spectra of TiO$_2$–N (10) after being used for oxidation, which demonstrates the appearance of Pb peaks. Such peaks were invisible in the spectra of TiO$_2$–N (10) before used for oxidation. Other supporting data is ascribed by increasing intensity of oxygen peak compared to the intensity of oxygen in the spectra of TiO$_2$–N (10) before used. The increase of the intensity assigned the increase of the oxygen amount that could be contributed by PbO$_2$ material. Furthermore, the intensities of Ti, N and C peaks found in TiO$_2$–N (10) after used were notably reduced, due to the incorporation of Pb and oxygen atoms into the photocatalyst.

The quantitative compositions of TiO$_2$–N (10) before and after being used for Pb(II) photo-oxidation that was derived from the EDX spectra by data processor in the machine are displayed in Table 2. Generally, the compositions were consistent with EDX spectra. The appearance of carbon element in both photocatalysts could be due to laboratory impurity and the residue of the incomplete decomposed urea, which are also observed in the respective FTIR spectra. The considerable quantity (5.20%) of Pb formed implied the effective photo-oxidation of TiO$_2$–N (10) under visible irradiation. Hence, there is obvious evidence that the photo-oxidation of Pb(II) yielded of PbO$_2$, by following the reaction in Equation (4). The formation of PbO$_2$ from Pb(II) photo-oxidation well agrees with the finding reported previously [8,9]. The formation of the solid PbO$_2$ is beneficial in terms of solid waste treatment due to the less toxic and handleable waste.

Table 2. The composition the doped photocatalyst based on the EDX data.

Element	The Content of the Element in TiO$_2$–N(10) (% Mole)	
	Before Photo-Oxidation	After Photo-Oxidation
Ti	27.10	21.40
O	47.40	52.30
N	15.30	11.40
C	10.20	7.60
Pb	-	5.20
Na	-	2.10

In addition to Pb, the new peak associated with Na element is also observable. The presence of Na could be originated from NaOH solution that was added to elevate pH into 8, from pH 4 of the solution during photo-oxidation. In the solution, NaOH was ionized into Na$^+$ and OH$^-$ and the positive ion of Na$^+$ adsorbed on the surface of TiO$_2$ photocatalyst having more electrons or negative charged surface. The Na was kept to attach on the surface of TiO$_2$–N (10) that was detectable by EDX analysis.

2.2.6. The Activity of the Doped-Photocatalyst with the Repetition Used

In order to know the activity of the N-doped TiO$_2$ photocatalyst after repetition use, in the present study the photo-oxidation of Pb(II) under optimal condition over TiO$_2$–N (10), after being used in several times, was observed. The results are presented as Figure 11 and Table 3, exhibiting a gradual decrease of photoactivity with more repetitive use.

The first use of TiO$_2$–N (10) photocatalyst resulted in 24.5 mg/g of the Pb(II) photo-oxidation, which was about 98% toward the initial Pb(II) concentration, referring to the high photoactivity of TiO$_2$–N (10). In the second use of the photocatalyst, an insignificant decrease of the Pb(II) photo-oxidation was obtained—22.63 mg/g, or about 91%. Such photo-oxidation result implied that the activity of TiO$_2$–N (10) photocatalyst remained high after being used once. Additionally, the very low decrease of the photo-oxidation could be effected by of 24.5 mg/g of the Pb occupation on the photocatalyst surface, which still could provide appreciable active surface of the photocatalyst. The appreciable surface active allowed the photocatalyst to have effective contact with visible light, to further produce suffice of OH radicals for the photo-oxidation.

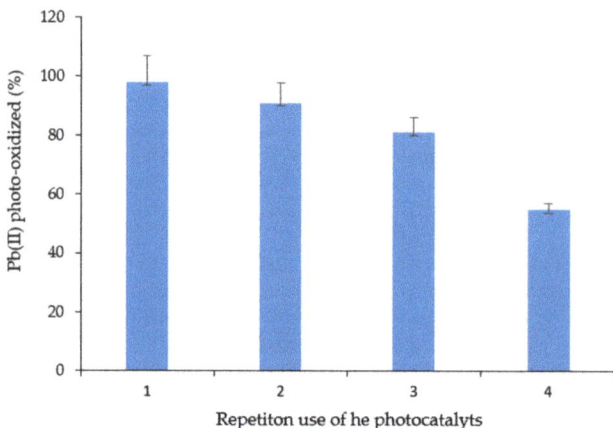

Figure 11. The activity of TiO$_2$–N (10) on the Pb(II) photo-oxidation with several repetitions.

Table 3. The activity of repeatedly used TiO$_2$–N (10) photocatalyst.

Repetition of TiO$_2$–N(10) Use	Pb Resulted from the Photo-Oxidation over TiO$_2$–N(10) (mg/g)	Pb(II) Photo-Oxidized (%)	Total Pb Resulted from the Photo-Oxidation Distributed over TiO$_2$–N(10) Surface (mg/g)
1st	24.53	98.12	24.53
2nd	22.63	90.50	47.16
3th	20.33	81.30	67.49
4th	13.76	55.02	81.25

The third repetitive use of the photocatalyst in the Pb(II) photo-oxidation yielded 20.30 mg/g or about 81 %, which was less than that that resulted from the second-use photocatalyst. The result of the photo-oxidation (81%) suggested the decent activity of the third-use photocatalyst. The slight depletion of the photo-oxidation could have been generated by the photocatalyst that was covered by 47.13 mg/g of the total Pb. This covering surface of the photocatalyst was believed to keep adequate surface area of the photocatalyst, which could effectively keep contact with the visible light to produce adequate number of OH for photo-oxidation.

When the photo- oxidation of Pb(II) process was conducted by applying the third-use photocatalyst through fourth process, the photo-oxidation drastically declined into 13.76 mg/g or around 55%. This low photo-oxidation result showed that the photocatalyst suffered from the activity. This significant decrease of the photo-activity was induced by smaller surface area of TiO$_2$–N (10) because of the covering by large Pb about 67.43 mg/g. Such noticeable covering obviously prevented the interaction between the photocatalyst with the visible light, which resulted in lack of OH radicals. From the data, it is clear that the repeatedly used photocatalyst up to three times showed significant activity in the Pb(II) photo-oxidation. Hence, the photocatalyst is believed to have promise in terms of application on an industrial scale.

3. Materials and Methods

3.1. Materials

The chemicals used in this research were TiO_2, urea, $Pb(NO_3)_2$, HCl and NaOH, which were purchased from E. Merck with analytical grade and were used without any purification.

3.2. Methods

3.2.1. Doping Process of N on TiO_2

Doping was performed by hydrothermal method reported previously [23] with small modification. TiO_2 powder (about 1 g) was dispersed in 200 mL of 1 g/L urea solution in water solvent. The mixture was placed in autoclave and then was heated at 150 °C for 24 h. The doped TiO_2 -N that resulted was dried at 100 °C for 30 min and continued by calcination at 400 °C for 2 h. The sample was kept for characterization and activity evaluation. With such an amount of urea, theoretically, N content in TiO_2 was about 5 % w. The same procedure was repeated for urea with 2 g/L and 3 g /L of the concentrations, giving approximately 10% w and 15% w of N content in the doped TiO_2 respectively. Therefore, the doped photocatalyst samples were coded as TiO_2–N (5), TiO_2–N (10) and TiO_2–N (15).

3.2.2. Characterization of N-Doped TiO_2

The doped photocatalysts obtained were characterized by using Pharmaspec UV-1700 Diffuse reflection ultra violet (DRUV) spectrophotometer to determine their band gap energies (Eg). The DRUV spectra were recorded in the wavelength (λ) range of 200–700 nm. The values of Eg were calculated based on the wavelengths of absorption edge by following relationship of Eg = 1240/λ [20]. The wavelengths of absorption edge were estimated based on the intersection of the straight lines of Y-axis with X-axis [20].

A Shimadzu 6000X-XRD machine with radiation source of Cu Kα (1,54056 Å) was operated at 30 mA of the current and 40 kV of the voltage to detect the crystallinity of the samples. The XRD patterns of the samples, having about 200 μm of the particle size, were scanned in the range of the 2 tetha of 5–80° with scanning rate was 5°/min. Fourier transform infrared (FT-IR) spectra with the wavenumber of 4000–400 cm^{-1} of the KBr pelleted samples were recorded on Prestige 21 Fourier-Transform Infrared spectrophotometer. From the FTIR spectra, the characteristic chemical bond vibrations could be found. In order to get the surface morphology of the samples, the SEM-EDX images of the samples were taken by Hitachi SU 3500 Scanning Electron Microscopy and Energy Dispersive X-ray (SEM-EDX) equipped with Coating Hitachi MC1000 ION SPUTTER 15 mA and 20 s. In this analysis, the samples were initially metalized by gold coating. All the instruments used are available at Gadjah Mada University, Yogyakarta, Indonesia.

3.2.3. Photo-Oxidation of Pb(II) in the Solution over TiO_2–N Photocatalyst

The photo-oxidation of Pb(II) was conducted by batch technique in the apparatus seen in Figure 12. The Pb(II) solution 15 mg//L 25 mL in a beaker glass was mixed with 15 mg of TiO_2–N (5), and the glass was put in the photocatalysis apparatus. Next, the beaker glass in the apparatus was irradiated with three wolfram lamps Philip 20 watt as source of visible light, accompanied by magnetically stirring with 200 rpm of the stirring rate for 30 min. The Pb(II) left in the solution was analyzed by Perkin-Elmer 110 AAS machine. The concentration of the Pb(II) was determined by extrapolation on the corresponding standard curve. The same procedure was copied for processes under dark condition, under UV irradiation emitted from 20 watt black light blue (BLB)-type UV lamps, for process with TiO_2–N containing N of 10% and 15%, and for processes with various irradiation times (5, 10, 20, 30, 45 and 60 min), photocatalyst weights (1, 5, 10, 15, 20 and 25 mg) and solution pH values (4, 6, 8, 10 and 12). When one parameter was varied, other parameters were kept to be constant. In addition, the repetitive use of the doped photocatalyst in the photo-oxidation was also preceded by following the same procedure. Each experiment was repeated three times, and the deviation of the photo-oxidation of Pb(II) results were found to be 5–10%.

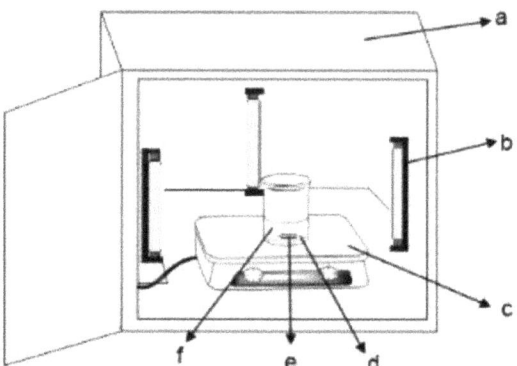

Figure 12. A set of apparatus used for Pb(II) photo-oxidation processes composed of: (**a**) melamine box, (**b**) visible or UV lamps, (**c**) magnetic stirrer plate, (**d**) photocatalyst powder, (**e**) magnetic stirrer bar and (**f**) sample solution.

4. Conclusions

The N-doped TiO_2 photocatalyst was successfully prepared using the hydrothermal method, which narrowed the band gap energy, allowing it to absorb visible light. It was found that doping N on TiO_2 structure could improve its activity in the Pb(II) photo-oxidation under visible light irradiation. The most effective Pb(II) photo-oxidation (98%) from Pb(II) 15 mg/L in 25 mL solution was reached by using TiO_2–N with 10% of N fraction, by applying the condition of 15 mg of the doped photocatalyst, 30 min and pH 8. It is also evident that PbO_2 is produced from the Pb(II) photo-oxidation. The doped TiO_2 photocatalyst with three repetitions demonstrated sufficient activity in the photo-oxidation. The low photocatalyst dose, short reaction time and neutral pH used to achieve the highest effective process are beneficial factors allowing the method to be applied for the treatment of industrial-Pb(II)-containing wastewater.

Author Contributions: Conceptualization, E.T.W.; methodology, A.S.; research experiment and data execution, T.R. and A.R.H.; data analysis, S.S.; writing—original draft preparation, T.R., A.R.H. and S.S.; writing—review and editing, E.T.W.; All authors have read and agreed to the published version of the manuscript.

Funding: This research was funded by Faculty of Mathematic and Natural Sciences Gadjah Mada University through a grant of Public Financial with the contract number 77/J01. 1.28/PL.06.02/2020.

Acknowledgments: Authors greatly thank to Faculty of Mathematic and Natural Sciences Gadjah Mada University.

Conflicts of Interest: The authors declare no conflict of interest. The funders had no role in the design of the study; in the collection, analyses or interpretation of data; in the writing of the manuscript; or in the decision to publish the results.

References

1. Pandey, S.; Fosso-Kankeu, E.; Spiro, M.; Waanders, F.; Kumar, N.; Ray, S.; Kim, J.; Kang, M. Equilibrium, kinetic, and thermodynamic studies of lead ion adsorption from mine wastewater onto MoS_2-clinoptilolite composite. *Mater. Today Chem.* **2020**, *18*, 100376. [CrossRef]
2. Soliman, A.M.; Elwy, H.M.; Thiemann, T.; Majedi, Y.; Labata, F.T.; Al-Rawashdeh, N.A. Removal of Pb(II) ions from aqueous solutions by sulphuric acid-treated palm tree leaves. *J. Taiwan Inst. Chem. Eng.* **2016**, *58*, 264–273. [CrossRef]

3. Asuquo, E.; Martin, A.; Nzerem, P.; Siperstein, F.; Fan, X. Adsorption of Cd(II) and Pb(II) ions from aqueous solutions using mesoporous activated carbon adsorbent: Equilibrium, kinetics and characterisation studies. *J. Environ. Chem. Eng.* **2017**, *5*, 679–698. [CrossRef]
4. Yousefi, T.; Mohsen, M.A.; Mahmudian, H.R.; Torab-Mostaedi, M.; Moosavian, M.A.; Hassan Aghaya, H. Removal of Pb(II) by modified natural adsorbent; thermodynamics and kinetics studies. *J. Water Environ. Nanotechnol.* **2018**, *3*, 265–272.
5. Pambudi, T.; Wahyuni, E.T.; Mudasir, M. Recoverable adsorbent of natural zeolite/Fe_3O_4 for removal of Pb(II) in water. *J. Mater. Environ. Sci.* **2020**, *11*, 69–78.
6. Mahmoud, M.E.; Abdou, A.E.H.; Ahmed, S.B. Conversion of waste styrofoam into engineered adsorbents for efficient removal of cadmium, lead and mercury from water. *ACS Sustain. Chem. Eng.* **2016**, *4*, 819–827. [CrossRef]
7. Wahyuni, E.T.; Siswanta, D.; Kunarti, E.S.; Supraba, D.; Budiraharjo, S. Removal of Pb(II) ions in the aqueous solution by photo-Fenton method. *Glob. Nest J.* **2019**, *21*, 180–186.
8. Wahyuni, E.T.; Aprilita, N.H.; Hatimah, H.; Wulandari, A.; Mudasir, M. Removal of Toxic Metal Ions in Water by Photocatalytic Method. *Am. Chem. Sci. J.* **2015**, *5*, 194–201. [CrossRef]
9. Ebraheim, G.; Karbassi, A.R.; Mehrdadi, N. Employing speciation of metals to assess photo-assisted electrochemical efficiency for improving rainwater quality in Tehran, Iran. *Int. J. Environ. Sci. Technol.* **2021**, 1–20. [CrossRef]
10. Pan, W.; Pan, C.; Bae, Y.; Giammar, D.E. Role of Manganese in Accelerating the Oxidation of Pb(II) Carbonate Solids to Pb(IV) Oxide at Drinking Water Conditions. *Environ. Sci. Technol.* **2019**, *53*, 6699–6707. [CrossRef] [PubMed]
11. Mills, A.; O'Rourke, C.; Moore, K. Powder semiconductor photocatalysis in aqueous solution: An overview of kinetics-based reaction mechanisms. *J. Photochem. Photobiol. A Chem.* **2015**, *310*, 66–105. [CrossRef]
12. Safari, G.; Hoseini, M.; Seyedsalehi, M.; Kamani, H.; Jaafari, J.; Mahvi, A.H. Photocatalytic degradation of tetracycline using nanosized titanium dioxide in aqueous solution. *Int. J. Environ. Sci. Technol.* **2014**, *12*, 603–616. [CrossRef]
13. Gautam, A.; Kshirsagar, A.; Biswas, R.; Banerjee, S.; Khanna, P.K. Photodegradation of organic dyes based on anatase and rutile TiO_2 nanoparticles. *RSC Adv.* **2016**, *6*, 2746–2759. [CrossRef]
14. Jariyanorasade, A.; Junyapoon, S. Factors affecting the degradation of linear alkylbenzene sulfonate by TiO_2 assisted photocatalysis and its kinetics. *Environ. Asia* **2018**, *11*, 45–60. [CrossRef]
15. Lin, Y.-H.; Hsueh, H.-T.; Chang, C.-W.; Chu, H. The visible light-driven photodegradation of dimethyl sulfide on S-doped TiO_2: Characterization, kinetics, and reaction pathways. *Appl. Catal. B Environ.* **2016**, *199*, 1–10. [CrossRef]
16. Li, H.; Hao, Y.; Lu, H.; Liang, L.; Wang, Y.; Qiu, J.; Shi, X.; Wang, Y.; Yao, J. A systematic study on visible-light N-doped TiO_2 photocatalyst obtained from ethylenediamine by sol-gel method. *Appl. Surf. Sci.* **2015**, *344*, 112–118. [CrossRef]
17. Kim, T.H.; Go, G.-M.; Cho, H.-B.; Song, Y.; Lee, C.-G.; Choa, Y.-H. A Novel Synthetic Method for N Doped TiO_2 Nanoparticles Through Plasma-Assisted Electrolysis and Photocatalytic Activity in the Visible Region. *Front. Chem.* **2018**, *6*, 458. [CrossRef] [PubMed]
18. Mahy, J.G.; Cerfontaine, V.; Poelman, D.; Devred, F.; Gaigneaux, E.M.; Heinrichs, B.; Lambert, S.D. Highly Efficient Low-Temperature N-Doped TiO_2 Catalysts for Visible Light Photocatalytic Applications. *Materials* **2018**, *11*, 584. [CrossRef]
19. Khan, T.T.; Bari, G.A.K.M.R.; Kang, H.-J.; Lee, T.-G.; Park, J.-W.; Hwang, H.J.; Hossain, S.M.; Mun, J.S.; Suzuki, N.; Fujishima, A.; et al. Synthesis of N-Doped TiO_2 for Efficient Photocatalytic Degradation of Atmospheric NO_x. *Catalysts* **2021**, *11*, 109. [CrossRef]
20. Xu, T.; Wang, M.; Tong, W. Effects of N doping on the microstructures and optical properties of TiO_2. *J. Wuhan Univ. Technol.-Mater. Sci. Ed.* **2019**, *34*, 55–63. [CrossRef]
21. Gomes, J.; Lincho, J.; Domingues, E.; Quinta-Ferreira, R.M.; Martins, R.C. N–TiO_2 photocatalysts: A Review of Their characteristics and capacity for emerging contaminants removal. *Water* **2019**, *11*, 373. [CrossRef]
22. Hua, L.; Yin, Z.; Cao, S. Recent Advances in Synthesis and Applications of Carbon-Doped TiO_2 Nanomaterials. *Catalysts* **2020**, *10*, 1431. [CrossRef]
23. Razali, M.H.; Ahmad-Fauzi, M.N.; Mohamed, A.R.; Sreekantan, S. Morphological, Structural and Optical Properties Study of Transition Metal Ions Doped TiO_2 Nanotubes Prepared by Hydrothermal Method. *Int. J. Mater. Mech. Manuf.* **2013**, *1*, 314–318. [CrossRef]
24. Wahyuni, E.; Istiningsi, I.; Suratman, A. Use of Visible Light for Photo Degradation of Linear Alkyl-benzene Sulfonate in Laundry Wastewater over Ag-doped TiO_2. *J. Environ. Sci. Technol.* **2020**, *13*, 124–130. [CrossRef]
25. Ghorbanpour, M.; Feizi, A. Iron-doped TiO_2 catalysts with photocatalytic activity. *J. Water Environ. Nanotechnol.* **2019**, *4*, 60–66. [CrossRef]
26. Pedroza-Herrera, G.; Medina-Ramírez, I.E.; Lozano-Álvarez, J.A.; Rodil, S.E. Evaluation of the photocatalytic cctivity of copper doped TiO_2 nanoparticles for the purification and/or disinfection of industrial effluents. *Catal. Today* **2020**, *341*, 37–48. [CrossRef]
27. Ansari, S.A.; Khan, M.M.; Ansari, M.O.; Cho, M.H. Nitrogen-doped titanium dioxide (N-doped TiO_2) for visible light photocatalysis. *New J. Chem.* **2016**, *40*, 3000–3009. [CrossRef]
28. Manivannan, M.; Rajendran, S. Investigation of inhibitive action of urea—Zn^{2+} system in the corrosion control of carbon steel in sea water. *Int. J. Eng. Sci. Tech.* **2011**, *3*, 8048–8060.

Article

Efficient N, Fe Co-Doped TiO₂ Active under Cost-Effective Visible LED Light: From Powders to Films

Sigrid Douven [1,*], Julien G. Mahy [1,2], Cédric Wolfs [1], Charles Reyserhove [1], Dirk Poelman [3], François Devred [2], Eric M. Gaigneaux [2] and Stéphanie D. Lambert [1]

1. Department of Chemical Engineering—Nanomaterials, Catalysis & Electrochemistry, University of Liège, B6a, Quartier Agora, Allée du six Août 11, 4000 Liège, Belgium; julien.mahy@uclouvain.be (J.G.M.); cedric.wolfs@uliege.be (C.W.); charles.reyserhove@gmail.com (C.R.); stephanie.lambert@uliege.be (S.D.L.)
2. Institute of Condensed Matter and Nanosciences—Molecular Chemistry, Materials and Catalysis (IMCN/MOST), Université catholique de Louvain, Place Louis Pasteur 1, Box L4.01.09, 1348 Louvain-La-Neuve, Belgium; francois.devred@uclouvain.be (F.D.); eric.gaigneaux@uclouvain.be (E.M.G.)
3. LumiLab, Department of Solid State Sciences, Ghent University, 9000 Gent, Belgium; dirk.poelman@ugent.be
* Correspondence: S.Douven@uliege.be; Tel.: +32-4-366-3563

Received: 6 April 2020; Accepted: 8 May 2020; Published: 14 May 2020

Abstract: An eco-friendly photocatalytic coating, active under a cost-effective near-visible LED system, was synthesized without any calcination step for the removal of organic pollutants. Three types of doping (Fe, N and Fe + N), with different dopant/Ti molar ratios, were investigated and compared with undoped TiO₂ and the commercial P25 photocatalyst. Nano-crystalline anatase-brookite particles were successfully produced with the aqueous sol-gel process, also at a larger scale. All samples displayed a higher visible absorption and specific surface area than P25. Photoactivity of the catalyst powders was evaluated through the degradation of *p*-nitrophenol in water under visible light (>400 nm). As intended, all samples were more performant than P25. The N-doping, the Fe-doping and their combination promoted the activity under visible light. Films, coated on three different substrates, were then compared. Finally, the photoactivity of a film, produced from the optimal N-Fe co-doped colloid, was evaluated on the degradation of (i) *p*-nitrophenol under UV-A light (365 nm) and (ii) rhodamine B under LED visible light (395 nm), and compared to undoped TiO₂ film. The higher enhancement is obtained under the longer wavelength (395 nm). The possibility of producing photocatalytic films without any calcination step and active under low-energy LED light constitutes a step forward for an industrial development.

Keywords: photocatalysis; Fe/N doping; titania; aqueous sol-gel process; LED visible light

1. Introduction

The last centuries have seen a steady increase in human activities, causing a remarkable technological development and soaring human populations. However, the industrial expansion has brought atmospheric, ground and water pollution, all harmful for humans and the environment [1]. Indeed, major pollution can cause human diseases like breathing problems, cardiovascular problems, cancers, neurobehavioral disorders, etc. It can also affect global warming, which worsens climate change, increases sea level rises and causes serious damage on e.g., animals and flora [1]. Major anthropogenic pollutants are aromatic compounds, pesticides, chlorinated compounds, SO_x, NO_x, heavy metals or petroleum hydrocarbons [1]. In order to decrease this emitted pollution, various chemical, physical and biological treatments exist [2,3]. Some molecules (like micropollutants and pharmaceuticals) are not eliminated or degraded by these processes and additional specific treatments are required to remove this small residual fraction of pollution. Among the possible methods, photocatalysis is a technique well-developed in the past years [4].

This technique consists of a set of redox reactions between organic pollutants and radicals or other active species. A semiconductor photocatalyst and any UV light source are required since the first active species are generated by the illumination of the photocatalyst [5,6]. This mechanism promotes the production of highly reactive species able to react and decompose organic molecules. In the best case, the final decomposition products are CO_2 and H_2O [5,6].

The most common photocatalyst is titanium dioxide (TiO_2) under the anatase phase [7–9]. The amount of energy required to activate anatase TiO_2 is high. Indeed, the width of its band gap (3.2 eV) corresponds to a light source with a wavelength inferior or equal to 388 nm [10]. This wavelength corresponds to UV radiation, which is more energetic than visible light.

The conventional light sources used for photocatalytic processes have several disadvantages like a high energy consumption, a high operational temperature (600–900 °C), difficulty in operation, and a short life span of between 500 and 2000 h [11–13]. In recent years [13–17], research has focused on the development of alternative lighting systems for photocatalytic water treatment using light-emitting diodes (LED). This lighting presents several advantages: it is cost-effective, eco-friendly, compact, with a narrow spectrum and a very long life span (> 50,000 h) [13–17]. However, UV LEDs have low efficiencies, with e.g., [18] taking pride in 20% external quantum efficiency for a 275-nm monochromatic LED. Decreasing the band gap of the photocatalyst, i.e., shifting toward visible LED, is thus very effective regarding the energy consumption.

The use of TiO_2 as photocatalyst has two main limitations [10]: (i) the fast charge recombination and (ii) the large band gap value. Indeed, if the recombination of the photo-generated species (e^- and h^+) is fast, the production of radicals is low and the degradation is less effective. Furthermore, as explained above, if the band gap is large, the energy required for the electron transfer is high and only UV radiation can be used.

To prevent these limitations, several studies have been carried out. Regarding the improvement of the recombination time, the major modification of TiO_2 materials is the addition of metallic nanoparticles or metallic ions as Ag [19,20], Au [19,21], Pt [22,23], Pd [20], Fe^{3+} [24–27] or Cu^{2+} [24,25,28]. In this case, the metallic nanoparticle or metallic ion plays a role of electron trap allowing to increase the recombination time [29]. Combination with other semiconductors has also been investigated [10] such as TiO_2/ZnO [30], TiO_2/CdS, TiO_2/Bi_2S_3 [10] or TiO_2/ZrO_2 [31]. In this case, synergetic effects lead to a better charge separation or an increased photostability. In order to extend the activity under visible light and therefore reduce the band gap, TiO_2 materials have been modified with different elements or molecules: (i) metallic ions [25,28,32] such as Cu [33] or Fe [26], (ii) dye molecules such as porphyrins [34–36] or (iii) non-metallic elements such as N [37–39], P [40,41], S [10,42] or F [10,43].

Depending on the application requirements, TiO_2 can be used as a powder or a film. Typically, the as-synthesized material is amorphous and has no photoactivity; it is subsequently calcined to increase its crystallinity and so, the photoactivity. For practical applications in water treatment, photocatalytic coatings are more convenient. Unlike powders, they do not require any filtration step. Nevertheless, increasing the amount of photocatalyst implies increasing the area of the coated surface, while it simply implies a higher concentration in the case of powders. In terms of design, increasing the coated surface is a challenging topic if one wants to fulfill space constraints and maximize illumination. Furthermore, calcination may be detrimental to the substrate. For example, stainless steel loses its anticorrosive properties after a high temperature treatment.

This study aims to develop an eco-friendly and efficient photocatalytic coating, active under a cost-effective near-visible LED system for the removal of organic pollutants. To reach this goal, the research was conducted in two steps: (i) crystalline TiO_2-based photocatalysts powders were prepared without any calcination step by a sol-gel process in water and the composition was optimized to extend photoactivity toward the visible range, then (ii) the best materials were coated on three different substrates (bare and brushed stainless steel and glass substrates) to assess its efficiency on organic pollutants removal under a near-visible LED light source.

TiO$_2$ was co-doped with Fe and N in order to extend its activity toward the visible region and reduce the charge recombination. Different amounts of Fe and N were tested. The corresponding pure and single-doped catalysts were also synthesized. The powder materials were characterized by nitrogen adsorption–desorption, X-ray diffraction, inductively coupled plasma–optical emission spectroscopy, diffuse reflectance and X-ray photo-electron spectroscopy. The photocatalytic activity of the powder samples was evaluated through the degradation of p-nitrophenol (PNP, $C_6H_5NO_3$) under conventional halogen visible light ($\lambda > 400$ nm). The degradation experiments highlighted the influence of the dopants on the visible activity. The optimized materials were deposited on bare and brushed stainless steel and glass substrates. The photoactivity of those samples were assessed on the basis of the degradation of (i) PNP under UV-A light (365 nm) and (ii) rhodamine B under near-visible LED light (395 nm). The degradation efficiency was compared to the one obtained with pure TiO$_2$ coatings. In order to demonstrate the feasibility of an industrial development, one of our photocatalysts was synthesized in a pilot-scale equipment of 10 L and compared to the corresponding material obtained at laboratory scale.

2. Results and Discussion

2.1. Sample Crystallographic Properties

The X-ray diffraction patterns (XRD) of pure and three doped TiO$_2$ catalysts are presented in Figure 1. Similar XRD patterns were obtained for the other samples.

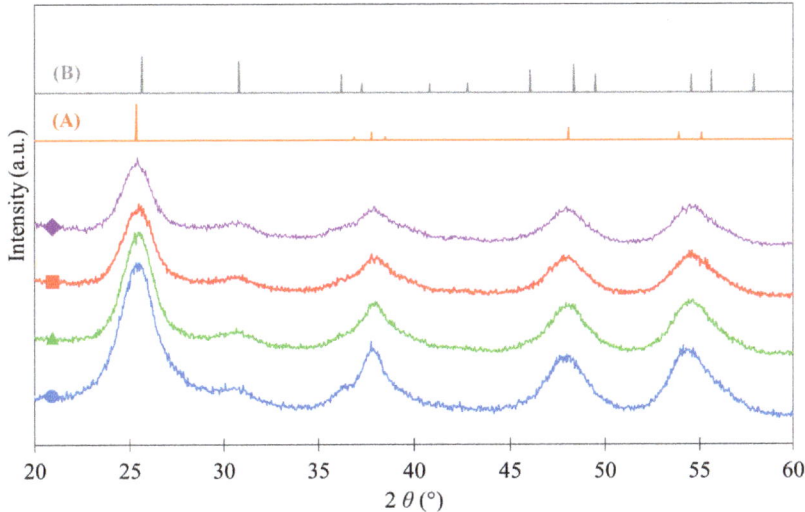

Figure 1. X-ray diffraction (XRD) patterns: (●) pure TiO$_2$, (▲) TiO$_2$/Fe0.5, (■) TiO$_2$/N43 and (♦) TiO$_2$/Fe0.5/N43. (A) Reference pattern of anatase and (B) reference pattern of brookite. Sample names and compositions are listed in Table 1.

Table 1. Textural and optical properties of TiO$_2$-based photocatalysts.

Sample	Fe Content mol/mol%	d_{XRD} nm ± 1	S_{BET} m^2 g^{-1} ± 5	V_{DR} cm^3 g^{-1} ± 0.01	V_P cm^3 g^{-1} ± 0.01	d_{BET} nm ± 1	$E_{g,direct}$ eV ± 0.01	$E_{g,indirect}$ eV ± 0.01
P25	–[1]	18[2]–8[3]	47	0.03	–[1]	31	3.45	3.05
Pure TiO$_2$	–[1]	5	180	0.09	0.10	9	3.25	2.90
TiO$_2$/Fe0.25	0.32	4	195	0.10	0.10	8	3.25	2.80
TiO$_2$/Fe0.5	0.56	5	180	0.10	0.10	9	3.20	2.75
TiO$_2$/N10	–[1]	5	185	0.10	0.10	8	3.15	2.90
TiO$_2$/N30	–[1]	4	185	0.09	0.09	8	3.15	2.90
TiO$_2$/N43	–[1]	5	220	0.11	0.11	7	3.20	2.90
TiO$_2$/N75	–[1]	4	210	0.11	0.11	7	3.20	2.90
TiO$_2$/Fe0.25/N10	0.28	4	155	0.08	0.08	10	3.10	2.80
TiO$_2$/Fe0.25/N30	0.27	4	200	0.10	0.10	8	3.10	2.75
TiO$_2$/Fe0.25/N43	0.27	4	240	0.12	0.12	6	3.10	2.80
TiO$_2$/Fe0.25/N75	0.31	4	235	0.12	0.12	7	3.15	2.70
TiO$_2$/Fe0.5/N10	0.54	5	185	0.10	0.10	8	3.05	2.70
TiO$_2$/Fe0.5/N30	0.54	4	210	0.11	0.11	7	3.1	2.65
TiO$_2$/Fe0.5/N43	0.55	4	230	0.12	0.12	7	3.05	2.65
TiO$_2$/Fe0.5/N75	0.53	4	220	0.11	0.11	7	3.05	2.65

[1] Not measured; [2] measured from anatase peak; [3] measured from rutile peak; S_{BET}: specific surface area estimated by the BET theory; V_{DR}: specific micropore volume estimated by the Dubinin–Raduskevitch method; d_{BET}: mean diameter of TiO$_2$ nanoparticles calculated from S_{BET} values; d_{XRD}: mean diameter of TiO$_2$ crystallites calculated using the Scherrer equation; $E_{g,direct}$: direct optical band gap values estimated with the transformed Kubelka–Munk function; $E_{g,indirect}$: indirect optical band gap values estimated with the transformed Kubelka–Munk function.

All samples are mainly composed of anatase with a small amount of brookite. The crystallite size, d_{XRD}, determined by the Scherrer equation (Equation (12) in Section 3.4) is 4–5 nm (Table 1) regardless of the nature or content of the dopant [33]. The calculated crystallite size could be slightly underestimated because of the presence of a small amount of brookite. The phase quantification is presented in Table 2. For all samples, the distribution between anatase and brookite phases was quite similar with ~90% of anatase and ~10% of brookite.

Table 2. Semi-quantitative analysis of sample crystallinity.

Sample	Anatase Content %	Brookite Content %	Rutile Content %
P25	80	–	20
Pure TiO$_2$	89	11	–
TiO$_2$/Fe0.25	90	10	–
TiO$_2$/Fe0.5	90	10	–
TiO$_2$/N10	91	9	–
TiO$_2$/N30	89	11	–
TiO$_2$/N43	91	9	–
TiO$_2$/N75	90	10	–
TiO$_2$/Fe0.25/N10	90	10	–
TiO$_2$/Fe0.25/N30	88	12	–
TiO$_2$/Fe0.25/N43	91	10	–
TiO$_2$/Fe0.25/N75	89	11	–
TiO$_2$/Fe0.5/N10	88	12	–
TiO$_2$/Fe0.5/N30	89	11	–
TiO$_2$/Fe0.5/N43	90	10	–
TiO$_2$/Fe0.5/N75	92	8	–

2.2. Textural Properties and Morphology

The nitrogen adsorption–desorption isotherms are presented in Figure 2 for four samples: pure TiO$_2$, TiO$_2$/Fe0.5, TiO$_2$/N43 and TiO$_2$/Fe0.5/N43. All other isotherms have a similar shape with a steep increase of the adsorbed volume at low pressure followed by a plateau. This corresponds to a microporous solid from the BDDT classification (type I isotherm) [44]. The corresponding specific surface area, S_{BET}, total pore volume, V_p, and micropore volume, V_{DR}, are reported in Table 1. The specific surface area varies between 155 and 240 m^2 g^{-1}, while the total pore volume varies between 0.08 and 0.12 cm^3 g^{-1}. The samples are essentially microporous, as the total pore volume is identical to the micropore volume. These surface properties are consistent with the ones usually found in literature for samples prepared with this peptization–precipitation method [33,45].

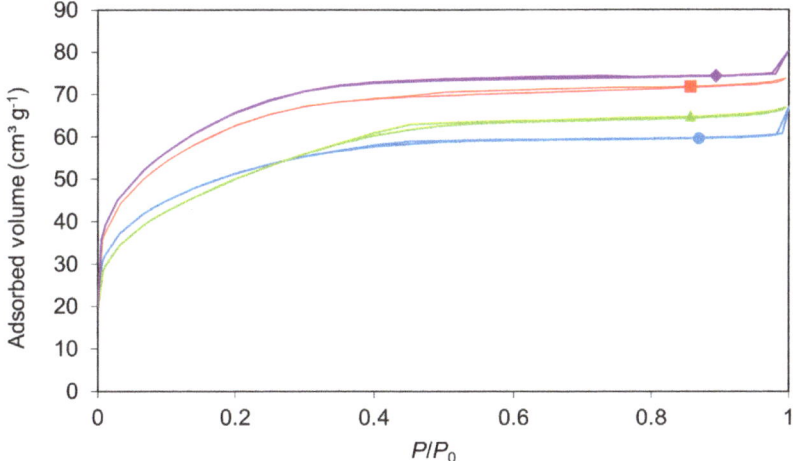

Figure 2. Nitrogen adsorption–desorption isotherms: (•) pure TiO$_2$, (▲) TiO$_2$/Fe0.5, (■) TiO$_2$/N43 and (♦) TiO$_2$/Fe0.5/N43.

A model developed earlier assumes that these materials are composed of spherical, non-porous TiO$_2$ nanoparticles between which small voids exist (i.e., < 2 nm, micropores) [26,33]. Thus, a particle size, d_{BET}, can be estimated from the S_{BET} value and Equation (11) (see Section 3.4). The values are reported in Table 1. For all samples, the d_{BET} is around 7–8 nm. The values are consistent with the ones determined by XRD and are independent of the nature and content of the dopant. Therefore, the assumption made before is realistic, and each spherical nanoparticle would correspond to one crystallite. This morphology is confirmed by transmission electron microscopy (TEM) observations from Figure 3 (see below).

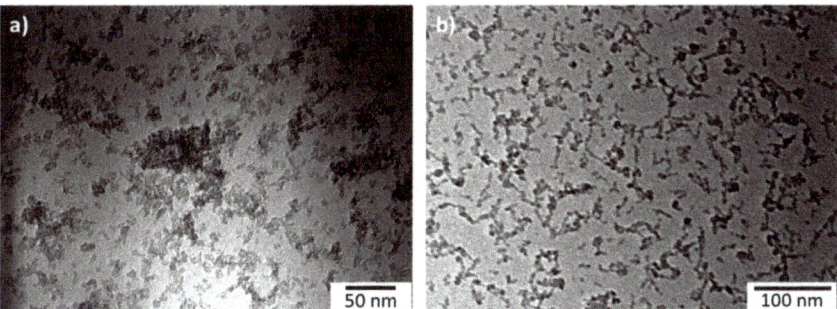

Figure 3. Transmission electron microscopy (TEM) pictures of (**a**) pure TiO$_2$ and (**b**) TiO$_2$/Fe0.5.

The isotherm of Evonik P25 (Figure S1 in Supplementary Materials) typically corresponds to a macroporous solid. At high pressure, the adsorbed volume increases asymptotically like in type II isotherm (macroporous solid) [44]. The S_{BET} and V_{DR} values are lower than pure and doped TiO_2 samples ones with values of 47 m^2 g^{-1} and 0.03 cm^3 g^{-1} respectively.

The morphology of the samples was visualized with a transmission electron microscope. The same morphology was observed for all samples with aggregates of spherical nanoparticles. The diameter of the nanoparticles is in the range of 5 to 8 nm. This size was similar to the crystallite size measured by XRD (Table 1). Pure TiO_2 and TiO_2/Fe0.5 TEM micrographs are presented in Figure 3 as examples.

2.3. Optical Properties

Optical properties were evaluated by diffuse reflectance spectroscopy. Figure 4 presents the evolution of the normalized Kubelka–Munk function $F(R\infty)$ with wavelength, λ, for the pure TiO_2, TiO_2/Fe0.5, TiO_2/N43, TiO_2/Fe0.5/N43 and Evonik P25.

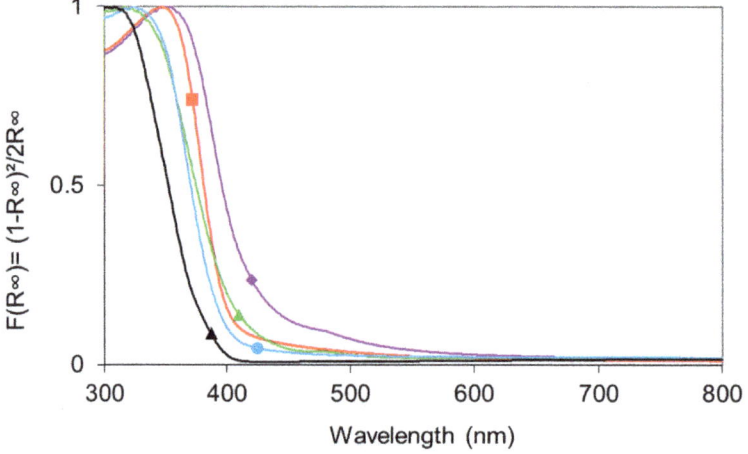

Figure 4. Normalized Kubelka–Munk function $F(R\infty)$ calculated from DR-UV-Vis absorbance spectra and transformed via the Kubelka–Munk function for (●) pure TiO_2, (▲) TiO_2/Fe0.5, (■) TiO_2/N43, (♦) TiO_2/Fe0.5/N43 and (▲) Evonik P25.

There is a clear shift of the curves towards longer wavelengths for all samples compared to Evonik P25, suggesting an activation under UV-Visible light. This result is in accordance with similar photocatalysts produced by different methods such as hydrothermal methods [46], sonochemical methods [47], and sol-gel methods in alcohol requiring calcination [48]. The pure TiO_2 is already shifted toward the visible compared to P25. This is probably due to nitrogen insertion during the synthesis [39]. Nitric acid is used as peptizing agent in the aqueous synthesis, leading to nitrogen insertion. The N-doping is confirmed by the XPS measurements (see Section 2.4). Adding iron to pure TiO_2 modifies the slope of the pure TiO_2 curve while adding nitrogen shifts it. The co-doped sample combines both effects (slope + translation) with, as a result, a higher shift than the single-doped samples.

The transformed Kulbelka–Munk function $(F(R\infty)h\nu)^{1/m}$ as a function of the energy enables to estimate the direct and indirect band gap values, taking m equal to $\frac{1}{2}$ and 2 respectively. The values are reported in Table 1. Both $E_{g,direct}$ and $E_{g,indirect}$ decrease with doping, suggesting a positive effect of iron and nitrogen on the activation under UV-Visible.

2.4. Sample Composition

The actual iron content, determined by inductively coupled plasma–optical emission spectroscopy (ICP-OES), is consistent with the intended content of iron introduced during the synthesis of the doped samples (Table 1).

The survey X-ray photo-electron spectroscopy (XPS) spectrum for the pure TiO_2 sample is presented in Figure S2 (Supplementary Materials). The different peaks (carbon, oxygen, nitrogen, titanium) are labelled. Similar spectra were obtained for all other samples. The intensity of the peaks representative of nitrogen is too low to make them visible on the general spectrum. Iron is not detected by XPS. This is probably due to the small amount and homogeneous distribution of iron in the sample. This has already been observed in literature [49].

Spectra specific to Ti $2p$, O $1s$, N $1s$ and C $1s$ are presented in Figure 5 for the TiO_2/Fe0.5/N43 sample.

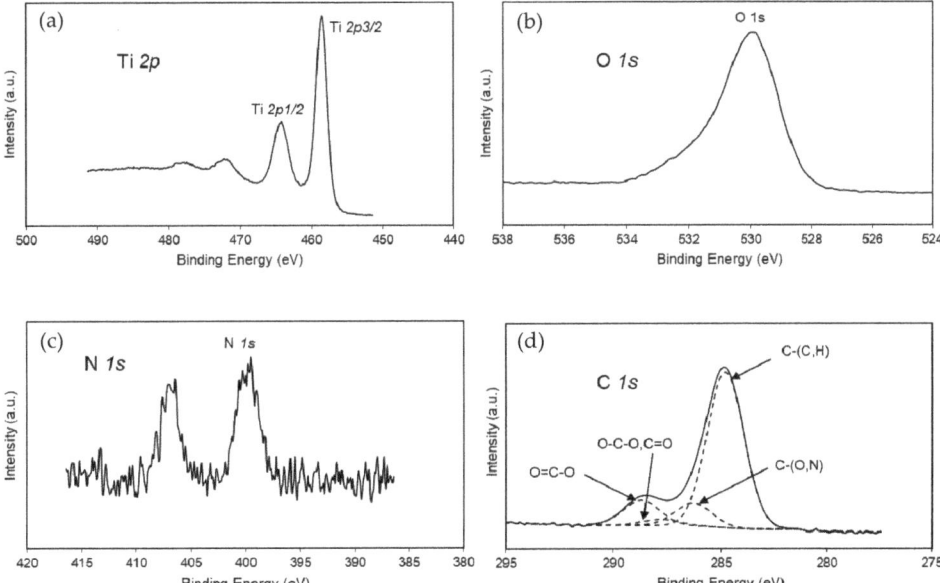

Figure 5. X-ray photo-electron spectroscopy (XPS) spectra of the TiO_2/Fe0.5/N43 sample: (**a**) Ti $2p$ region, (**b**) O $1s$ region, (**c**) N $1s$ region and (**d**) C $1s$ region.

The Ti $2p$ spectrum (Figure 5a) shows the Ti $2p1/2$ and Ti $2p3/2$ peaks at 464 and 459 eV respectively. They are attributed to Ti^{4+} species in TiO_2 [39,50].

As for the O $1s$ spectrum (Figure 5b), the sample's peak at 530 eV is associated to Ti–O in TiO_2. The tail at higher binding energy in the O $1s$ peak is difficult to exploit because of the presence of a significant amount of oxygen caused by carbonaceous contamination accumulating at the surface of the sample (inherent to XPS and unavoidable) or bonded to nitrogen. The O $1s$/Ti $2p$ ratio is around 2.5 which is close to the stoichiometric ratio in TiO_2. The gap between 2 and 2.5 is due to this oxygen involved in the surface carbonaceous contamination.

The C $1s$ contribution is divided into four components (Figure 5d). The C–(C,H) contribution at 284.8 eV is a classical aliphatic carbon contamination used to calibrate the measurements. In our standard routine procedure for decomposing the C $1s$ contribution, we define the contribution of carbon involved in simple bond with O or N at 1.5 eV higher (286.3 eV). The signals at a binding energy of around 288 eV are attributed to the contribution of C multiple-bonded to O. The other samples show comparable decomposition of the C signal.

In the case of the N $1s$ spectrum (Figure 5c), two peaks are visible for pure TiO_2: one centered on 400 eV and one around 407 eV. According to literature, a N $1s$ peak around 400 eV may correspond to interstitial Ti–O–N [50,51] leading to visible absorption. This activation under UV/Visible light is consistent with the diffuse reflectance measurements (Figure 4). Indeed, a higher visible absorption is observed in N-doped samples compared to Evonik P25. Other impurities such as ammonium ions also show an XPS peak at about 400 eV [52]. Since ammonium chloride was used for the N doping, the presence of NH_4^+ could lead to a misleading interpretation of XPS results. This is the reason why samples were washed several times. The second peak at 407 eV may be linked to residual nitrate due to the residual nitric acid from the synthesis as shown in [39].

The two peaks related to nitrogen are present in all XPS spectra but there is no clear difference between the samples and the N/Ti ratios are quite similar for all samples. This has already been observed in the literature [53,54]. Even if increasing the amount of nitrogen during the synthesis does not seem to have an impact on the surface composition of the samples (XPS), all samples are actually different, as shown by diffuse reflectance tests (Section 2.3) and photocatalytic activity tests (Section 2.5). One possible explanation is the inhomogeneous repartition of nitrogen along the depth of the samples. Furthermore, even if the presence of photoactive N dopant is revealed by other techniques (optical absorption, Electron Paramagnetic Resonance), the signal sometimes seems to escape XPS detection, as pointed out by [52]. In Smirniotis et al. [53], the relative atomic concentrations of N in N-doped TiO_2 (aerosol synthesis) do not vary significantly between samples obtained with different concentration of nitric acid. Furthermore, the relative atomic concentrations obtained by XPS differ from the ones obtained by energy-dispersive X-ray (EDX) spectroscopy. In Gil et al. [54], nitrogen is not detected.

Let us mention that the small changes in the XRD patterns and in the N-signal in XPS show that the amount of nitrogen that is actually incorporated in the TiO_2 lattice is probably small, and clearly lower than 10 to 75%. So, in the label of the samples, the N content corresponds to the starting one.

2.5. Photocatalytic Activity of the Powders

Adsorption tests performed in the dark showed a mean adsorption of PNP of 3%. Results of the degradation tests are presented in Figure 6 for all samples. All samples have a better activity under visible light than P25 (~12% of PNP degradation), even for the pure TiO_2 (~25% of PNP degradation). As previously explained, due to the synthesis method and the use of nitric acid, all the samples are doped with nitrogen, even the pure TiO_2 sample. The best sample, TiO_2/Fe0.5/N43, reaches a PNP degradation of 67%.

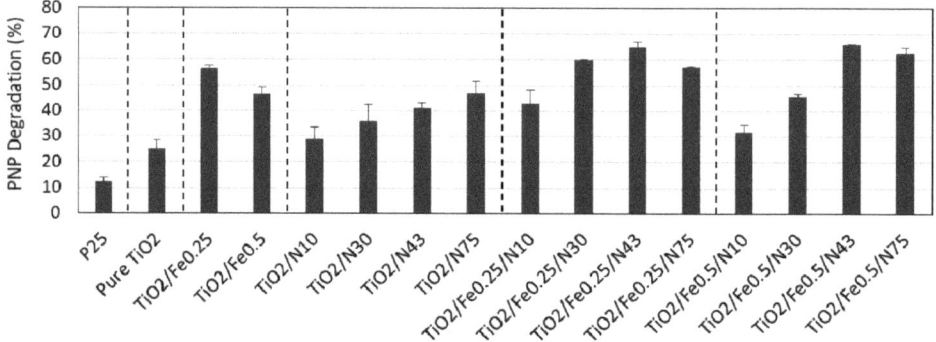

Figure 6. *p*-Nitrophenol (PNP) degradation (%) under visible light (λ > 400 nm) for all samples after 24 h of reaction time.

Both dopants have a positive effect on the degradation of PNP under the conditions of the catalytic experiments. For Fe single-doped TiO_2, 0.25 mol% of Fe enhances the photocatalytic activity, but increasing its content beyond that value becomes detrimental to the degradation of PNP [26].

During the peptization, the metal ion dopant species (M^{n+}) are hydrolyzed, leading to the formation of $[M_x(OH)_y]^{(nx-y)+}$. Hydroxylated species associated to Fe^{3+} dopant are known to be particularly stable, therefore increasing the –OH density at the surface of the catalyst [26]. This could explain the enhancement of the photocatalytic activity of the Fe^{3+}-doped samples owing to a higher production of hydroxyl radicals and a higher hydrophilicity. Furthermore, the reduction of the band gap value for Fe-doped samples (Table 1) suggests the presence of an intra-band gap level, thus reducing the energy required to activate the samples, i.e., shifting the activity toward visible.

The mechanism of organic pollutant degradation by pure TiO_2 is well-established [55]. By absorbing the energy brought by the light, free electrons (e^-) and holes (h^+) are generated and can react with O_2 and H_2O respectively to generate the corresponding oxidative species $O_2^{-\bullet}$ and $^\bullet OH$.

The use of Fe^{3+} induces the photo-Fenton effect, i.e. the photo-reduction of ferric ions into ferrous ones by reacting with water in the presence of light:

$$Fe^{3+} + H_2O + h\nu \rightarrow Fe^{2+} + {}^\bullet OH + H^+ \tag{1}$$

where h is the Planck constant (6.63×10^{-34} J.s) and ν is the light frequency (Hz).

There is therefore an increased production of hydroxyl radicals $^\bullet OH$ which can degrade organic pollutants.

Furthermore, Fe^{3+} can act as charge trapping sites, hindering the e^--h^+ recombination process. The photo-generated holes and electrons trapped by Fe^{3+} can easily be transferred to surface –OH groups and adsorbed oxygen at the surface of TiO_2 to produce $^\bullet OH$ and O_2^- radicals (see Equations (2) to (8)) [25,28,33]. This induces an enhanced photocatalytic activity.

$$Fe^{3+} + h^+ \rightarrow Fe^{4+} \tag{2}$$

$$Fe^{3+} + e^- \rightarrow Fe^{2+} \tag{3}$$

$$Fe^{2+} + O_2(ads) \rightarrow Fe^{3+} + O_2^- \tag{4}$$

$$Fe^{2+} + Ti^{4+} \rightarrow Fe^{3+} + Ti^{3+} \tag{5}$$

$$Ti^{3+} + O_2(ads) \rightarrow Ti^{4+} + O_2^- \tag{6}$$

$$Fe^{4+} + e^- \rightarrow Fe^{3+} \tag{7}$$

$$Fe^{4+} + OH^-(ads) \rightarrow Fe^{3+} + {}^\bullet OH(ads) \tag{8}$$

However, a high Fe^{3+} concentration is detrimental to the photocatalytic activity because Fe^{3+} acts as e^--h^+ recombination center through cyclic redox reactions without the generation of active radicals available for the photocatalytic degradation process.

Adding N to TiO_2 has a positive effect on the degradation of PNP. The higher the amount of N, the higher the degradation of PNP. The molar percentages, included in the label of the samples (Table 1), correspond to the amounts of nitrogen introduced during the synthesis. Samples containing nitrogen were washed three times in order to remove the excess of salts and therefore to eliminate the residual NH_4^+ and Cl^-. In fact, the NH_4^+ ions show an XPS peak at about 400 eV, the same binding energy as the N photoactive species in TiO_2/N.

The reasons for the enhancement of the photoactivity with N-doping are still not clear, regarding the chemical nature of doping centers and the modification of the band structure. Now authors mostly agree on the fact that the presence of N leads to a band gap narrowing, with an interband gap some tenths of electronvolts over the valence band. The nitrogen species can be incorporated to the TiO_2 lattice in a substitutional or interstitial position. In the substitutional position, a nitrogen atom substitutes an oxygen at a regular lattice site while in the interstitial position the nitrogen atom is chemically bound to a lattice oxygen. Substitutional N-doping can be excluded in our study, since the corresponding peak around 396-7 eV is never present. This is consistent with literature, where the substitutional nitrogen is usually only observed for materials prepared by physical methods [56]. At this stage, even if interstitial N-doping is generally favored in sol-gel chemistry and consistent with the 400 eV peak [38], the decomposition of the N peak does not allow to claim the presence of Ti–O–N interstitial species only.

The highest photocatalytic activity is obtained for co-doped samples, assuming a combination of positive effects of both dopants.

The obtained photo-efficiency can be compared with literature. However, the large range of operating conditions reported in the photocatalytic experiments makes the comparison difficult. Indeed, the pollutant, its concentration, the illumination, or the catalyst amount were often different from one study to another.

In Gil et al. [54], sol-gel N-doped TiO_2 photocatalysts were synthesized and tested on the degradation of different emergent pollutants as caffeine, diclofenac, ibuprofen and salicylic acid. The photoactivity was evaluated under UV radiation with very low pollutant concentrations (1–15 ppm) and a catalyst concentration of 0.1 to 1 g/L. The degradation was faster than in our study, only 2 h, due to the higher energetic light used (UV radiation vs. visible light).

In Smirniotis et al. [53], N-doped TiO_2 is produced by a flame aerosol method to enhance visible activity. The degradation of phenol was conducted for 2 h and the comparison with commercial Evonik P25 was performed. The catalyst concentration was 1 g/L and the phenol concentration was 5×10^{-4} M. The best N-doped catalyst degraded 50% of phenol while P25 degraded only less than 5%. Our model pollutant, PNP, is a more complex molecule than phenol. The reaction time was therefore higher (24 h). However, when comparing our best sample with P25, the gain in photoactivity was similar to Smirniotis et al. [53].

In Suwannaruang et al. [57], Fe/N co-doped TiO_2 was synthesized via sol-gel hydrothermal method. The photoactivity was evaluated on the degradation of ciprofloxacin (20 mg/L corresponding to 6×10^{-5} M) under LED illumination (no information available about the wavelength) and a catalyst concentration of 1 g/L. The best catalyst (2.5 % N–1.5 % Fe) degraded 67% of ciprofloxacin after 6 h. Our best catalyst, TiO_2/Fe0.5/N43, degraded 67% of PNP after 24 h with a concentration of 10^{-4} M.

In Aba-Guevara et al. [49], Fe/N co-doped TiO_2 was also synthesized via sol-gel and microwave methods. The photoactivity was evaluated on the degradation of amoxicillin and streptomycin with a concentration of 30 mg/L (~ 8×10^{-5} M and 5×10^{-5} M) under visible light (> 400 nm). The catalyst concentration was 1 g/L. After 5 h of illumination, the best sample (the sol-gel one with 0.7% Fe/0.5% N) degraded 42% of amoxicillin and 26% of streptomycin. A relevant comparison is difficult because those molecules are different from PNP.

A previous study of our laboratory reported the photoactivity obtained, in similar operating conditions, but with TiO_2/ZrO_2. When comparing the results, a better activity is reached with our TiO_2/Fe0.5/N43 sample than the TiO_2/ZrO_2 sample (67 vs. 40% of PNP degradation in same conditions [31]).

2.6. Film Crystallinity and Thickness

Films were deposited on three substrates with different colloids presented in the first part of this study: the TiO_2/Fe0.5 and the TiO_2/Fe0.5/N43 colloids. Coatings with pure TiO_2 were also made for comparative purposes.

The films were washed with water and dried before characterization to ensure the adhesion of the coating to the substrate. For all samples, the film remains unmodified after washing; the surface is slightly iridescent.

The crystallinity of the TiO_2 coatings was assessed using grazing incidence X-ray diffraction (GIXRD). Similar patterns were obtained for all samples; only the TiO_2/Fe0.25/N43 pattern deposited on brushed steel is represented in Figure 7. The broad peak at 25°, associated to anatase, is clearly observed. Regarding the peak shape, it is similar to the one obtained with the powder (Figure 1). Indeed, the broadness of the peak is caused by the very low value of the crystallite sizes (around 4 nm, Table 1). Some additional peaks are observed around 32, 43 and 51° corresponding to the brushed stainless steel substrate (chromium oxide PDF 00-059-0308 and chromium PDF 01-088-2323 phases).

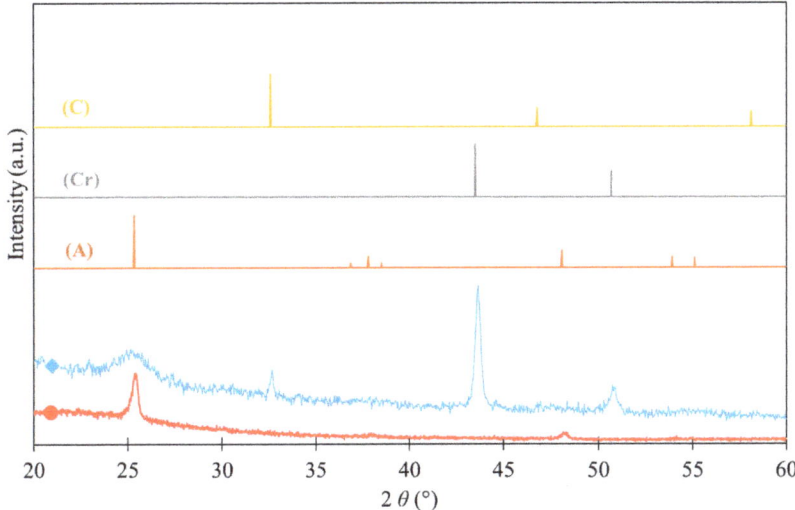

Figure 7. Grazing incidence X-ray diffraction (GIXRD) pattern of (♦) a TiO$_2$/Fe0.25/N43 film on steel, (●) a calcined TiO$_2$ film on glass, (A) reference pattern of anatase, (Cr) reference pattern of chromium, and (C) reference pattern of chromium oxide.

The presence of an additional layer between the substrate and the TiO$_2$ coatings was also investigated in order to point out the versatility of deposition. The GIXRD pattern of the calcined TiO$_2$ sublayer deposited on glass is presented in Figure 7. In this case, a highly crystalline TiO$_2$ sublayer was obtained.

Films produced by an aqueous sol-gel method typically have low thicknesses. As shown in a previous study [58], with the same deposition parameters, the thickness of a film deposited on a glass substrate is similar to the one obtained on steel for this type of sol-gel synthesis. The estimation gives a value of about 80 ± 10 nm on a glass substrate. Therefore, direct measurement of the thickness of the layer on steel is not possible by mechanical profilometry, since the roughness of the steel substrate is at the half-micron scale, or by optical profilometry, since the layer is transparent.

2.7. Photocatalytic Activity of the Films

The films were tested in two different setups in order to illustrate their photocatalytic potential in different situations. In the first case, a pesticide residue (PNP) was eliminated by the films under UV-A light. In the second case, LED light was used to degrade a dye (Rhodamine B).

The TiO$_2$/Fe0.5 colloid was deposited on different surfaces. The photoactivity was evaluated on PNP degradation. The results are presented in Table 3 after 72 h of illumination. A photocatalytic activity was maintained on each surface, highlighting its versatility of deposition. Generally, when a sublayer of anatase TiO$_2$ was present, the activity was increased due to this highly crystalline sublayer (Figure 7). Deposition on brushed steel gave higher PNP degradation than on bare steel. It can be assumed that the higher roughness of the brushed steel increased the surface of photocatalysts exposed to the light source. This substrate was therefore chosen for further PNP photocatalytic experiments with the optimized colloid composition identified in the first part of this study: the TiO$_2$/Fe0.5/N43 colloid (Section 2.5).

Table 3. PNP degradation (%) under UV-A light (λ = 365 nm) after 72 h for the TiO$_2$/Fe0.5 colloid deposited on various surfaces.

Sample	PNP Degradation after 72 h (%) ± 3
Brushed steel	26
Brushed steel + SiO$_2$ sublayer	19
Brushed steel + TiO$_2$ sublayer	32
Bare steel	18
Bare steel + SiO$_2$ sublayer	15
Bare steel + TiO$_2$ sublayer	12
Glass	26
Glass + TiO$_2$ sublayer	68

The photocatalysts are intended to be used for water depollution. Therefore, leaching experiments were performed with all samples from Table 3 in order to check the adherence of the films and the absence of TiO$_2$ or Fe leaching in water. The results are presented in Table S1, no leaching was detected whatever the sample.

Similar PNP degradation experiments were then performed with pure TiO$_2$ and TiO$_2$/Fe0.5/N43 coatings on brushed steel substrates. The evolution of the PNP degradation is presented in Figure 8.

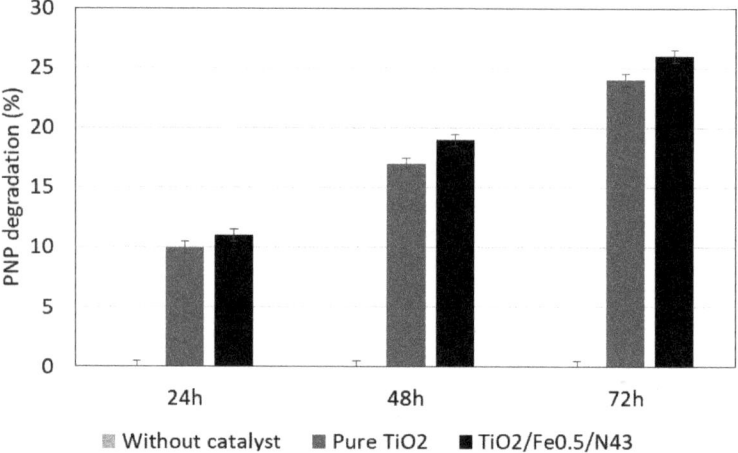

Figure 8. *p*-Nitrophenol (PNP) degradation (%) for pure TiO$_2$ and TiO$_2$/Fe0.5/N43 films under UV-A light (λ = 365 nm) after 24–48–72 h of illumination. No degradation took place without catalyst (error bars still shown).

The degradation of PNP increases with time (Figure 8). There is only a slight difference between pure TiO$_2$ and TiO$_2$/Fe0.5/N43 upon illumination at 365 nm. This is expected, since the desired effect of Fe- and N-doping is to shift the band gap energy to lower values. Thus, doped TiO$_2$ is barely at an advantage under UV-A light. Both photocatalysts are active at this wavelength as shown by Figure 4.

Nevertheless, when comparing absorbances (not shown here), values of 0.82 and 0.87 are obtained for pure TiO$_2$ and TiO$_2$/Fe0.5/N43 respectively with a ratio of 1.06 between the pure and the doped sample.

By contrast, while using LED light of wavelength 395 nm and RB, the degradation is much more significant for TiO$_2$/Fe0.5/N43 than for pure TiO$_2$. This difference in performance cannot be attributed to adsorption phenomena, which reach equilibrium after 20 min (not shown here). Figure 9 presents the variation of the concentration (normalized by the initial concentration) as a function of time.

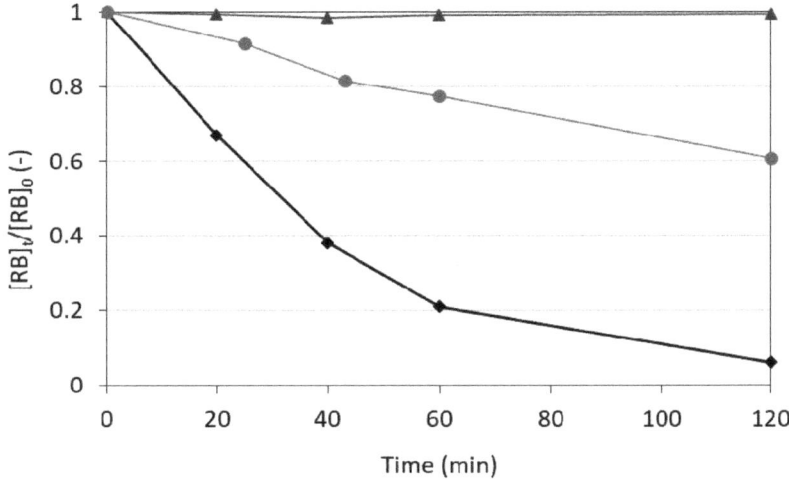

Figure 9. RB degradation for (•) pure TiO$_2$, (♦) TiO$_2$/Fe0.5/N43 films and (▲) without catalyst under LED light (λ = 395 nm) as a function of time.

Unlike in the previous case, RB can be sensitized by 395 nm light and contribute to its own degradation by interacting with TiO$_2$. This might explain the low, but nevertheless significant, degradation potential of pure TiO$_2$, alongside with the involuntary N-doping described in Section 2.4.

Upon this wavelength, pure TiO$_2$ has a low absorbance (0.59) and is less efficient than TiO$_2$/Fe0.5/N43 (0.79), with an absorbance ratio of 1.34 between the pure and the doped sample.

The longer the wavelength, the more efficient the degradation (energy-wise), since less energy is used for the creation of each e$^-$-h$^+$ pair. Practically, it is particularly true when shifting from UV-A light to visible light, as visible LEDs are much more efficient than their UV-A counterparts. It confirms the assumption of the positive effect of the doping on the shift of the activity toward the visible region.

As reported in numerous papers, dye degradation usually follows a first-order reaction rate. An excellent fit of our data was obtained with the first-order rate equation:

$$-kt = \ln\left(\frac{C}{C_0}\right) \quad (9)$$

where k is the rate constant, t is the time, and C and C_0 are the concentrations of pollutants at time t and initial time t_0, respectively. The kinetic constants were estimated using the least squares method (Table 4). As k depends on the number of active sites per unit volume of solution treated, a second rate constant k′ can be calculated.

$$k\prime = k/\left(\frac{S}{V}\right) \quad (10)$$

where S is the surface of the photocatalyst and V is the volume of the solution. It is more representative of the efficiency of the photocatalyst.

Table 4. Kinetic rate constants of reactions with TiO$_2$ films.

	TiO$_2$/Fe0.5–PNP	TiO$_2$/Fe0.5/N43–RB
k (h^{-1})	6.52 × 10^{-3}	1.43
k′ (h^{-1} m$^3{}_{solution}$/m$^2{}_{cata}$)	9.73 × 10^{-5}	2.13 × 10^{-2}
R^2 (correlation coefficient)	0.9996	0.9948

As mentioned earlier, a quantitative comparison to literature is hard to carry out. In this case, it would be limited to batch systems using photocatalytic films to degrade either pollutant, at similar pH and not involving any additional chemical, as all these parameters influence the process. The observed kinetic constant k should be reported against the ratio S/V (otherwise, the constant ignores the amount of catalyst). Other parameters could be cited. Literature usually includes all information necessary to calculate the kinetic constant, but its value is rarely transferrable to other works.

In Lu et al. [59], kinetic constants range from 12.5×10^{-3} to 225×10^{-3} h^{-1} for PNP degradation, but the authors do not report the photocatalyst surface. In Kothavale et al. [60], a spray pyrolysis method is used in order to create N-doped TiO$_2$ films. In this case, the kinetic constant can be estimated to be around 0.168 h^{-1} for RB, but even though the area of the film is known, the total volume of solution is not. In Gastello et al. [61], TiO$_2$ films on CoFe$_2$O$_4$ particles are synthesized. The value of k' can be calculated and is around 6.9×10^{-2} h^{-1} m^3$_{solution}$/m^2$_{cata}$ for RB. The value is in the same order of magnitude, but higher. However, it must be noted that the authors use an optimal pH (3.5) and a powerful 500W-lamp.

2.8. Large-Scale Synthesis of TiO$_2$/Fe0.5

In order to point out a possible industrial production of our catalysts, the TiO$_2$/Fe0.5 material was produced at pilot scale in a 10 L glass reactor. This sample was chosen as it was a good compromise between enhanced activity under low energy light, compared to the pure sample, and the amount of dopant added. A picture of the 10 L colloid and the pilot reactor are presented in Figure S3 in the Supplementary Materials. This colloid was similar to the corresponding colloid obtained at laboratory scale, i.e., a blue-orange stable suspension.

The XRD pattern of the pilot scale TiO$_2$/Fe0.5 (Figure S4 in Supplementary Materials) is similar to the XRD pattern of the laboratory TiO$_2$/Fe0.5 (Figure 1) with the main peaks of anatase and a small fraction of brookite. The PNP degradation with the pilot scale TiO$_2$/Fe0.5 was 42% ± 3% after 8 h of illumination, which is similar to the degradation reached with the laboratory TiO$_2$/Fe0.5 (Figure 6). This comparison highlighted the efficient scale-up from laboratory to pilot scale, going one step further to a pre-industrial implementation.

3. Materials and Methods

3.1. Pure TiO$_2$ Synthesis

The undoped TiO$_2$ catalyst was prepared using titanium (IV) tetraisopropoxide (TTIP, >97%, Sigma-Aldrich, St. Louis, MO, USA), nitric acid (HNO$_3$, 65%, Merck, Suprapur, Fort Kennerworth, NJ, USA), isopropanol (IsoP, 99.5%, Acros Organics, Fisher Scientific, Hampton, NH, USA), and distilled water as starting materials.

First, 250 mL of distilled water was acidified with HNO$_3$ to reach a pH of 1. In a second vessel, 35 mL of TTIP was mixed with 15 mL of IsoP for 15 min. Then, the TTIP-IsoP mixture was added to the acidified water under stirring at 80 °C. After 24 h, a light blue sol was obtained [45]. Part of the sol was used directly to produce thin films on stainless steel. The remaining sol was dried under ambient air to obtain a white powder. The undoped TiO$_2$ sample is labelled pure TiO$_2$.

Evonik P25 TiO$_2$ material was used as reference material and is denoted P25.

3.2. Doped and Co-Doped TiO$_2$ Synthesis

Several Fe, N single-doped and co-doped TiO$_2$ catalysts were prepared. Two different Fe/Ti molar percentages were studied: 0.25 mol% and 0.5 mol% starting from ferric nitrate (Fe(NO$_3$)$_3$·9H$_2$O, Merck). The N content varied between 10 and 75 mol% starting from ammonium chloride (NH$_4$Cl, UCB, Leuven, Belgium). The dopant(s) was (were) dissolved in the acidified water before addition to the TTIP-IsoP mixture. After 24 h, white to light blue stable sols were obtained. Samples were then dried under ambient air at room temperature. Samples containing nitrogen were washed three times with water to remove the excess of nitrogen salt. Finally, all samples were dried at 100 °C for 12 h. Prior to the drying, part of the sol was deposited on various substrates.

Samples are denoted as TiO$_2$/FeX/NY, with X and Y corresponding to the starting content (in mol%) of Fe and N respectively.

3.3. Film Deposition

Films were produced by bar-coating with an Elcometer 4340 Automatic Film Applicator (Elcometer Limited, Edge Lane, Manchester, UK) with a bar-speed of 1 cm s^{-1}. Films were coated on three substrates: bare and brushed stainless steel substrates (7.5 cm × 2.5 cm × 0.8 cm, 316L stainless steel, Mecanic Systems, Braine-l'Alleud, Belgium) and glass microscope slides (AF32ECO, Schott AG, Mainz, Germany). Films were dried at 100 °C for 30 s, washed with distilled water, dried with compressed air, and finally dried again at 100 °C for 30 s to get a perfectly clean surface and to remove non-adherent TiO$_2$ material.

Three suspensions were used for film application: pure TiO$_2$, TiO$_2$/Fe0.5 and TiO$_2$/Fe0.5/N43.

Films were also deposited on different sublayers to evaluate the versatility of deposition of the aqueous TiO$_2$ colloids. For bare and brush stainless steel substrates, two sublayers were tested: a sublayer of anatase TiO$_2$ synthesized as [29,41], calcined 1 h at 500 °C, and a sublayer of sol-gel SiO$_2$ synthesized according to [62]. For glass substrate, only anatase TiO$_2$ sublayer was evaluated.

3.4. Powder Characterization

The actual amount of iron in the doped TiO$_2$ samples was determined by inductively coupled plasma–optical emission spectroscopy (ICP–OES, Varian Inc., Palo Alto, CA, USA) with a Varian Liberty Series II. Solutions for analysis were prepared as follows: (i) 0.2 g of sample was digested in 12 mL of HCl (37%) + 6 mL of HNO$_3$ (65%) + 2 mL of HF (40%); (ii) the solution was then transferred into a 200 mL calibrated flask, finally filled with deionized water. The solution was then analyzed using the ICP-OES device.

Nitrogen adsorption–desorption isotherms were measured at 77K on a multi-sampler Micromeritics ASAP 2420 (Micromeritics, Norcross, GA, USA). Those isotherms provide information on the textural properties of the samples, such as the specific surface area, S_{BET}, by the Brunauer-Emmett- Teller (BET) method, the specific micropore volume, V_{DR}, by the Dubinin-Radushkevich theory and the pore volume, V_p, calculated from the adsorbed volume at saturation. Pore volume can be determined precisely enough for non-macroporous materials, which is the case in this study, except for P25.

An average particle size, d_{BET}, was estimated from S_{BET} values by assuming spherical and non-porous TiO$_2$ anatase nanoparticles using the following formula [33]:

$$\frac{d_{BET}}{6} = \frac{\frac{1}{\rho_{Anatase}}}{S_{BET}} \quad (11)$$

where $\rho_{Anatase}$ is the apparent density of TiO$_2$ anatase, estimated to be equal to 3.89 × 10^3 kg m^{-3} [20].

The crystallographic properties were determined using X-ray diffraction (XRD). The powder patterns were recorded with a Bruker D8 Twin-Twin powder diffractometer (Bruker, Billerica, MA, USA) using Cu-K$_\alpha$ radiation. The size of the TiO$_2$ crystallites, d_{XRD}, was determined via the Scherrer formula (Equation (12)):

$$d_{XRD} = 0.9 \frac{\lambda}{(B \cos \theta)} \quad (12)$$

where d_{XRD} is the crystallite size (nm), B the peak full-width at half maximum after correction of the instrumental broadening (rad), λ the wavelength (nm), and θ the Bragg angle (rad).

The semi-quantitative analysis of phases was performed using the Profex software [63]. This program is based on Rietveld refinement method [64,65].

The samples were observed by transmission electron microscopy (TEM) with a Phillips CM 100 device with an accelerating voltage of 200 kV.

The optical properties were evaluated by diffuse reflectance in the 250–800 nm region with a Perkin Elmer Lambda 1050 S UV/VIS/NIR spectrophotometer, equipped with a spectralon coated integrating sphere (150 mm InGaAs Int. Sphere from PerkinElmer, Waltham, MA, USA) and using Al_2O_3 as reference. The absorbance spectra were transformed using the Kubelka–Munk function [33,66,67] to produce a signal, normalized for comparison between samples. The values of the band gaps ($E_{g,direct}$ and $E_{g,indirect}$) were estimated for all samples. The details of this method are widely described elsewhere [26,55].

X-ray photoelectron spectra were obtained with a SSI-X-probe (SSX-100/206) Fisons spectrometer (Surface Science Instruments, Mountain View, CA, USA) equipped with a monochromatized microfocused Al X-ray source (1486.6 eV), operating at 10 kV and 20 mA. The analysis chamber, in which samples were placed, was under a pressure of 10^{-6} Pa. Sample charging was adjusted using flood gun energy at 8 eV and a fine-meshed nickel grid placed 3 mm above the sample surface [68]. The pass energy was set to 150 eV and the spot size was 1.4 mm^2. The normal to the surface of the sample and the direction of electron collection formed an angle of 55°. Under these conditions, the mid-height width (FWHM) of the Au $4f7/2$ peak photo-peak measured on a standard sample of clean gold was about 1.6 eV. The following sequence of spectra was recorded: survey spectrum, C $1s$, O $1s$, N $1s$, Fe $2p$ and Ti $2p$ and again C $1s$ to check the stability of charge compensation with time and the stability of the samples over time.

The C–(C, H) component of the carbon C $1s$ peak was set to 284.8 eV in order to calibrate the scale in binding energy. Three other components of the carbon peak (C–(O, N), C=O or O–C–O and O–C=O) were resolved, giving insight on how much oxygen was present because of carbon contamination. Data was processed by using the CasaXPS software (Casa Software Ltd, Teignmouth, UK). Some spectra were decomposed using the Gaussian and Lorentzian function product model (least squares fitting) after subtraction of a nonlinear Shirley baseline [69].

3.5. Film Characterization

The crystallinity of the films was characterized by grazing incidence X-ray diffraction (GIXRD) in a Bruker D8 diffractometer using Cu K_α radiation and operating at 40 kV and 40 mA. The incidence beam angle was 0.25°.

The film thickness was estimated by profilometry (Veeco Dektak 8 Stylus Profiler, Bruker, Billerica, MA, USA) on a film deposited on glass (Marienfeld Superior – 25 mm × 75 mm × 1 mm, Paul Marienfeld GmbH & Co. KG, Lauda-Königshofen, Germany) in the same deposition conditions as for steel.

Leaching experiments were performed with the coated substrates to evaluate their integrity in water. For each sample, the coated slide was placed in 50 mL of ultrapure water under stirring. After 48 h, the water was analyzed by ICP-OES to detect Ti and Fe. The detection limit was 0.03 and 0.01 ppm for Fe and Ti respectively.

3.6. Photocatalytic Activity of the Powders under Visible Light

The photocatalytic activity of all catalysts was determined through the degradation of a model pollutant, *p*-nitrophenol (PNP) for 24 h. The experimental set-up was described in a previous article [33,45]. For each catalyst, sealed batch reactors were exposed to a halogen lamp with a continuous spectrum from 300 to 800 nm (300 W, 220 V) measured with a Mini-Spectrometer TM-UV/vis C10082MD (Hamamatsu, Japan). A UV filter placed on the lamp eliminated wavelengths shorter than 400 nm. The temperature of the lamp and the reactors was maintained at 20 °C by a cooling system with recirculating water. The catalyst powder was dispersed in PNP solution ($C_0 = 10^{-4}$ kmol m^{-3}) in order to reach 1 kg m^{-3} in each batch reactor. A constant mixing was maintained by magnetic stirrers. Three reactors were used for each catalyst in order to study the reproducibility of the results. Furthermore, for each photocatalyst, dark tests (catalyst + PNP without light) were performed in order to take into account the potential adsorption of the pollutant onto the catalyst. One additional flask was exposed to light without catalyst to evaluate PNP natural decomposition under visible light.

No spontaneous break down of the PNP in the absence of catalyst was detected, so that any decrease of PNP concentration could be attributed to the catalytic degradation. The entire system was isolated in the dark in order to prevent interaction with the room lighting.

The degradation of PNP was determined by measuring the absorbance of PNP by UV/Vis spectroscopy (GENESYS 10S UV–Vis from Thermo Scientific, Waltham, MA, USA) at 317 nm. The degradation percentage of PNP, D_{PNP}, was calculated by Equation 13 for each batch reactor:

$$D_{PNP}(\%) = \left(1 - \frac{[PNP]_{24h}}{[PNP]_0}\right) \times 100\ \% \tag{13}$$

where $[PNP]_{24h}$ represents the residual concentration of PNP at time t = 24 h and $[PNP]_0$ represents the initial concentration of PNP at time t = 0 h.

3.7. Photocatalytic Activity of the Films

The photocatalytic activity of TiO_2 films was evaluated (i) by monitoring the degradation of *p*-nitrophenol (PNP) under ultraviolet light (UV-A, Osram Sylvania, Blacklight-Bleu Lamp, F 18W/BLB-T8, OSRAM GmbH, Munich, Germany) and (ii) by monitoring the degradation of rhodamine B (RB) under visible LED light (LED Würth Elektronic WL-SUMW SMT Ultraviolet Ceramic Waterclear, Würth Elektronik GmbH & Co. KG, Waldenburg, Germany) in two similar photocatalytic experiments.

The spectra of the lamps were measured with a Mini-Spectrometer TM-UV/vis C10082MD from Hamamatsu. UV-A and LED lights can be considered as quasi monochromatic with a wavelength of 365 nm and 395 nm respectively.

The experimental procedure is very similar for both pollutants. Indeed, each coated steel slide was placed in a Petri dish with 25 mL of PNP solution (10^{-4} kmol m^{-3}) or RB solution (2.5×10^{-6} kmol m^{-3}). The Petri dish was closed with a lid in order to avoid evaporation. The degradations of PNP or RB were evaluated from absorbance measurements with a Genesys 10S UV-Vis spectrophotometer (Thermo Scientific, Waltham, MA, USA) at 317 nm or 554 nm respectively. Adsorption tests were performed in the dark (dark tests) to determine whether PNP or RB was adsorbed by the films or the substrates. Blank tests, consisting of irradiating the pollutant solution in a Petri dish without any catalyst or support, were carried out to estimate the decomposition of PNP or RB under the corresponding light. The Petri dishes with catalyst and pollutant were stirred on orbital shakers at 80 rpm. Aliquots of PNP or RB were sampled along the experiment and put back in the Petri dishes after absorbance measurements to keep the volume constant. The photocatalytic degradation is equal to the total degradation of PNP or RB taking the catalyst adsorption (dark test) into account. Each photocatalytic measurement was triplicated to assess the reproducibility of data.

Regarding the PNP degradation, different substrates (bare steel, brushed steel and glass) and sublayers (TiO_2, SiO_2) were tested with the TiO_2/Fe0.5 coating. Then only pure TiO_2 and TiO_2/Fe0.5/N43 deposited on brushed stainless steel were evaluated through the degradation of PNP.

Regarding the RB degradation, only the coatings deposited on brushed steel without sublayer were evaluated, starting from pure TiO_2 and TiO_2/Fe0.5/N43 colloids.

3.8. Large-Scale Synthesis of TiO_2/Fe0.5

The large-scale synthesis was carried out in a 10 L pilot glass reactor (Figure S3 in Supplementary Materials) with a jacket warming system. First, 7.2 L of distilled water was acidified by HNO_3 to reach a pH of 1. Then, 1 L of TTIP was added to 300 mL of IsoP and 6.5 g of $Fe(NO_3)_3 \cdot 9H_2O$. The mixture was stirred at room temperature for 30 min. The mixture was then added to the acidified water and stirred at 250 rpm. The reaction took place for 4 h at 80°C. A blue-orange translucent sol was obtained, and a small fraction (50 mL) was dried under ambient air to obtain a powder.

The corresponding powder was analyzed by XRD (Section 3.4), and its photoactivity was evaluated on the degradation of PNP (Section 3.6).

4. Conclusions

In this work, an aqueous sol-gel process was successfully applied to produce Fe, N single-doped and co-doped TiO_2 photocatalysts at low temperature without any calcination step. Different *dopant/Ti* molar ratios were tested: 0.25 or 0.5 mol% of Fe and 10 to 75 mol% of N (starting concentrations). The corresponding undoped TiO_2 catalyst was synthesized for comparison.

The physico-chemical characterizations showed that catalysts are composed of nano-crystalline anatase-brookite particles, with higher specific surface areas than P25 (~ 200 m^2 g^{-1} vs. 47 m^2 g^{-1}). All samples presented a higher visible absorption than P25. The XPS spectra showed that all the samples were doped with nitrogen, leading to a shift of the activity toward visible. Both dopants extend the TiO_2 photoactivity toward the visible region with the best result for a co-doped sample. The optimal starting composition is TiO_2 doped with 43 mol% of nitrogen and 0.5 mol% of iron.

Thin films were produced on stainless steel with the undoped TiO_2 and the best N-Fe co-doped sample. The photoactivity was estimated under 365 nm UV-A light and 395 nm visible LED light. The doped sample had a better activity than the pure one, with a clear enhancement under the longer wavelength (395 nm). This confirms the positive influence of the dopants on the shift of activity toward the visible region. The versatility of the photocatalysts was demonstrated on the degradation of two molecules, *p*-nitrophenol and rhodamine B, and on the deposition on different surfaces. A colloid synthesis was efficiently scaled-up from laboratory to 10 L pilot scale. Those results highlight the possibility of producing photocatalytic films without any calcination step and active under LED light, whose energy consumption is lowest in the visible range. This constitutes a step forward for an industrial development.

Supplementary Materials: The following are available online at http://www.mdpi.com/2073-4344/10/5/547/s1, Figure S1: Nitrogen adsorption–desorption isotherm of Evonik P25, Figure S2: XPS general spectrum of the pure TiO_2 sample, Figure S3: Large-scale suspension of TiO_2/Fe0.5 and the pilot reactor used for its synthesis, Figure S4: XRD patterns of the large-scale TiO_2/Fe0.5 sample. (A) Reference pattern of anatase and (B) Reference pattern of brookite, Table S1. Ti and Fe leaching for the TiO_2/Fe0.5 colloid deposited on various surfaces after 48 h in water.

Author Contributions: Conceptualization, methodology, writing, S.D.; investigation and analysis, S.D., J.G.M. and C.R.; writing–original draft preparation, S.D., J.M. and C.W.; DR-UV-Vis absorbance, D.P.; X-ray photoelectron spectroscopy, F.D. and E.M.G.; supervision, funding acquisition and project administration, S.D.L.; correction of the paper before submission: all authors. All authors have read and agreed to the published version of the manuscript.

Funding: Authors are grateful to the Département de la Recherche et du Développement technologique DGO6 of the Service Public de Wallonie and GreenWin, Competitivity Cluster of Wallonia, for financial support under Grant n°7744 (BlueV project).

Conflicts of Interest: The authors declare no conflict of interest. The funders had no role in the design of the study; in the collection, analyses, or interpretation of data; in the writing of the manuscript, or in the decision to publish the results.

References

1. Khan, M.A.; Ghouri, A.M. Environmental Pollution: Its Effects on Life and Its Remedies. *Res. World J. Arts Sci. Commer.* **2011**, *2*, 276–285.
2. Pignatello, J.J.; Oliveros, E.; MacKay, A. Advanced Oxidation Processes for Organic Contaminant Destruction Based on the Fenton Reaction and Related Chemistry. *Crit. Rev. Environ. Sci. Technol.* **2006**, *36*, 1–84. [CrossRef]
3. Kuyukina, M.S.; Ivshina, I.B. Application of Rhodococcus in Bioremediation of Contaminated Environments. In *Biology of Rhodococcus*; Springer: Berlin/Heidelberg, Germany, 2010; pp. 231–262. [CrossRef]
4. Mills, A.; Le Hunte, S. An overview of semiconductor photocatalysis. *J. Photochem. Photobiol. A Chem.* **1997**, *108*, 1–35. [CrossRef]
5. Di Paola, A.; García-López, E.; Marcì, G.; Palmisano, L. A survey of photocatalytic materials for environmental remediation. *J. Hazard. Mater.* **2012**, *211–212*, 3–29. [CrossRef] [PubMed]

6. Rauf, M.A.; Ashraf, S.S. Fundamental principles and application of heterogeneous photocatalytic degradation of dyes in solution. *Chem. Eng. J.* **2009**, *151*, 10–18. [CrossRef]
7. Fujishima, A.; Hashimoto, K.; Watanabe, T. *TiO$_2$ Photocatalysis: Fundamentals and Applications*; Bkc: Tokyo, Japan, 1999; ISBN 493905103X 9784939051036.
8. Hoffmann, M.R.; Martin, S.T.; Choi, W.; Bahnemann, D.W. Environmental Applications of Semiconductor Photocatalysis. *Chem. Rev.* **1995**, *95*, 69–96. [CrossRef]
9. Linsebigler, A.L.; Lu, G.; Yates, J.T. Photocatalysis on TiO$_2$ Surfaces: Principles, Mechanisms, and Selected Results. *Chem. Rev.* **1995**, *95*, 735–758. [CrossRef]
10. Pelaez, M.; Nolan, N.T.; Pillai, S.C.; Seery, M.K.; Falaras, P.; Kontos, A.G.; Dunlop, P.S.M.; Hamilton, J.W.J.; Byrne, J.A.; O'Shea, K.; et al. A review on the visible light active titanium dioxide photocatalysts for environmental applications. *Appl. Catal. B Environ.* **2012**, *125*, 331–349. [CrossRef]
11. Jo, W.-K.; Tayade, R.J. New Generation Energy-Efficient Light Source for Photocatalysis: LEDs for Environmental Applications. *Ind. Eng. Chem. Res.* **2014**, *53*, 2073–2084. [CrossRef]
12. Subagio, D.P.; Srinivasan, M.; Lim, M.; Lim, T.-T. Photocatalytic degradation of bisphenol-A by nitrogen-doped TiO$_2$ hollow sphere in a vis-LED photoreactor. *Appl. Catal. B Environ.* **2010**, *95*, 414–422. [CrossRef]
13. Oseghe, E.O.; Ofomaja, A.E. Study on light emission diode/carbon modified TiO$_2$ system for tetracycline hydrochloride degradation. *J. Photochem. Photobiol. A Chem.* **2018**, *360*, 242–248. [CrossRef]
14. Casado, C.; Timmers, R.; Sergejevs, A.; Clarke, C.T.; Allsopp, D.W.E.; Bowen, C.R.; van Grieken, R.; Marugán, J. Design and validation of a LED-based high intensity photocatalytic reactor for quantifying activity measurements. *Chem. Eng. J.* **2017**, *327*, 1043–1055. [CrossRef]
15. Hossaini, H.; Moussavi, G.; Farrokhi, M. Oxidation of diazinon in cns-ZnO/LED photocatalytic process: Catalyst preparation, photocatalytic examination, and toxicity bioassay of oxidation by-products. *Sep. Purif. Technol.* **2017**, *174*, 320–330. [CrossRef]
16. Chevremont, A.-C.; Farnet, A.-M.; Coulomb, B.; Boudenne, J.-L. Effect of coupled UV-A and UV-C LEDs on both microbiological and chemical pollution of urban wastewaters. *Sci. Total Environ.* **2012**, *426*, 304–310. [CrossRef]
17. Ghosh, J.P.; Langford, C.H.; Achari, G. Characterization of an LED Based Photoreactor to Degrade 4-Chlorophenol in an Aqueous Medium Using Coumarin (C-343) Sensitized TiO$_2$. *J. Phys. Chem. A* **2008**, *112*, 10310–10314. [CrossRef]
18. Takano, T.; Mino, T.; Sakai, J.; Noguchi, N.; Tsubaki, K.; Hirayama, H. Deep-ultraviolet light-emitting diodes with external quantum efficiency higher than 20% at 275 nm achieved by improving light-extraction efficiency. *Appl. Phys. Express* **2017**, *10*, 031002. [CrossRef]
19. Epifani, M.; Giannini, C.; Tapfer, L.; Vasanelli, L. Sol–Gel Synthesis and Characterization of Ag and Au Nanoparticles in SiO$_2$, TiO$_2$, and ZrO$_2$ Thin Films. *J. Am. Ceram. Soc.* **2000**, *83*, 2385–2393. [CrossRef]
20. Espino-Estévez, M.R.; Fernández-Rodríguez, C.; González-Díaz, O.M.; Araña, J.; Espinós, J.P.; Ortega-Méndez, J.A.; Doña-Rodríguez, J.M. Effect of TiO2-Pd and TiO$_2$-Ag on the photocatalytic oxidation of diclofenac, isoproturon and phenol. *Chem. Eng. J.* **2016**, *298*, 82–95. [CrossRef]
21. Vaiano, V.; Iervolino, G.; Sannino, D.; Murcia, J.J.; Hidalgo, M.C.; Ciambelli, P.; Navío, J.A. Photocatalytic removal of patent blue V dye on Au-TiO$_2$ and Pt-TiO$_2$ catalysts. *Appl. Catal. B Environ.* **2016**, *188*, 134–146. [CrossRef]
22. Ofiarska, A.; Pieczyńska, A.; Fiszka Borzyszkowska, A.; Stepnowski, P.; Siedlecka, E.M. Pt–TiO$_2$-assisted photocatalytic degradation of the cytostatic drugs ifosfamide and cyclophosphamide under artificial sunlight. *Chem. Eng. J.* **2016**, *285*, 417–427. [CrossRef]
23. Semlali, S.; Pigot, T.; Flahaut, D.; Allouche, J.; Lacombe, S.; Nicole, L. Mesoporous Pt-TiO$_2$ thin films: Photocatalytic efficiency under UV and visible light. *Appl. Catal. B Environ.* **2014**, *150–151*, 656–662. [CrossRef]
24. Di Paola, A.; Marcì, G.; Palmisano, L.; Schiavello, M.; Uosaki, K.; Ikeda, S.; Ohtani, B. Preparation of Polycrystalline TiO$_2$ Photocatalysts Impregnated with Various Transition Metal Ions: Characterization and Photocatalytic Activity for the Degradation of 4-Nitrophenol. *J. Phys. Chem. B* **2002**, *106*, 637–645. [CrossRef]
25. Rauf, M.A.; Meetani, M.A.; Hisaindee, S. An overview on the photocatalytic degradation of azo dyes in the presence of TiO$_2$ doped with selective transition metals. *Desalination* **2011**, *276*, 13–27. [CrossRef]

26. Malengreaux, C.M.; Pirard, S.L.; Léonard, G.; Mahy, J.G.; Herlitschke, M.; Klobes, B.; Hermann, R.; Heinrichs, B.; Bartlett, J.R. Study of the photocatalytic activity of Fe3+, Cr3+, La3+ and Eu3+ single-doped and co-doped TiO$_2$ catalysts produced by aqueous sol-gel processing. *J. Alloys Compd.* **2017**, *691*, 726–738. [CrossRef]
27. Carp, O.; Huisman, C.L.; Reller, A. Photoinduced reactivity of titanium dioxide. *Prog. Solid State Chem.* **2004**, *32*, 33–177. [CrossRef]
28. Litter, M.I. Heterogeneous photocatalysis: Transition metal ions in photocatalytic systems. *Appl. Catal. B Environ.* **1999**, *23*, 89–114. [CrossRef]
29. Léonard, G.L.-M.; Malengreaux, C.M.; Mélotte, Q.; Lambert, S.D.; Bruneel, E.; Van Driessche, I.; Heinrichs, B. Doped sol–gel films vs. powders TiO$_2$: On the positive effect induced by the presence of a substrate. *J. Environ. Chem. Eng.* **2016**, *4*, 449–459. [CrossRef]
30. Léonard, G.L.-M.; Pàez, C.A.; Ramírez, A.E.; Mahy, J.G.; Heinrichs, B. Interactions between Zn2+ or ZnO with TiO$_2$ to produce an efficient photocatalytic, superhydrophilic and aesthetic glass. *J. Photochem. Photobiol. A Chem.* **2018**, *350*, 32–43. [CrossRef]
31. Mahy, J.G.; Lambert, S.D.; Tilkin, R.G.; Wolfs, C.; Poelman, D.; Devred, F.; Gaigneaux, E.M.; Douven, S. Ambient temperature ZrO$_2$-doped TiO$_2$ crystalline photocatalysts: Highly efficient powders and films for water depollution. *Mater. Today Energy* **2019**, *13*, 312–322. [CrossRef]
32. Papadimitriou, V.C.; Stefanopoulos, V.G.; Romanias, M.N.; Papagiannakopoulos, P.; Sambani, K.; Tudose, V.; Kiriakidis, G. Determination of photo-catalytic activity of un-doped and Mn-doped TiO$_2$ anatase powders on acetaldehyde under UV and visible light. *Thin Solid Films* **2011**, *520*, 1195–1201. [CrossRef]
33. Malengreaux, C.M.; Douven, S.; Poelman, D.; Heinrichs, B.; Bartlett, J.R. An ambient temperature aqueous sol-gel processing of efficient nanocrystalline doped TiO$_2$-based photocatalysts for the degradation of organic pollutants. *J. Sol-Gel Sci. Technol.* **2014**, *71*, 557–570. [CrossRef]
34. Granados, G.; Martínez, F.; Páez-Mozo, E.A. Photocatalytic degradation of phenol on TiO$_2$ and TiO$_2$/Pt sensitized with metallophthalocyanines. *Catal. Today* **2005**, *107–108*, 589–594. [CrossRef]
35. Mahy, J.G.; Paez, C.A.; Carcel, C.; Bied, C.; Tatton, A.S.; Damblon, C.; Heinrichs, B.; Man, M.W.C.; Lambert, S.D. Porphyrin-based hybrid silica-titania as a visible-light photocatalyst. *J. Photochem. Photobiol. A Chem.* **2019**, *373*, 66–76. [CrossRef]
36. Tasseroul, L.; Pirard, S.L.; Lambert, S.D.; Páez, C.A.; Poelman, D.; Pirard, J.-P.; Heinrichs, B. Kinetic study of p-nitrophenol photodegradation with modified TiO$_2$ xerogels. *Chem. Eng. J.* **2012**, *191*, 441–450. [CrossRef]
37. Hu, L.; Wang, J.; Zhang, J.; Zhang, Q.; Liu, Z. An N-doped anatase/rutile TiO$_2$ hybrid from low-temperature direct nitridization: Enhanced photoactivity under UV-/visible-light. *RSC Adv.* **2014**, *4*, 420–427. [CrossRef]
38. Di Valentin, C.; Pacchioni, G.; Selloni, A.; Livraghi, S.; Giamello, E. Characterization of Paramagnetic Species in N-Doped TiO$_2$ Powders by EPR Spectroscopy and DFT Calculations. *J. Phys. Chem. B* **2005**, *109*, 11414–11419. [CrossRef]
39. Mahy, J.G.; Cerfontaine, V.; Poelman, D.; Devred, F.; Gaigneaux, E.M.; Heinrichs, B.; Lambert, S.D. Highly efficient low-temperature N-doped TiO$_2$ catalysts for visible light photocatalytic applications. *Materials (Basel)* **2018**, *11*, 584. [CrossRef]
40. Iwase, M.; Yamada, K.; Kurisaki, T.; Prieto-Mahaney, O.O.; Ohtani, B.; Wakita, H. Visible-light photocatalysis with phosphorus-doped titanium(IV) oxide particles prepared using a phosphide compound. *Appl. Catal. B Environ.* **2013**, *132–133*, 39–44. [CrossRef]
41. Bodson, C.J.; Heinrichs, B.; Tasseroul, L.; Bied, C.; Mahy, J.G.; Man, M.W.C.; Lambert, S.D. Efficient P- and Ag-doped titania for the photocatalytic degradation of waste water organic pollutants. *J. Alloys Compd.* **2016**, *682*, 144–153. [CrossRef]
42. Cheng, X.; Liu, H.; Chen, Q.; Li, J.; Wang, P. Construction of N, S codoped TiO$_2$ NCs decorated TiO$_2$ nano-tube array photoelectrode and its enhanced visible light photocatalytic mechanism. *Electrochim. Acta* **2013**, *103*, 134–142. [CrossRef]
43. Di Valentin, C.; Pacchioni, G. Trends in non-metal doping of anatase TiO$_2$: B, C, N and F. *Catal. Today* **2013**, *206*, 12–18. [CrossRef]
44. Lecloux, A. Exploitation des isothermes d'adsorption et de désorption d'azote pour l'étude de la texture des solides poreux. *Mém. Soc. R. Des Sci. Liège 6ème série* **1971**, *1*, 169–209.

45. Mahy, J.G.; Léonard, G.L.-M.; Pirard, S.; Wicky, D.; Daniel, A.; Archambeau, C.; Liquet, D.; Heinrichs, B. Aqueous sol–gel synthesis and film deposition methods for the large-scale manufacture of coated steel with self-cleaning properties. *J. Sol-Gel Sci. Technol.* **2017**, *81*, 27–35. [CrossRef]
46. Li, X.; Chen, Z.; Shi, Y.; Liu, Y. Preparation of N, Fe co-doped TiO_2 with visible light response. *Powder Technol.* **2011**, *207*, 165–169. [CrossRef]
47. Kim, T.-H.; Rodríguez-González, V.; Gyawali, G.; Cho, S.-H.; Sekino, T.; Lee, S.-W. Synthesis of solar light responsive Fe, N co-doped TiO_2 photocatalyst by sonochemical method. *Catal. Today* **2013**, *212*, 75–80. [CrossRef]
48. Su, Y.; Xiao, Y.; Li, Y.; Du, Y.; Zhang, Y. Preparation, photocatalytic performance and electronic structures of visible-light-driven Fe–N-codoped TiO_2 nanoparticles. *Mater. Chem. Phys.* **2011**, *126*, 761–768. [CrossRef]
49. Aba-Guevara, C.G.; Medina-Ramírez, I.E.; Hernández-Ramírez, A.; Jáuregui-Rincón, J.; Lozano-Álvarez, J.A.; Rodríguez-López, J.L. Comparison of two synthesis methods on the preparation of Fe, N-Co-doped TiO_2 materials for degradation of pharmaceutical compounds under visible light. *Ceram. Int.* **2017**, *43*, 5068–5079. [CrossRef]
50. Azouani, R.; Tieng, S.; Chhor, K.; Bocquet, J.-F.; Eloy, P.; Gaigneaux, E.M.; Klementiev, K.; Kanaev, A.V. TiO_2 doping by hydroxyurea at the nucleation stage: Towards a new photocatalyst in the visible spectral range. *Phys. Chem. Chem. Phys.* **2010**, *12*, 11325–11334. [CrossRef]
51. Bittencourt, C.; Rutar, M.; Umek, P.; Mrzel, A.; Vozel, K.; Arčon, D.; Henzler, K.; Krüger, P.; Guttmann, P. Molecular nitrogen in N-doped TiO_2 nanoribbons. *RSC Adv.* **2015**, *5*, 23350–23356. [CrossRef]
52. Livraghi, S.; Chierotti, M.R.; Giamello, E.; Magnacca, G.; Paganini, M.C.; Cappelletti, G.; Bianchi, C.L. Nitrogen-Doped Titanium Dioxide Active in Photocatalytic Reactions with Visible Light: A Multi-Technique Characterization of Differently Prepared Materials. *J. Phys. Chem. C* **2008**, *112*, 17244–17252. [CrossRef]
53. Smirniotis, P.G.; Boningari, T.; Damma, D.; Inturi, S.N.R. Single-step rapid aerosol synthesis of N-doped TiO_2 for enhanced visible light photocatalytic activity. *Catal. Commun.* **2018**, *113*, 1–5. [CrossRef]
54. Gil, A.; García, A.M.; Fernández, M.; Vicente, M.A.; González-Rodríguez, B.; Rives, V.; Korili, S.A. Effect of dopants on the structure of titanium oxide used as a photocatalyst for the removal of emergent contaminants. *J. Ind. Eng. Chem.* **2017**, *53*, 183–191. [CrossRef]
55. Mahy, J.G.; Lambert, S.D.; Léonard, G.L.-M.; Zubiaur, A.; Olu, P.-Y.; Mahmoud, A.; Boschini, F.; Heinrichs, B. Towards a large scale aqueous sol-gel synthesis of doped TiO_2: Study of various metallic dopings for the photocatalytic degradation of p-nitrophenol. *J. Photochem. Photobiol. A Chem.* **2016**, *329*, 189–202. [CrossRef]
56. El Koura, Z.; Cazzanelli, M.; Bazzanella, N.; Patel, N.; Fernandes, R.; Arnaoutakis, G.E.; Gakamsky, A.; Dick, A.; Quaranta, A.; Miotello, A. Synthesis and Characterization of Cu and N Codoped RF-Sputtered TiO_2 Films: Photoluminescence Dynamics of Charge Carriers Relevant for Water Splitting. *J. Phys. Chem. C* **2016**, *120*, 12042–12050. [CrossRef]
57. Suwannaruang, T.; Hildebrand, J.P.; Taffa, D.H.; Wark, M.; Kamonsuangkasem, K.; Chirawatkul, P.; Wantala, K. Visible light-induced degradation of antibiotic ciprofloxacin over Fe–N–TiO_2 mesoporous photocatalyst with anatase/rutile/brookite nanocrystal mixture. *J. Photochem. Photobiol. A Chem.* **2020**, *391*, 112371. [CrossRef]
58. Mahy, J.G.; Wolfs, C.; Mertes, A.; Vreuls, C.; Drot, S.; Smeets, S.; Dircks, S.; Boergers, A.; Tuerk, J.; Lambert, S.D. Advanced photocatalytic oxidation processes for micropollutant elimination from municipal and industrial water. *J. Environ. Manag.* **2019**, *250*, 109561. [CrossRef]
59. Lu, Y.; Xu, Y.; Wu, Q.; Yu, H.; Zhao, Y.; Qu, J.; Huo, M.; Yuan, X. Synthesis of Cu_2O nanocrystals/TiO_2 photonic crystal composite for efficient p-nitrophenol removal. *Colloids Surf. A Physicochem. Eng. Asp.* **2018**, *539*, 291–300. [CrossRef]
60. Kothavale, V.P.; Patil, T.S.; Patil, P.B.; Bhosale, C.H. Photoelectrocatalytic degradation of Rhodamine B using N doped TiO_2 thin Films. *Mater. Today Proc.* **2020**, *23*, 382–388. [CrossRef]
61. Gastello, E.; Estrada, D.; Estrada, W.; Luyo, C.; Espinoza, J.; Ponce, S.; de Oca, J.M.; Rodriguez, J.M. TiO_2 films on $CoFe_2O_4$ nanoparticles for the Photocatalytic oxidation of Rhodamine B: Influence of the alcoholic solutions. *Proc. SPIE* **2019**, *11371*, 113710C. [CrossRef]
62. Hao, S.; Lin, T.; Ning, S.; Qi, Y.; Deng, Z.; Wang, Y. Research on cracking of SiO_2 nanofilms prepared by the sol-gel method. *Mater. Sci. Semicond. Process.* **2019**, *91*, 181–187. [CrossRef]
63. Doebelin, N.; Kleeberg, R. Profex: A graphical user interface for the Rietveld refinement program BGMN. *J. Appl. Crystallogr.* **2015**, *48*, 1573–1580. [CrossRef] [PubMed]

64. Tobaldi, D.M.; Piccirillo, C.; Rozman, N.; Pullar, R.C.; Seabra, M.P.; Škapin, A.S.; Castro, P.M.L.; Labrincha, J.A. Effects of Cu, Zn and Cu-Zn addition on the microstructure and antibacterial and photocatalytic functional properties of Cu-Zn modified TiO_2 nano-heterostructures. *J. Photochem. Photobiol. A Chem.* **2016**, *330*, 44–54. [CrossRef]
65. Lopes, D.; Daniel-da-Silva, A.L.; Sarabando, A.R.; Arias-Serrano, B.I.; Rodríguez-Aguado, E.; Rodríguez-Castellón, E.; Trindade, T.; Frade, J.R.; Kovalevsky, A.V. Design of Multifunctional Titania-Based Photocatalysts by Controlled Redox Reactions. *Materials* **2020**, *13*, 758. [CrossRef] [PubMed]
66. Kubelka, P.; Munk, F. Ein Beitrag zur Optik der Farban striche. *Z. Tech. Phys.* **1931**, *12*, 593–601.
67. Kubelka, P. New contributions to the optics of intensely light-scattering materials. *J. Opt. Soc. Am.* **1948**, *38*, 448–457. [CrossRef]
68. Bryson, C.E. Surface potential control in XPS. *Surf. Sci.* **1987**, *189–190*, 50–58. [CrossRef]
69. Shirley, D.A. High-Resolution X-Ray Photoemission Spectrum of the Valence Bands of Gold. *Phys. Rev. B* **1972**, *5*, 4709–4714. [CrossRef]

© 2020 by the authors. Licensee MDPI, Basel, Switzerland. This article is an open access article distributed under the terms and conditions of the Creative Commons Attribution (CC BY) license (http://creativecommons.org/licenses/by/4.0/).

Article

Synthesis of Thin Titania Coatings onto the Inner Surface of Quartz Tubes and Their Photoactivity in Decomposition of Methylene Blue and Rhodamine B

Stanislav D. Svetlov [1], Dmitry A. Sladkovskiy [2], Kirill V. Semikin [2], Alexander V. Utemov [2], Rufat Sh. Abiev [1] and Evgeny V. Rebrov [1,3,4,*]

1. Department of Optimization of Chemical and Biotechnological Equipment, St. Petersburg State Institute of Technology (Technical University), St. Petersburg 190013, Russia; svetlovstanislav@gmail.com (S.D.S.); ohba@lti-gti.ru (R.S.A.)
2. Resource-Saving Department, St. Petersburg State Institute of Technology (Technical University), St. Petersburg 190013, Russia; dmitry.sla@gmail.com (D.A.S.); kirrse@gmail.com (K.V.S.); avutemov@rambler.ru (A.V.U.)
3. School of Engineering, University of Warwick, Coventry CV4 7AL, UK
4. Department of Chemical Engineering and Chemistry, Eindhoven University of Technology, P.O. Box 513, 5600 MB Eindhoven, The Netherlands
* Correspondence: E.Rebrov@warwick.ac.uk

Abstract: An evaporation-deposition coating method for coating the inner surface of long (>1 m) quartz tubes of small diameter has been studied by the introduction of two-phase (gas-liquid) flow with the gas core flowing in the middle and a thin liquid film of synthesis sol flowing near the hot tube wall. The operational window for the deposition of continuous titania coatings has been obtained. The temperature range for the deposition of continuous titania coatings is limited to 105–120 °C and the gas flow rate is limited to the range of 0.4–1.0 L min^{-1}. The liquid flow rate in the annular flow regime allows to control the coating thickness between 3 and 10 micron and the coating porosity between 10% and 20%. By increasing the liquid flow rate, the coating porosity can be substantially reduced. The coatings were characterized by X-ray diffraction, N_2 chemisorption, thermogravimetric analysis, and scanning electron microscopy. The coatings were tested in the photocatalytic decomposition of methylene blue and rhodamine B under UV-light and their activity was similar to that of a commercial P25 titania catalyst.

Keywords: titania coatings; gas-liquid flow; sol-gel method; methylene blue; rhodamine B

1. Introduction

Semiconductor catalysts, in particular titania, were widely applied in wastewater treatment [1,2], environmental applications [3], energy storage [4], biological applications [5], and in the production of fuels and chemicals [6]. Titania absorbs only UV-light and the position of the absorption band depends on the phase composition [7]. Titania has two polymorphs, anatase and rutile. The transformation of anatase to rutile starts at 400 °C [8]. Anatase has a higher photocatalytic activity in the case of relatively thick films (on the order of several microns) [9]. However, anatase dissolves in acidic solutions at a faster rate than the rutile phase and it suffers degradation under accelerated photocatalytic cycles, and therefore its durability is often compromised. To stabilize the catalytic activity, often a mixture of anatase and rutile is desirable.

Thin titania coatings [10,11], titania nanoparticles (NPs) [12–15], and supported titania catalysts [16] were widely employed in the decomposition of organic dyes. Among them, the suspension of colloidal NPs demonstrated the highest productivity, because it provides a good contact between the titania and the organic pollutants. However, the handling of colloidal suspensions often causes clogging, and it requires an expensive post-treatment filtration process. Several coating methods to deposit catalysts onto structured substrates were reviewed by Meille [17]. Briefly, they include suspension sedimentation [18–20], sol-gel synthesis [21–24], a hybrid sol-gel method with additional pre-processing or post-processing steps [25,26], and electrochemical sedimentation [27]. The addition of active metals is often performed by an additional impregnation step followed by calcination [28,29]. The suspension and sol-gel methods are most widely used due to their simplicity and wide range of coating thickness and catalyst porosity that they could offer. The coating thickness can be varied from 300 nm to 100 μm by the suspension methods and it is determined by the size of the particles used [17]. The sol-gel method gives a thickness starting from 100 nm [30,31]. In general, the sol composition, the type of surfactant, and solvent evaporation conditions determine the coating porosity and thickness [32]. The solvent removal from open surfaces, such as flat plates and the outer surfaces of tubes, often happens at moderate heating just above room temperature. However, this method is not applicable for solvent removal from the inner surface of a tube. Bravo et al. introduced a gas flow to displace the liquid in a tube [33]. Their method was applied for highly viscous liquids (0.15–0.25 Pa·s). However, the extension of the method to other synthesis sols did not provide continuous coatings. Previously a combustion-evaporation method was developed, where a tube filled with a sol was slowly moving into a tubular oven maintained far above the boiling temperature of the solvent [34]. The method is similar to the static coating methods but uses elevated temperature and introduces an additional control parameter, the tube displacement speed, that allows to control the coating thickness. When an inert gas was added to the liquid flow, coatings with a very high adhesion to the wall were obtained. The addition of a gas eliminates fouling, the major problem in slurry reactors. The coatings obtained allow for stable operation for a very long time on stream, often for several hundreds of hours [35]. The high heat transfer rate allows fast cooling of reaction mixture therefore highly exothermic hydrogenation reactions can be performed under solvent-free conditions. In this way, Pd-Bi/TiO$_2$ coatings were obtained and tested in hydrogenation reaction under flow conditions [36]. The catalyst remained stable for 100 h of continuous operation with a high selectivity to the desired product (98%) in the continuous mode. More recently, the hydrogenation of an imine (N-Cyclohexyl-(benzylidene) imine) into a secondary amine in the continuous flow was demonstrated [37]. The long-term coating stability allowed reaching a turnover number (TON) of 150,000, an unprecedented value under operation with coated catalysts.

In the boiling-decomposition method, the coating thickness and its morphology are determined by boiling conditions inside the tube. There are four modes of two-phase flow in tubes of small diameter: bubbly, slug, annular and churn [38]. For channels of large diameter, there exists also stratified and parallel flow regime. Slug flow refers to the phenomenon whereby gas-liquid flow is present in a tube over a wide range of intermediate gas and liquid velocities. It was observed that the addition of a non-reactive gas can create the slug flow regime, increasing mixing and associated heat and mass transfer rates [39]. However, a slug flow regime may often result in the formation of solid plugs, while boiling under annular flow regime allows to obtain a uniform coating [35]. The boiling heat transfer was studied for two-phase slug and annular flow in microchannels [40–43]. However, most results from the past experimental studies display a substantial disagreement on the influence of the flow conditions and mechanisms on the heat transfer rate. Partially, this can be explained by the fact that the wall microstructures make use of the capillary forces to evenly distribute the liquid fuel over the wall, so that the appearance of uncontrolled dry patches can be avoided in the channels of small (below 2 mm) inner diameter. There is limited number of experimental studies describing boiling heat transfer in channels of small diameter. Peterson and Ma [43] investigated the maximum heat flux to the flow which allowed to estimate the minimum furnace temperature required for coating deposition. Helbig et al. [44] studied the flow in a channel with grooved walls. They concluded that before forming a dry spot, the liquid in the grooves begins to behave in an unstable manner and breaks up into droplets, which may result in discontinuous coatings. Sibiryakov et al.

presented a numerical solution for liquid boiling in triangular channels [45] and in channels with smooth and grooved walls [46]. Warrier et al. [47] proposed correlations for heat transfer coefficient under two-phase boiling conditions previously reported in [48–50]. Similar to the single-phase flow, the heat transfer coefficient increases with an increase in the flow rate. The presence of solutes affects the surface tension, density, and boiling rate of the liquid [51]. Therefore, the boiling rate of a solution can differ significantly from that of water. Recently, Wang et al. described the boiling of liquids in the presence of additives [52].

In this work, the boiling-decomposition method was investigated under two-phase flow conditions with the introduction of a non-reacting gas flow to the synthesis sol flow. An operational window resulting in the formation of stable continuous coatings was experimentally studied. The effect of gas and liquid flow rate, the oven temperature, and the tube displacement speed on the titania morphology and coating thickness was studied. The effect of an additional post-processing annealing step on phase composition was also investigated.

2. Results and Discussion

An annular gas-liquid flow regime was chosen for coating deposition. In this regime, the liquid flows as a thin film near the inner channel wall, while gas flows in the center of the tube.

Such flow conditions provide a good thermal contact between the liquid and the channel wall, so the boiling rate can be controlled by the thickness of the flowing liquid film and the temperature excess (the difference between the actual temperature and the boiling temperature), similar to a single-phase flow. The presence of the gas core prevents the formation of solid plugs which were previously observed when large liquid slugs were present in the channel.

The effect of gas and liquid flow rates and the oven temperature on the average thickness of the titania coatings was studied in three series of experiments. In series A, the effect of temperature was studied at a fixed gas and liquid velocity of 2.3 m s^{-1} and 1.2 mm s^{-1}, respectively (Figure 1a). The preheater temperature was set to prevent boiling before the furnace. Increasing the temperature increases the boiling rate and leads to the formation of more dense titania coatings. Therefore, the mean coating thickness monotonously decreases with temperature. In this temperature range, vaporization of the liquid film is promoted by the lower latent heat of vaporization of solvent. On the contrary, very porous (foam-like) coatings were produced in the temperature range above 150 °C. These coatings were rather fragile and a considerable amount of material was detached from the surface in the subsequent calcination step. Thus, the resulting thickness of the coatings obtained at elevated evaporator temperatures is less than in the case of moderate temperatures. The solvent was partially decomposed, and the coatings were gray in color after deposition due to the presence of carbon deposits. Therefore, the maximum temperature was fixed to 150 °C in the subsequent optimization experiments.

In series B, the effect of liquid velocity on the coating thickness was studied at a constant temperature and a constant gas flow rate (Figure 1b). The range of flow rates between 0.6 and 1.8 mm s^{-1} was chosen based on the results of our previous study [34]. Previously, contrasting trends were obtained, with the coating thickness either decreasing or increasing by increase of liquid mass flow rate. The increase of boiling rate can be attributed to a coalescence of gas bubbles, which increases the thermal flux and therefore the boiling rate. The increasing bubble nucleation frequency induced by the higher flow rate promotes bubble detachment from the wall due to increased drag force. Figure 1b shows that an increase of liquid flow rate enhances the heat transfer performance and the boiling rate, as a consequence of the thinner liquid film and thus higher evaporation and coating deposition rates. The higher coating mass in this flow range can be explained by the increased precursor evaporation rate [53].

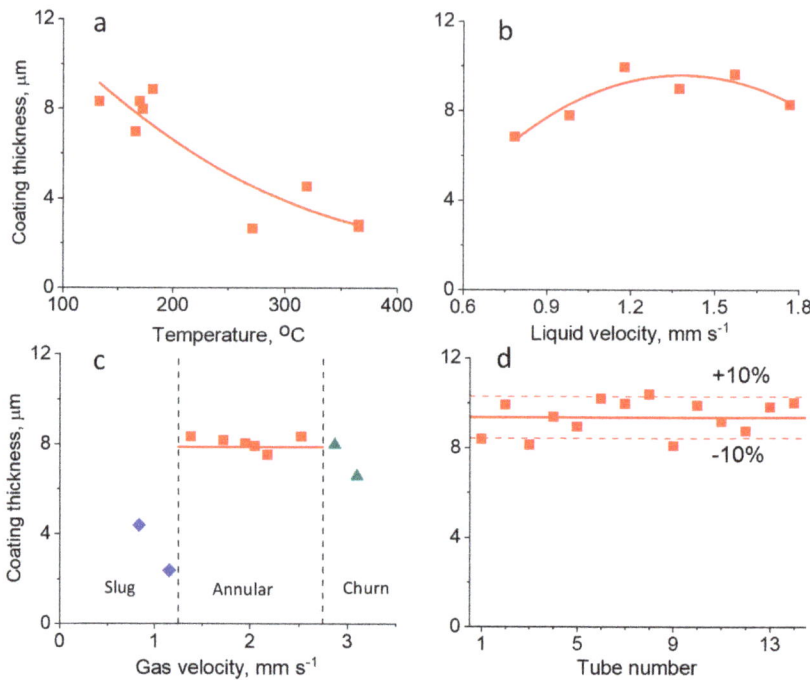

Figure 1. Effect of (**a**) oven temperature, (**b**) liquid velocity; (**c**) gas velocity onto the mean coating thickness in as-synthesized samples, the two vertical dashed lines show flow regime transitions; (**d**) reproducibility experiments performed at optimised conditions (liquid flow rate: 1.2 mm s^{-1}, gas flow rate: 2.0 m s^{-1}, temperature: 115 °C).

The liquid also contains a large number of vapor bubbles, and rather porous coatings were formed with a larger thickness, as shown schematically in Figure 2a. The respective optical images are shown in Figure 3a–d. The white color of the coatings is due to multiple light reflections in their porous structure. However, an opposite trend was found in the higher liquid flow range, above 1.5 mm s^{-1}. In this range, the coating thickness decreases as a result of higher bubble nucleation frequency and the formation of a continuous gas layer near the hot channel wall [54]. The bubbles prevent an efficient heat transfer to the liquid film and therefore the coating mass decreases (Figure 1b). Moreover, some droplets can be carried away from the liquid film by the gas flow (Figure 3e) and they do not contribute to the formation of coating. Nucleate boiling does not occur at these conditions and the coatings formed are semi-transparent films without a developed pore structure (Figure 3e,f). A schematic mechanism for their formation in shown in Figure 2b. A similar mechanism was observed at higher process temperatures corresponding to high heat fluxes.

The gas velocity has a minor effect on the coating thickness under annular flow regime (Figure 1c). With increasing flow rate above 2.0 m s^{-1}, a significant amount of liquid becomes transferred from the annular film to the gas core. At a gas velocity of 2.8 m s^{-1}, a transition to mist flow regime occurs where all of the liquid is entrained in the gas flow. Due to the large drop in heat-transfer coefficient that accompanies tube wall dryout in the mist flow, the evaporation rate decreases, and this leads to a decrease in the coating thickness (Figure 1c). On the other side, a transition to slug flow occurs below a gas velocity of 1.3 m s^{-1}. In the periodic passage of elongated bubbles and liquid slugs which is characteristic of the slug flow pattern, the maximum heat transfer rate is achieved when the elongated bubbles are formed at relatively low gas flow rates (below 1 m s^{-1}).

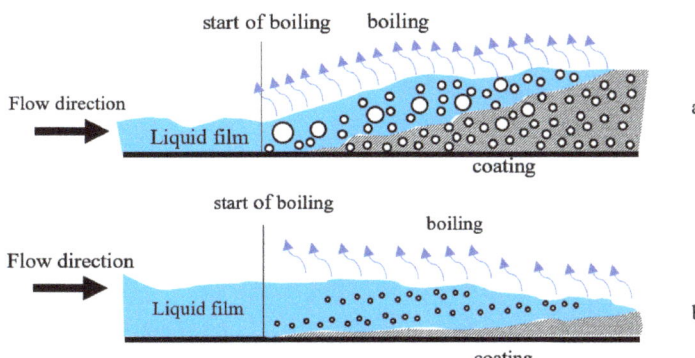

Figure 2. Schematic mechanism of coating formation. (**a**) formation of porous coatings at liquid flow rate of 0.8–1.5 mm s^{-1}, (**b**) formation of dense semi-transparent coatings at liquid flow rate of 1.5–1.8 mm s^{-1}.

Figure 3. Images of coated tubes obtained at different liquid flow velocities: (**a**) 0.8, (**b**) 1.0, (**c**) 1.2, (**d**) 1.4, (**e**) 1.6, (**f**) 1.8 mm s^{-1}. Temperature: 115 °C, gas flow rate: 2.0 m s^{-1}.

The XRD diffractogram of the coatings obtained in the slug and annular flow regimes are shown in Figure 4. Due to a rather low coating thickness, the scanning range was limited to 24–31° 2-thetas to reduce the beam time, however no impurities of other phases were observed when a wider range of angles was analyzed. Due to a sharp curvature of the coated tubes, the measurements in a wider range are very time consuming as they result in a large scattering and a very low signal to noise ratio requiring very low XRD scanning rates. The presence of other reflections could be hindered by a preferred orientation of the crystals onto the tube wall. The main phase was rutile, as confirmed by its strongest peak at 27.0° 2-theta while small amounts of anatase were also present in the samples obtained in the annular flow regime. The positions of the strongest XRD peaks were consistent with the standard XRD data of the anatase and rutile TiO$_2$ phase (JCPDS no. 21-1272 and JCPDS no. 21-1276, respectively).

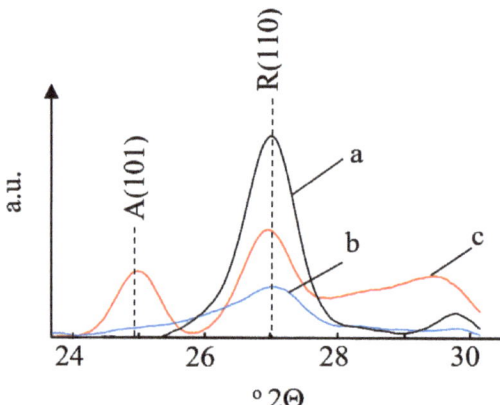

Figure 4. XRD diffractogram of as-synthesized coatings obtained at different gas flow rates. Abbreviations: A—anatase, R—rutile. Gas flow rate: (a) 0.6 m s^{-1}, (b) 1.0 m s^{-1}, (c) 1.4 m s^{-1}. Liquid flow rate: 1.2 mm s^{-1}. Temperature: 115 °C.

The formation of anatase occurs in a lower temperature range as compared to that of rutile. This allows to conclude that the surface temperature was lower in the annular flow regime due to the much higher heat transfer rate, which is in line with the previous discussion. It should be mentioned that the temperature of anatase to rutile transition depends also on the particle size, shape of the nanostructure, and presence of dopants and structural defects [55,56]. However, these parameters were very similar in all samples. Therefore, the phase composition of the resulting coatings is mainly determined by the hydrodynamics of two-phase flow and the related heat transfer rates. The sample obtained in the annular regime demonstrated an apparent density of 3.37 kg m^{-3}, which was increased to 3.85 kg m^{-3} after calcination. Moreover, the anatase phase completely disappeared after the additional calcination step (Figure 4). A non-porous rutile has a density of 4.24 g cm^{-3}. Thus, a simple estimation shows that the porosity of coatings decreases from 20.5% to 9.2% after calcination. Based on the above data, it can be concluded that optimal conditions correspond to a temperature of 115–120 °C, a liquid velocity of 1.2 mm s^{-1}, and a gas velocity of 1.4–2.6 m s^{-1}.

Figure 5 shows SEM images of coating produced in the annular flow regime under optimized process conditions. It can be seen that the mean coating thickness is about 9 µm, and it reduced to 7 µm after an additional calcination step at 400 °C. These data are in good agreement with the data obtained from the gravimetric analysis. The respective N_2 adsorption-desorption isotherms and the pore-size distribution are shown in Figure 6. The specific surface area of the coating was 9.2 m^{2}g^{-1} with a mean pore size of 5.1 nm. The mean crystallite size obtained from the XRD analysis is 10 nm. This corresponds to a low end of the range of particle sizes observed in P25 titania catalysts. The P25 titania can be seen as benchmarking for the coatings produced in this study. It has a wide particle size distribution between 10 and 40 nm and a typical surface area of 50 m^{2}g^{-1}, while the coatings obtained in this study demonstrated a value five times smaller. It appears that a large part of the coatings has no porosity and therefore it is not accessible for N_2 adsorption. As no structure directing agents were used in the synthesis sol (see experimental section), the formation of a partially non-porous coating cannot be excluded. A coating porosity in the 10–20% range also supports this conclusion.

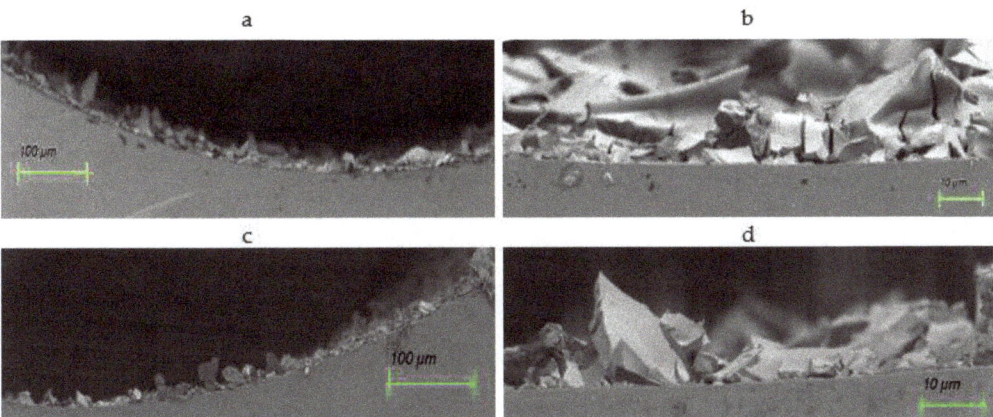

Figure 5. SEM images of optimised coatings obtained at a liquid flow rate of 1.2 mm s^{-1} and a gas flow rate of 2.2 m s^{-1} in the annular flow regime. (**a**,**b**) as-synthesized, (**c**,**d**) after calcination at 400 °C.

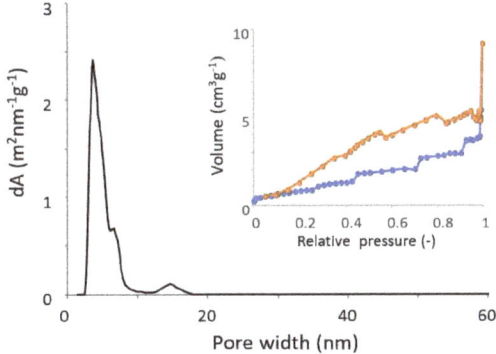

Figure 6. Pore size distribution and nitrogen adsorption/desorption isotherms over the coating obtained at a liquid flow rate of 1.2 mm s^{-1} and a gas flow rate of 2.2 m s^{-1}.

The reproducibility of the coating method was studied by coating fourteen tubes under optimized process conditions. It can be seen in Figure 1d that the relative standard deviation was 12%, demonstrating the rather good reproducibility of the method.

The operational window of the method is shown in Figure 7. The minimum temperature of deposition is obtained from the energy balance in the system when the heat transfer rate is equal to the rate of solvent evaporation.

$$hA(T_b - T_{sat}) = \dot{m}_L \Delta H, \tag{1}$$

where h is the heat transfer coefficient, A is the inner surface area of the tube, T_b is the wall temperature, T_{sat} is the temperature of evaporation, $\dot{m}_L$ is the liquid mass flow rate, and ΔH is the enthalpy of solvent evaporation. This energy balance can be assumed based on the fact that as the liquid is already preheated to the evaporation temperature in the preheater section, the heating of the gas phase above the boiling temperature can be neglected. The rearrangement of Equation (1) allows to estimate the minimum boiling temperature as a function of liquid flow rate:

$$T_b = \frac{\Delta H \cdot \dot{m}_L}{h \cdot A} + T_{sat} \tag{2}$$

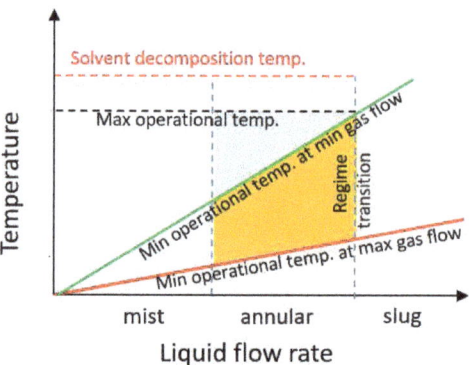

Figure 7. Operational window for coating deposition. The blue color shows the operational range at a minimum gas flow rate of 0.4 L min^{-1}, the orange color shows the extension of operational range towards lower temperatures at a maximum gas flow rate of 1.0 L min^{-1}.

The heat transfer coefficient increases at high gas flow rate and therefore the slope of the minimum temperature line (shown in red in Figure 7) decreases. The two vertical lines show the transitions to the slug and churn flow regimes where the operation should be avoided. Finally, the upper temperature range, shown by a horizontal maximum operational temperature line in Figure 7, should always be below the solvent decomposition temperature. The exact difference depends on the solvent type and usually stays in the range of 10–30 K. In this study, the maximum operational temperature corresponding to the formation of continuous coatings is limited to 135 °C. It can further be decreased to 115 °C (minimum operational temperature) by increasing the gas flow rate from 1.4 to 2.6 m s^{-1}.

The coatings obtained were tested in the decomposition of two organic compounds at 20 °C. The kinetic data were corrected by subtracting the rate of non-catalytic reaction measured in a blank experiment with a non-coated tube. The $\ln(C_0/C)$ values versus time provided a straight line (Figure 8). Therefore, the photocatalytic reaction rate is described by a first order kinetics with a rate constant of 0.0120 min^{-1} for MB and 0.0253 min^{-1} for RhB. The value reported for MB is in a very good agreement with that (0.014 min^{-1}) reported over a P25 titania catalyst [57]. The rate constant for the decomposition of RhB exceeds this value (0.0194 min^{-1}) reported over a TiO$_2$/SiC catalyst [16].

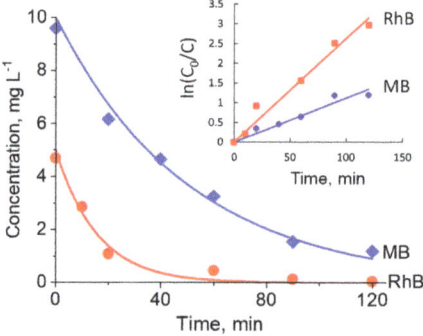

Figure 8. Concentration of methylene blue (MB) and rhodamine B (RhB) as a function of time in stop flow conditions over a TiO$_2$ coated tube. The coating was produced under the same conditions as those reported in Figure 5. The data are corrected based on the rate over non-coated tubes measured in blank experiments. The insert shows the fitting curves based on a first order kinetics.

3. Materials and Methods

The initial sol was prepared from titanium isopropoxide (97.0 wt.%, Sigma-Aldrich, St. Louis, MO, USA), isopropanol (99.5 wt.%, Sigma-Aldrich, St. Louis, MO, USA) and HNO$_3$ (65 wt%, Sigma-Aldrich). First, a solution of nitric acid (0.1 mL) was added to isopropanol (8.45 g) to obtain a solution A. Then, titanium isopropoxide was added dropwise to solution A to obtain the final solution. The mixture was preheated to 60 °C and stirred for 2 h. Figure 9 shows the experimental set-up employed for coating deposition. The nitrogen flow and the flow of synthesis mixture were mixed in a coaxial mixer and fed to a tubular furnace via a preheater section. The quartz tube (i.d. 3.0 mm, o.d. 4.0 mm) was positioned vertically in a furnace with a 50 mm length. The liquid solution was fed with a syringe pump and the gas flow was fed with a mass flow controller. In this study, the liquid flow rate was varied from 0.33 to 0.75 mL min^{-1} and the gas flow rate was varied from 0.43 to 1.27 L min^{-1}. The temperature in the preheater section was set just below the boiling temperature of the synthesis mixture, while the temperature in the furnace was varied in the 115–350 °C range. The images of the gas-liquid flow were recorded using a high-speed videocamera. In the beginning of each experiment, a gas-liquid flow was fed in the tube. Once the temperature of the tubular furnace reached the setpoint, the quartz tube was fed to the furnace using a stepper motor. The tube displacement speed was fixed at 1.6 mm s^{-1}. At the end of each deposition run, the coated tube was detached from the connecting lines and annealed in an oven at 400 °C for 3 h with a heating rate of 1 K min^{-1}.

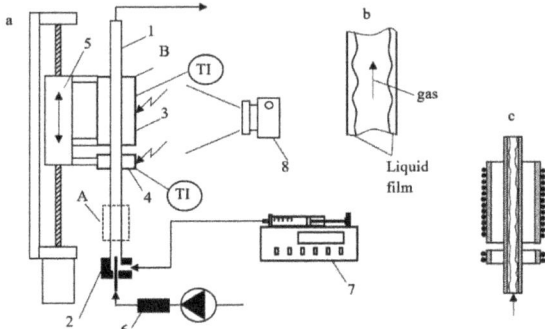

Figure 9. Schematic view of the experimental set-up for coating deposition (**a**) general view: (1) quartz tube, (2) T-mixer, (3) tubular furnace, (4) electrical preheater section, (5) stepper motor, (6) mass flow controller, (7) syringe pump, (8) high speed camera, TI is the temperature indicator, (**b**) schematic view of annular flow regime, (**c**) enlarged view of the furnace and the preheater section.

For analysis, the tubes were cut into short sections with a length of 15 mm. The XRD patterns were recorded with a Shimadzu XRD-7000 diffractometer in the 23–32° 2θ range using a CuKα irradiation. The scanning rate was 2.0° 2-theta min^{-1}. The specific surface area and the pore volume were obtained from N$_2$ adsorption-desorption isotherms obtained on a Quantachrome Autosorb-6iSA apparatus. For these measurements, the coatings obtained were mechanically removed from the inner tube wall. Before the measurements, the samples were degassed at 250 °C for 1 h. The specific surface area was calculated using a multi-point BET method from the adsorption isotherm. The mean coating thickness (δ) was determined by Equation (3).

$$\delta = \frac{\Delta m}{\rho_c \cdot \pi \cdot d \cdot L_t} \quad (3)$$

where Δm is the weight change after coating deposition, ρ_c is the apparent coating density determined from the sample porosity, d is the tube inner diameter, and L_t is the coated length. In this method, the coating (geometrical) volume was approximated by the product of $\pi \cdot d \cdot \delta \cdot L_t$. The apparent coating density was calculated by Equation (4):

$$\rho_c = \frac{m}{V_{TiO2} + V_p} \tag{4}$$

where V_{TiO2} is the specific volume of titania, V_p is the specific pore volume.

The decomposition of methylene blue ($C_{16}H_{18}ClN_3S$, MB) and rhodamine B ($C_{28}H_{31}ClN_2O_3$, RhB) was studied in the coated tubes at 20 °C under stop flow conditions under UV light (power: 26 W in the range of 320–370 nm) illuminated from a distance of 2 cm. The solution volume in the tube was 0.87 mL. The size of the light source was considerably larger than the illuminated reactor area, providing a uniform intensity over the entire length.

RhB and MB dyes (purity $\geq$ 99.0 wt.%) were obtained from Fluka. The initial concentrations of MB and RhB were 10.0 and 5.0 mg L^{-1}, respectively. After the initial adsorption in the dark, the concentration decreased by 4%. The concentration was measured with a UV-VIS spectrometer (UV-1800, Shimadzu, Kyoto, Japan) using calibration curves. A quartz cuvette with an optical path of 10 mm was used. The samples were diluted 10 times in distilled water for analysis. Blank experiments were also carried out with non-coated quartz tubes to obtain the rate of non-catalytic decomposition at the same experimental conditions.

4. Conclusions

An operational window for the evaporation-deposition method for the controlled deposition of micrometer-thick titania coatings on the inner surface of long quartz tubes has been studied. The liquid flow rate and the oven temperature were the most important parameters that control coating morphology. In particular, the effect of liquid velocity in the range of 0.8–1.8 mm s^{-1} and the boiling temperature in the range of 115–350 °C was systematically investigated. The operation in the annular flow regime allows to reduce the coating deposition temperature from 135 to 115 °C by increasing the gas flow rate from 1.4 to 2.6 mm s^{-1}. This allows uniform solvent removal leading to the formation of continuous coatings with an average thickness in the range between 3 and 10 mm. The porosity of the coatings decreases with increasing liquid flow rate. The preheating of the precursor mixture to the temperature just below its boiling point was an important factor to increase the reproducibility of the method. A mean standard deviation in the coating thickness of 12% was obtained under optimised conditions. The coatings obtained were active in the reactions of photocatalytic decomposition of methylene blue and rhodamine B with no observed catalyst deactivation. The photocatalytic reaction rates were comparable to those previously reported over P25 titania catalysts.

Author Contributions: Conceptualization, E.V.R. and S.D.S.; methodology, R.S.A.; formal analysis, S.D.S.; investigation, S.D.S., D.A.S., K.V.S. and A.V.U.; writing—original draft preparation, S.D.S.; review and editing, E.V.R.; supervision, R.S.A. and E.V.R.; funding acquisition, E.V.R. All authors have read and agreed to the published version of the manuscript.

Funding: This research was funded by the Russian Science Foundation, grant number 20-69-46041.

Data Availability Statement: Not applicable.

Acknowledgments: The authors would like to thank Yuyan Gong from the University of Warwick for SEM analysis.

Conflicts of Interest: The authors declare no conflict of interest.

References

1. Sacco, O.; Vaiano, V.; Rizzo, L.; Sannino, D. Photocatalytic activity of a visible light active structured photocatalyst developed for municipal wastewater treatment. *J. Clean. Prod.* **2018**, *175*, 38–49. [CrossRef]
2. Yang, H.; Yang, J. Photocatalytic degradation of rhodamine B catalyzed by TiO_2 films on a capillary column. *RSC Adv.* **2018**, *8*, 11921–11929. [CrossRef]
3. Zhang, Q.; Fu, Y.; Wu, Y.; Zhang, Y.N.; Zuo, T. Low-Cost Y-Doped TiO_2 Nanosheets film with highly reactive {001} facets from CRT waste and enhanced photocatalytic removal of Cr(VI) and methyl orange. *ACS Sustain. Chem. Eng.* **2016**, *4*, 1794–1803. [CrossRef]
4. Zhao, J.; Yang, Y.; Li, Y.; Zhao, L.; Wang, H.; Song, G.; Tang, G. Microencapsulated phase change materials with TiO_2-doped PMMA shell for thermal energy storage and UV-shielding. *Sol. Energy Mater. Sol. Cells* **2017**, *168*, 62–68. [CrossRef]
5. Endres, P.J.; Paunesku, T.; Vogt, S.; Meade, T.J.; Woloschak, G.E. DNA-TiO_2 nanoconjugates labeled with magnetic resonance contrast agents. *J. Am. Chem. Soc.* **2007**, *129*, 15760–15761. [CrossRef] [PubMed]
6. Fitra, M.; Daut, I.; Irwanto, M.; Gomesh, N.; Irwan, Y.M. Effect of TiO_2 thickness dye solar cell on charge generation. *Energy Procedia* **2013**, *36*, 278–286. [CrossRef]
7. Yang, H.; Zhu, S.; Pan, N. Studying the mechanisms of titanium dioxide as ultraviolet-blocking additive for films and fabrics by an improved scheme. *J. Appl. Polym. Sci.* **2004**, *92*, 3201–3210. [CrossRef]
8. Cardoso, B.N.; Kohlrausch, E.C.; Laranjo, M.T.; Benvenutti, E.V.; Balzaretti, N.M.; Arenas, L.T.; Santos, M.J.L.; Costa, T.M.H. Tuning anatase-rutile phase transition temperature: TiO_2/SiO_2 nanoparticles applied in dye-sensitized solar cells. *Int. J. Photoenergy* **2019**, *2019*, 7183978. [CrossRef]
9. Luttrell, T.; Halpegamage, S.; Tao, J.; Kramer, A.; Sutter, E.; Batzill, M. Why is anatase a better photocatalyst than rutile?—Model studies on epitaxial TiO_2 films. *Sci. Rep.* **2015**, *4*, 4043. [CrossRef]
10. Wu, J.M. Photodegradation of rhodamine B in water assisted by titania nanorod thin films subjected to various thermal treatments. *Environ. Sci. Technol.* **2007**, *41*, 1723–1728. [CrossRef]
11. Crişan, M.; Mardare, D.; Ianculescu, A.; Drăgan, N.; Niţoi, I.; Crişan, D.; Voicescu, M.; Todan, L.; Oancea, P.; Adomniţei, C.; et al. Iron doped TiO_2 films and their photoactivity in nitrobenzene removal from water. *Appl. Surf. Sci.* **2018**, *455*, 201–215. [CrossRef]
12. Bisen, N.; Shrivastava, P.; Hariprasad, N.; Anju, S.G.; Yesodharan, E.P.; Suguna, Y.; Gaya, U.I.; Abdullah, A.H.; Tseng, T.K.; Lin, Y.S.; et al. Sunlight induced removal of Rhodamine B from water through Semiconductor Photocatalysis: Effects of adsorption, reaction conditions and additives. *Int. J. Mol. Sci.* **2013**, *11*, 2336–2361.
13. Van Viet, P.; Sang, T.T.; Hien, N.Q.; Thi, C.M.; Hieu, L. Synthesis of a silver/TiO_2 nanotube nanocomposite by gamma irradiation for enhanced photocatalytic activity under sunlight. *Nucl. Instrum. Methods Phys. Res. Sect. B Beam Interact. Mater. Atoms* **2018**, *429*, 14–18. [CrossRef]
14. Sanzone, G.; Zimbone, M.; Cacciato, G.; Ruffino, F.; Carles, R.; Privitera, V.; Grimaldi, M.G. Ag/TiO_2 nanocomposite for visible light-driven photocatalysis. *Superlattices Microstruct.* **2018**, *123*, 394–402. [CrossRef]
15. Hafizah, N.; Sopyan, I. Cement bonded sol-gel TiO_2 powder photocatalysis for phenol removal. *Appl. Mech. Mater.* **2015**, *776*, 271–276. [CrossRef]
16. Allé, P.H.; Fanou, G.D.; Robert, D.; Adouby, K.; Drogui, P. Photocatalytic degradation of Rhodamine B dye with TiO_2 immobilized on SiC foam using full factorial design. *Appl. Water Sci.* **2020**, *10*, 207. [CrossRef]
17. Meille, V. Review on methods to deposit catalysts on structured surfaces. *Appl. Catal. A* **2006**, *315*, 1–17. [CrossRef]
18. Agrafiotis, C.; Tsetsekou, A.; Stournaras, C.J.; Julbe, A.; Dalmazio, L.; Guizard, C. Evaluation of sol-gel methods for the synthesis of doped-ceria environmental catalysis systems. Part I: Preparation of coatings. *J. Eur. Ceram. Soc.* **2002**, *22*, 15–25. [CrossRef]
19. McCarty, J.G. Kinetics of PdO combustion catalysis. *Catal. Today* **1995**, *26*, 283–293. [CrossRef]
20. Rice, C.V.; Raftery, D. Photocatalytic oxidation of trichloroethylene using TiO_2 coated optical microfibers. *Chem. Commun.* **1999**, *10*, 895–896. [CrossRef]
21. Belochapkine, S.; Shaw, J.; Wenn, D.; Ross, J.R.H. The synthesis by deposition-precipitation of porous γ-alumina catalyst supports on glass substrates compatible with microreactor geometries. *Catal. Today* **2005**, *110*, 53–57. [CrossRef]
22. Cini, P.; Blaha, S.R.; Harold, M.P.; Venkataraman, K. Preparation and characterization of modified tubular ceramic membranes for use as catalyst supports. *J. Memb. Sci.* **1991**, *55*, 199–225. [CrossRef]
23. Xiaoding, X.; Vonk, H.; Cybulski, A.; Moulijn, J.A. Alumina washcoating and metal deposition of ceramic monoliths. *Stud. Surf. Sci. Catal.* **1995**, *91*, 1069–1078. [CrossRef]
24. Rebrov, E.V.; Klinger, E.A.; Berenguer-Murcia, A.; Sulman, E.M.; Schouten, J.C. Selective hydrogenation of 2-methyl-3-butyne-2-ol in a wall-coated capillary microreactor with a $Pd_{25}Zn_{75}/TiO_2$ catalyst. *Org. Process Res. Dev.* **2009**, *13*, 991–998. [CrossRef]
25. Rebrov, E.V. Sol-gel synthesis of zeolite coatings and their application in catalytic microstructured reactors. *Catal. Ind.* **2009**, *1*, 322–347. [CrossRef]
26. Jiang, P.; Lu, G.; Guo, Y.; Guo, Y.; Zhang, S.; Wang, X. Preparation and properties of a γ-Al_2O_3 washcoat deposited on a ceramic honeycomb. *Surf. Coat. Technol.* **2005**, *190*, 314–320. [CrossRef]
27. Stefanov, P.; Stoychev, D.; Valov, I.; Kakanakova-Georgieva, A.; Marinova, T. Electrochemical deposition of thin zirconia films on stainless steel 316 L. *Mater. Chem. Phys.* **2000**, *65*, 222–225. [CrossRef]
28. Matatov-Meytal, Y.; Barelko, V.; Yuranov, I.; Sheintuch, M. Cloth catalysts in water denitrification. I. Pd on glass fibers. *Appl. Catal. B* **2000**, *27*, 127–135. [CrossRef]

29. Ismagilov, Z.R.; Matus, E.V.; Yakutova, A.M.; Protasova, L.N.; Ismagilov, I.Z.; Kerzhentsev, M.A.; Rebrov, E.V.; Schouten, J.C. Design of Pt-Sn catalysts on mesoporous titania films for microreactor application. *Catal. Today* **2009**, *147*, S81–S86. [CrossRef]
30. Giornelli, T.; Löfberg, A.; Bordes-Richard, E. Grafting of VO_x/TiO_2 catalyst on anodized aluminum plates for structured catalytic reactors. *Thin Solid Films* **2005**, *479*, 64–72. [CrossRef]
31. Protasova, L.N.; Rebrov, E.V.; Glazneva, T.S.; Berenguer-Murcia, A.; Ismagilov, Z.R.; Schouten, J.C. Control of the thickness of mesoporous titania films for application in multiphase catalytic microreactors. *J. Catal.* **2010**, *271*, 161–169. [CrossRef]
32. Pan, J.H.; Zhao, X.S.; Lee, W.I. Block copolymer-templated synthesis of highly organized mesoporous TiO_2-based films and their photoelectrochemical applications. *Chem. Eng. J.* **2011**, *170*, 363–380. [CrossRef]
33. Bravo, J.; Karim, A.; Conant, T.; Lopez, G.P.; Datye, A. Wall coating of a $CuO/ZnO/Al_2O_3$ methanol steam reforming catalyst for micro-channel reformers. *Chem. Eng. J.* **2004**, *101*, 113–121. [CrossRef]
34. Cherkasov, N.; Ibhadon, A.O.; Rebrov, E.V. Novel synthesis of thick wall coatings of titania supported Bi poisoned Pd catalysts and application in selective hydrogenation of acetylene alcohols in capillary microreactors. *Lab Chip* **2015**, *15*, 1952–1960. [CrossRef] [PubMed]
35. Rebrov, E.V.; Cherkasov, N. Disruptive technology for fine chemicals synthesis with catalyst-coated tube reactors. *Chim. Oggi/Chem. Today* **2018**, *36*, 17–20. [CrossRef]
36. Cherkasov, N.; Ibhadon, A.O.; Rebrov, E.V. Solvent-free semihydrogenation of acetylene alcohols in a capillary reactor coated with a $Pd-Bi/TiO_2$ catalyst. *Appl. Catal. A* **2016**, *515*, 108–115. [CrossRef]
37. Exposito, A.J.; Bai, Y.; Tchabanenko, K.; Rebrov, E.V.; Cherkasov, N. Process intensification of continuous-flow imine hydrogenation in catalyst-coated tube reactors. *Ind. Eng. Chem. Res.* **2019**, *58*, 4433–4442. [CrossRef]
38. Rebrov, E.V. Two-phase flow regimes in microchannels. *Theor. Found. Chem. Eng.* **2010**, *44*, 355–367. [CrossRef]
39. Warnier, M.J.F. *Taylor Flow Hydrodynamics in Gas-Liquid-Solid Micro Reactors*; Technische Universiteit Eindhoven: Eindhoven, The Netherlands, 2009.
40. Qu, W.; Mudawar, I. Flow boiling heat transfer in two-phase micro-channel heat sinks—II. Annular two-phase flow model. *Int. J. Heat Mass Transf.* **2003**, *46*, 2773–2784. [CrossRef]
41. Yang, F.; Dai, X.; Peles, Y.; Cheng, P.; Khan, J.; Li, C. Flow boiling phenomena in a single annular flow regime in microchannels (I): Characterization of flow boiling heat transfer. *Int. J. Heat Mass Transf.* **2014**, *68*, 703–715. [CrossRef]
42. Abiev, R.S. Hydrodynamics and heat transfer of circulating Two-phase Taylor flow in microchannel heat pipe: Experimental study and Mathematical Model. *Ind. Eng. Chem. Res.* **2020**, *59*, 3687–3701. [CrossRef]
43. Peterson, G.P.; Ma, H.B. Theoretical analysis of the maximum heat transport in triangular grooves: A study of idealized micro heat pipes. *J. Heat Transf.* **1996**, *118*, 731–739. [CrossRef]
44. Helbig, K.; Alexeev, A.; Gambaryan-Roisman, T.; Stephan, P. Evaporation of falling and shear-driven thin films on smooth and grooved surfaces. *Flow, Turbul. Combust.* **2005**, *75*, 85–104. [CrossRef]
45. Sibiryakov, N.; Kabov, O.; Belosludtsev, V. Numerical simulation of flow in triangular minichannel. *EPJ Web Conf.* **2019**, *196*, 00051. [CrossRef]
46. Sibiryakov, N.; Kabov, O. Numerical simulation of flow with evaporation in triangular grooves. *J. Phys. Conf. Ser.* **2019**, *1369*, 012060. [CrossRef]
47. Warrier, G.R.; Dhir, V.K.; Momoda, L.A. Heat transfer and pressure drop in narrow rectangular channels. *Exp. Therm. Fluid Sci.* **2002**, *26*, 53–64. [CrossRef]
48. Lazarek, G.M.; Black, S.H. Evaporative heat transfer, pressure drop and critical heat flux in a small vertical tube with R-113. *Int. J. Heat Mass Transf.* **1982**, *25*, 945–960. [CrossRef]
49. Kandlikar, S.G. A general correlation for saturated two-phase flow boiling heat transfer inside horizontal and vertical tubes. *J. Heat Transfer* **1990**, *112*, 219–228. [CrossRef]
50. Liu, Z.; Winterton, R.H.S. A general correlation for saturated and subcooled flow boiling in tubes and annuli, based on a nucleate pool boiling equation. *Int. J. Heat Mass Transf.* **1991**, *34*, 2759–2766. [CrossRef]
51. Nayar, K.G.; Panchanathan, D.; McKinley, G.H.; Lienhard, J.H. Surface Tension of Seawater. *J. Phys. Chem. Ref. Data* **2014**, *43*, 043103. [CrossRef]
52. Wang, Z.; Karapetsas, G.; Valluri, P.; Sefiane, K.; Williams, A.; Takata, Y. Dynamics of hygroscopic aqueous solution droplets undergoing evaporation or vapour absorption. *J. Fluid Mech.* **2021**, *912*, 1–30. [CrossRef]
53. Fukano, T.; Furukawa, T. Prediction of the effects of liquid viscosity on interfacial shear stress and frictional pressure drop in vertical upward gas-liquid annular flow. *Int. J. Multiph. Flow* **1998**, *24*, 587–603. [CrossRef]
54. Magnini, M.; Thome, J.R. A CFD study of the parameters influencing heat transfer in microchannel slug flow boiling. *Int. J. Therm. Sci.* **2016**, *110*, 119–136. [CrossRef]
55. Zhang, H.; Chen, B.; Banfield, J.F. The size dependence of the surface free energy of titania nanocrystals. *Phys. Chem. Chem. Phys.* **2009**, *11*, 2553–2558. [CrossRef] [PubMed]
56. Zhang, H.; Banfield, J.F. Understanding polymorphic phase transformation behavior during growth of nanocrystalline aggregates: Insights from TiO_2. *J. Phys. Chem. B* **2000**, *104*, 3481–3487. [CrossRef]
57. Le, H.A.; Linh, L.T.; Chin, S.; Jurng, J. Photocatalytic degradation of methylene blue by a combination of TiO_2-anatase and coconut shell activated carbon. *Powder Technol.* **2012**, *225*, 167–175. [CrossRef]

Article

Crystalline ZnO Photocatalysts Prepared at Ambient Temperature: Influence of Morphology on *p*-Nitrophenol Degradation in Water

Julien G. Mahy [1,*,†], Louise Lejeune [1,†], Tommy Haynes [1], Nathalie Body [1], Simon De Kreijger [1], Benjamin Elias [1], Raphael Henrique Marques Marcilli [2], Charles-André Fustin [2] and Sophie Hermans [1,*]

1. Molecular Chemistry, Materials and Catalysis (MOST), Institute of Condensed Matter and Nanosciences (IMCN), Université Catholique de Louvain, Place Louis Pasteur 1, 1348 Louvain La Neuve, Belgium; l.lejeune@student.uclouvain.be (L.L.); tommy.haynes@uclouvain.be (T.H.); nathalie.body@uclouvain.be (N.B.); simon.dekreijger@uclouvain.be (S.D.K.); benjamin.elias@uclouvain.be (B.E.)
2. Bio and Soft Matter Division (BSMA), Institute of Condensed Matter and Nanosciences (IMCN), Université Catholique de Louvain, Place Louis Pasteur 1, 1348 Louvain La Neuve, Belgium; raphael.marques@uclouvain.be (R.H.M.M.); charles-andre.fustin@uclouvain.be (C.-A.F.)
* Correspondence: julien.mahy@uclouvain.be (J.G.M.); sophie.hermans@uclouvain.be (S.H.); Tel.: +32-4-3664771 (J.G.M.); +32-10-47-28-10 (S.H.)
† These two authors participated equally to the research.

Citation: Mahy, J.G.; Lejeune, L.; Haynes, T.; Body, N.; De Kreijger, S.; Elias, B.; Marcilli, R.H.M.; Fustin, C.-A.; Hermans, S. Crystalline ZnO Photocatalysts Prepared at Ambient Temperature: Influence of Morphology on *p*-Nitrophenol Degradation in Water. *Catalysts* **2021**, *11*, 1182. https://doi.org/10.3390/catal11101182

Academic Editor: Magdalena Janus

Received: 20 August 2021
Accepted: 27 September 2021
Published: 28 September 2021

Publisher's Note: MDPI stays neutral with regard to jurisdictional claims in published maps and institutional affiliations.

Copyright: © 2021 by the authors. Licensee MDPI, Basel, Switzerland. This article is an open access article distributed under the terms and conditions of the Creative Commons Attribution (CC BY) license (https://creativecommons.org/licenses/by/4.0/).

Abstract: Since the Industrial Revolution, technological advances have generated enormous emissions of various pollutants affecting all ecosystems. The detection and degradation of pollutants has therefore become a critical issue. More than 59 different remediation technologies have already been developed, such as biological remediation, and physicochemical and electrochemical methods. Among these techniques, advanced oxidation processes (AOPs) have been popularized in the treatment of wastewater. The use of ZnO as a photocatalyst for water remediation has been developing fast in recent years. In this work, the goals are to produce ZnO photocatalysts with different morphologies, by using a green sol-gel process, and to study both the influence of the synthesis parameters on the resulting morphology, and the influence of these different morphologies on the photocatalytic activity, for the degradation of an organic pollutant in water. Multiple morphologies were produced (nanotubes, nanorods, nanospheres), with the same crystalline phase (wurtzite). The most important parameter controlling the shape and size was found to be pH. The photoactivity study on a model of pollutant degradation shows that the resulting activity is mainly governed by the specific surface area of the material. A comparison with a commercial TiO_2 photocatalyst (Evonik P25) showed that the best ZnO produced with this green process can reach similar photoactivity without a calcination step.

Keywords: aqueous sol-gel process; ZnO; photocatalysis; pollutant degradation

1. Introduction

Although too often poorly appreciated and considered as an almost inexhaustible resource, water is becoming a scarce commodity [1]. Its excessive and disproportionate use in some regions of the world, combined with the overall increase in the population, adds increasing pressure on water reserves and increases the general level of aqueous pollution. To face these problems, treating and decontaminating wastewater for reuse appears to be a promising solution [2–4].

In general, it is possible to distinguish the following three main families of water contaminants: chemical contaminants (organic and inorganic), microbial contaminants, such as viruses and bacteria, and, finally, radiological contaminants. Depending on the type and quantity of pollutants present, as well as the volume of water to be treated, various treatment methods can be used [5]. This work will focus on organic pollutants and the associated degradation processes.

Recently, innovative water pollution control techniques have emerged, as a result of water quality legislation strengthening. Among these, advanced oxidation processes (AOPs) are attracting growing interest. These processes constitute promising alternatives for the degradation of organic pollutants, which are non-biodegradable and refractory to conventional treatments [6].

All AOP technologies are based on the production and use of hydroxyl radicals (OH•), which represent the most powerful oxidizing species that can be used in the field of water and industrial effluent treatment. The advantage of AOPs lies in their ability to degrade almost all organic molecules, by reacting with –C=C– double bonds and by attacking aromatic nuclei, which are major constituents of refractory pollutants. Due to their ability to break down the most recalcitrant compounds into biologically degradable molecules and/or mineral compounds (CO_2 and H_2O), they can be used in addition to conventional techniques, such as adsorption on activated carbon, reverse osmosis, or biological treatments [6].

Among AOPs, photocatalysis relies on the activation of a semiconductor-type photocatalyst with light energy. When the photons meet the surface of the photocatalyst, they are absorbed by the material, which allows the production of highly reactive oxidizing and reducing species on the surface of the semiconductor photocatalyst. These radicals, generated near the catalyst surface, from water and dissolved oxygen, are then able to attack chemical bonds and induce the total or partial destruction of a wide variety of organic compounds [6–8].

Among the different semiconductors that can be used in photocatalysis, ZnO has drawn a lot of attention in the photocatalytic remediation of wastewater fields, due to its high free-exciton binding energy (60 meV), high electrical conductivity, and strong redox ability with valence (VB) and conduction (CB) band positions [7,9]. Moreover, ZnO presents chemical and thermal stability [10].

ZnO is a semiconductor with a band gap of around 3.37 eV [7], so it is activated by UV light. As for the archetypal TiO_2, many studies have been conducted to modify ZnO light absorption properties, in order to shift it in the visible range or to increase its photoactivity in the UV range [11–13]. Different preparation methods are found to synthesize ZnO photocatalysts, such as the sol-gel method, precipitation, microwave-assisted methods, or thermal oxidation [9,10,14–16]. Sol-gel methods present the advantages of occurring under soft conditions (i.e., at a low temperature and low pressure), producing liquid sol or solid gel, to obtain materials in different shapes, such as coatings, powders, or monoliths, and this process is also often compatible with water as solvent, reducing the environmental impact of the preparative steps [8,17–19]. The sol-gel process is based on the hydrolysis and condensation of metal alkoxides, to produce metal oxide materials [18]. By playing on different parameters, such as the pH, the catalyst, or the time of reaction, fine tuning of the metal oxide material characteristics (nanostructure, morphology, or surface properties) can occur [17,20,21].

To date, the photocatalytic applications of ZnO nanostructures have been investigated by numerous researchers. However, relatively little is known about the performance of ZnO catalysts in relation to their morphologies, in a systematic comparative manner [7].

In this work, the goals will be to produce ZnO photocatalysts with different morphologies, by using a green sol-gel process, and to study both the influence of the synthesis parameters on the resulting morphology, and the influence of these different morphologies on the photocatalytic activity for the degradation of an organic pollutant in water. To reach these goals, an aqueous sol-gel synthesis of ZnO will be studied, and the impact of three synthesis parameters (pH, stirring, time of reaction) will be analyzed using a design of experiment (DoE) plan, implemented with JMP® Pro 15 software. All the ZnO photocatalysts will be characterized by PXRD (powder X-ray diffraction), TEM (transmission electron microscopy), XPS (X-ray photoelectron spectroscopy), nitrogen adsorption–desorption measurements, and DRUVS (diffuse reflectance UV–visible spectroscopy). In the last part of this study, the photocatalytic activities of ZnO materials will be assessed on the degradation of a water model pollutant that is commonly found in the pesticide *p*-nitrophenol (PNP). The resulting photoactivities will be compared with the well-known commercial Evonik Aeroxide P25 TiO_2 photocatalyst. This commercial product is synthesized by a high-temperature aerosol process [22].

2. Results and Discussion

As explained in Section 3.1, different protocols were followed, in order to obtain ZnO nanoparticles (NPs) with different morphologies. Two different synthesis protocols were used, named syntheses 1 and 2 in the following, to facilitate reading. Synthesis 1 is a sol-gel method carried out at room temperature, using NaOH as basic titrant, and is adapted from [9], while synthesis 2 is a sol-gel method performed at 60 °C, using KOH as the basic titrant and absolute EtOH as a solvent, and is adapted from [10].

2.1. Synthesis 1: A Study of Three Reaction Parameters

First, ZnO NPs were prepared by the synthesis 1 protocol, as detailed in Section 3.1.1. The following three main reaction parameters were studied: the pH, which was varied from 8 to 12.5, the stirring (or not) of the solution, and the reaction duration (varied between 1 and 7 days). In order to accurately reveal the joint impact of these three factors on the size response, an experimental plan was designed using JMP® Pro 15 software. The tested conditions are detailed in Table 1.

Table 1. Experimental plan designed by JMP® 15 software.

Table	Code Name	pH	Time (days)	Stirring
1	Z1	10.25	7	No
2	Z2	10.25	1	Yes
3	Z3	12.5	4	No
4	Z4	8	4	Yes
5	Z5	12.5	1	No
6	Z6	8	7	Yes
7	Z7	12.5	1	Yes
8	Z8	8	7	No
9	Z9	12.5	7	Yes
10	Z10	8	1	No
11	Z11	12.5	1	No
12	Z12	8	7	Yes
13	Z13	12.5	7	Yes
14	Z14	8	1	No
15	Z15	12.5	7	No
16	Z16	8	1	Yes
17	Z17	10.25	4	Yes
18	Z18	10.25	4	No

2.1.1. Phase Composition

The crystalline ZnO phase present in all samples was identified by means of PXRD, as can be observed in Figure 1, which shows the diffraction pattern of one of the 18 ZnO samples (Z13). The position of the recorded diffraction peaks corresponds to that of the ZnO bulk diffraction spectrum, which can be indexed as hexagonal wurtzite (JCPDS 36-1451). The (004) and (202) plane peaks are less visible than the reference peaks because of the background noise. All the other 17 samples exhibit the same XRD patterns, with various peak widths, indicative of different crystallite sizes. Some XRD patterns (Figure S1 as an example) show additional peaks in the 2θ range of 5°–25°. After an extra washing with deionized water, these peaks disappear, which suggests that they are relative to a zinc acetate residue and not to another ZnO phase. Thanks to the Scherrer formula (Equation (1)), it is possible to calculate the crystallite size (d_{XRD}) of the 18 ZnO samples from their diffraction patterns (Table 2, d_{XRD} (nm)). The full width at half maximum (FWHM, named β) is calculated for the (102) plane (fourth peak), and θ is the angle corresponding to this diffraction plane. β and θ were taken directly from the EVA software that was provided with the PXRD instrument.

$$d_{XRD} = 0.9 \frac{\lambda}{\beta \cos(\theta)} \quad (1)$$

where β is the full width of the peak at half maximum, after correction of the instrumental broadening (rad), λ the wavelength (nm), and θ the Bragg angle (rad).

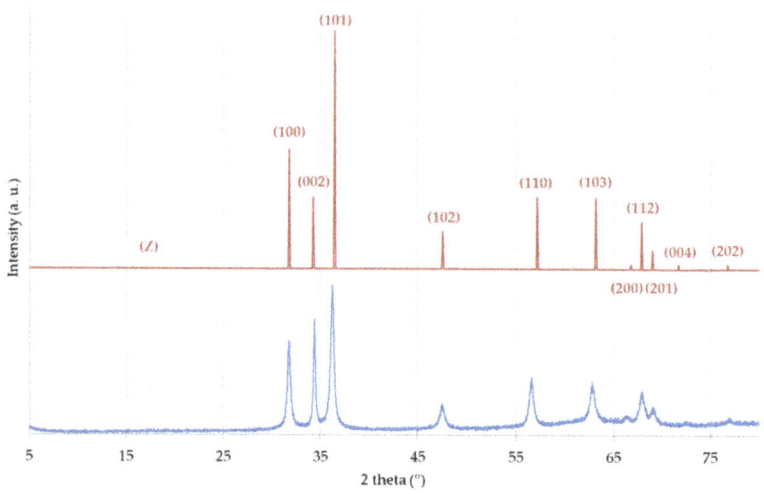

Figure 1. XRD diffraction pattern obtained for a ZnO sample prepared by Synthesis 1 (Z13—blue line), as compared to a wurtzite reference (JCPDS 36-1451) diffraction spectrum (Z—red line).

Table 2. Overview of the crystallite size (d_{XRD}) calculated from diffraction patterns and average size of NPs and of their agglomerates (d_{TEM}), and the related standard deviations (σ_{TEM}) measured from TEM images for the 18 syntheses included in the design of experimental plan. The diameter d_{TEM} for non-spherical NPs and agglomerates corresponds to the height, the length and the diameter of the circumscribed circle of triangular or trapezoidal, rectangular and regular hexagonal morphologies, respectively.

JMP N°	Code Name	d_{XRD} (nm)	d_{TEM} Nanoparticle (nm)	σ_{TEM} (nm)	d_{TEM} Agglomerate (nm)	σ_{TEM} (nm)	Morphology Nanoparticle	Morphology Agglomerate
1	Z1	32	41	13	138	34	rectangular	geometrical
2	Z2	52	29	8	109	36	elongated	triangular and spherical
3	Z3	23	30	8	109	41	elongated	triangular and spherical
4	Z4	109	28	15	717	115	spherical	hexagonal
5	Z5	21	25	10	91	25	elongated	triangular and spherical
6	Z6	124	125	53	414	166	geometrical	spherical and geometrical
7	Z7	18	20	5	100	29	spherical	triangular and spherical
8	Z8	71	48	12	1015	458	spherical	geometrical
9	Z9	19	28	10	104	30	elongated	triangular and spherical

Table 2. Cont.

JMP N°	Code Name	d_{XRD} (nm)	d_{TEM} Nanoparticle (nm)	σ_{TEM} (nm)	d_{TEM} Agglomerate (nm)	σ_{TEM} (nm)	Morphology Nanoparticle	Morphology Agglomerate
10	Z10	105	-	-	110	26	tubular	tubular
11	Z11	23	25	7	96	31	spherical	triangular and spherical
12	Z12	110	126	-	451	130	spherical	geometrical
13	Z13	19	20	7	91	25	elongated	triangular and spherical
14	Z14	104	30	15	771	97	spherical	hexagonal
15	Z15	19	21	8	98	32	elongated	triangular and spherical
16	Z16	112	37	5	696	134	spherical	hexagonal
17	Z17	46	9	5	203	53	spherical	spherical
18	Z18	48	37	19	158	57	rectangular	geometrical

- = not measured.

All the samples were also analyzed by XPS, which confirmed the ZnO composition. All the samples presented the same XPS spectra (an example is given in Supplementary Materials, Figure S2).

2.1.2. Morphology and Size

The ZnO NPs morphology of the 18 samples was investigated by TEM analysis. As is shown for nine samples in Figure 2, ZnO NPs can present the following variety of morphologies: spherical, elongated, or geometrical (rectangular, trapezoidal). Most of them were agglomerated in bigger particles. The morphology of the agglomerates is even more significantly different; they either have perfect geometry, such as spheres or regular hexagons, or smaller and more irregular shapes.

Table 2. Concerning NPs and agglomerates exhibiting a non-spherical morphology, the measured d_{TEM} corresponds to the height for triangular or trapezoidal shapes, the length for the rectangular or elongated shapes and the diameter of the circumscribed circle for the regular hexagonal shapes.

Differences between the diameters calculated from diffraction patterns and those measured from TEM images are observable. It is difficult to indiscriminately compare the sizes measured from XRD and TEM analyses. First of all, the sizes measured by XRD and TEM do not have the same meaning. Indeed, by means of the Scherrer formula (Equation (1)), an average crystallite size is determined, while from TEM images, the size of NPs or agglomerates of a specific zone of the sample can be measured. This is a fact that is important to keep in mind when the d_{XRD} is smaller than the d_{TEM}, which means that several crystallites constitute a single NP. If the comparison of d_{XRD} and d_{TEM} gives the same numbers, it shows that the observed nanoparticles are mainly composed of only one crystallite.

Secondly, an important approximation of the Scherrer equation is that all the crystallites are considered as being spherical. As observed in Figure 2, it is clear that not all the morphologies are spherical.

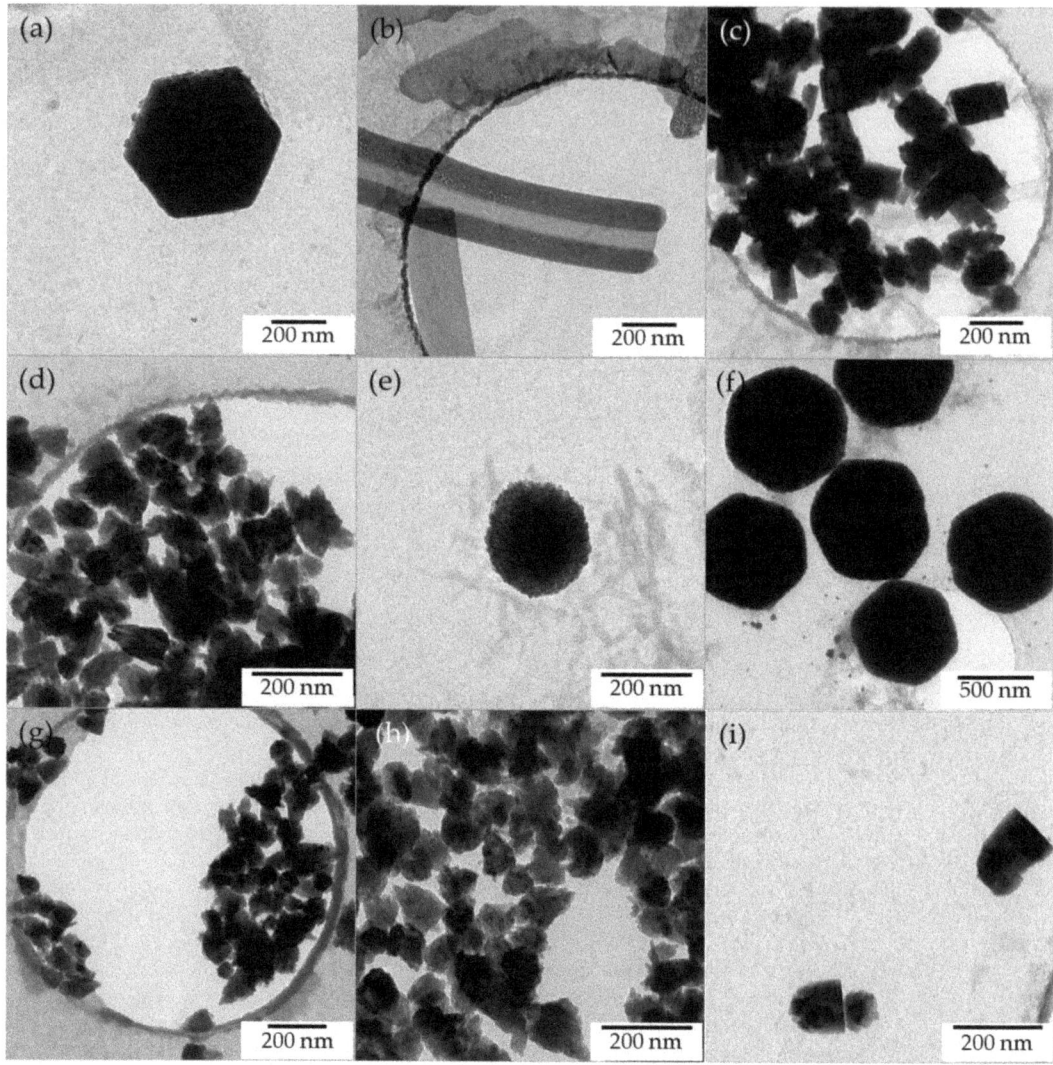

Figure 2. TEM images of nine ZnO samples (**a**) Z4 (**b**) Z10 (**c**) Z1 (**d**) Z15 (**e**) Z17 (**f**) Z14 (**g**) Z2, (**h**) Z13 and (**i**) Z18 in which different morphologies of the obtained NPs and agglomerates can be clearly appreciated. The different code names correspond to the different syntheses (see Table 2).

Moreover, a bias between the observed size, as compared to the real size, occurs for both the size determination techniques employed [23]. Since a TEM image is a 2D projection of the sample, the latter could be slightly distorted, depending on its orientation, and thus the size on the projection is not exactly the same as the actual object. Concerning XRD, for the Scherrer calculations to be meaningful, the diffraction peak width should be purely that of the material itself, and should consequently be free from side effects. The following several factors, other than NP sizes, could contribute to the width of the peaks: instrumental broadening or the use of a non-monochromatic X-ray source, inhomogeneous strain or crystal lattice imperfections, temperature factors, etc. [24,25]. Furthermore, crystallite size broadening is more important at a large value of 2θ, while instrumental width and microstrain broadening are also the largest at a high 2θ value. The asymmetry of the peaks is more pronounced at a 2θ angle lower than 30° [24,25]. As a compromise, the (102) diffraction plane located at a 2θ value of 48° was chosen to be used in the Scherrer calculations. Thus, the d_{XRD} determined is the apparent size of the crystallites in the direction perpendicular to the (102) plane.

Finally, as previously mentioned, the NPs prepared with synthesis 1 are agglomerated in bigger particles of various sizes and shapes. Therefore, the precise measurement of a single NP is intricated in TEM images. It is noteworthy that the standard deviations of the mean sizes of both NPs and agglomerates measured by TEM are really high, whereas XRD provided a single average value of crystallite size, without any estimation of the size distribution. A more complex variant of the Scherrer equation can be developed, to take into account the crystallite size distribution [25], but it was not considered in this work. The Scherrer equation is thus useful to compare several samples with each other, rather than to precisely quantify the absolute size.

On the basis of the values gathered in Table 2, the d_{XRD} and d_{TEM} values are compared and found to correspond, to a certain extent. Nevertheless, concerning the reactions performed at pH 8 (samples Z4, Z10, Z8, Z6, Z12, Z14, and Z16), TEM and XRD analyses exhibit the biggest differences. Reactions at a low pH seem to be less reproducible than reactions at a higher pH. It was not possible to measure an isolated NP on the TEM images of the Z10 sample. Indeed, as observed in Figure 2b, only long tubular forms were visible.

2.1.3. Statistical Analysis of the Three Tested Parameters

In order to study the influence of the three tested parameters (pH, stirring, and reaction time) in this synthesis 1 protocol (Table 1), a statistical analysis was performed using the JMP® Pro 15 software. It was chosen to work with crystallite sizes that were determined from XRD patterns, rather than size measured on TEM images. Indeed, the approximations of the Scherrer equation, as described in Equation (1), would have impacted all the size values in the same way, while it was difficult to precisely measure the NP size on TEM images, due to their agglomeration in bigger particles, and to anisotropy. To visualize the relationships between the three tested parameters and the d_{XRD} response, a scatter plot was drawn (Figure 3) for the 18 ZnO experiments. A marked correlation between pH and d_{XRD} is appreciated in Figure 3 (middle), while stirring and reaction time do not seem to influence the d_{XRD} obtained.

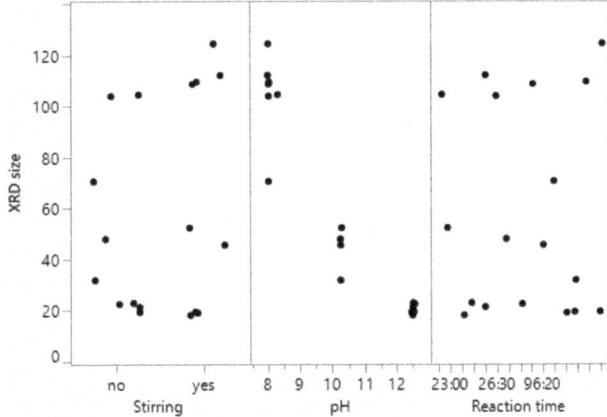

Figure 3. Scatter plot of d_{XRD} values in function of stirring, pH and reaction time. A correlation between d_{XRD} and pH is clearly observable.

To investigate the percentage of correlation between the pH and d_{XRD}, further statistical analyses were performed. The actual pH of the 18 launched experiments varied by a few tenths of a unit from the expected pH of the designed JMP® plan, because of the accuracy of the pH meter. For the statistical analyses, all the pH values were standardized to the theoretical values of 8.0, 10.25, and 12.5, in order to reduce the repeatability error. Then, a statistical model was built as follows: using a model of a degree equal to two, a standard least squares method, and a screening report.

Figure 4 is a plot of the obtained d_{XRD}, using the Scherrer formula to identify the d_{XRD} values expected by the built model. The adjustment summary is gathered in Table 3. The R^2 is equal to 0.97, meaning that 97% is explained by the built model, which is therefore relevant. The analysis of variance report (Table 4) provides the calculations for comparing the fitted model to a model where all the predicted values are equal to the response mean (58.5578 nm). The degree of freedom is the number of parameters implemented to fit the model. This degree is equal to six for the model, for the following parameters: pH, stirring, reaction time, pH * pH, pH * stirring, and pH * reaction time. The parameters (stirring * reaction time) and (reaction time * reaction time) were not implemented here. An analysis including these two additional parameters was also performed (see Figures S3 and S4, and Table S1 in Supplementary Materials). Nevertheless, the LogWorth of the pH was lower and the error sum of squares was higher, which means that the model explained less than the analysis performed with less parameters.

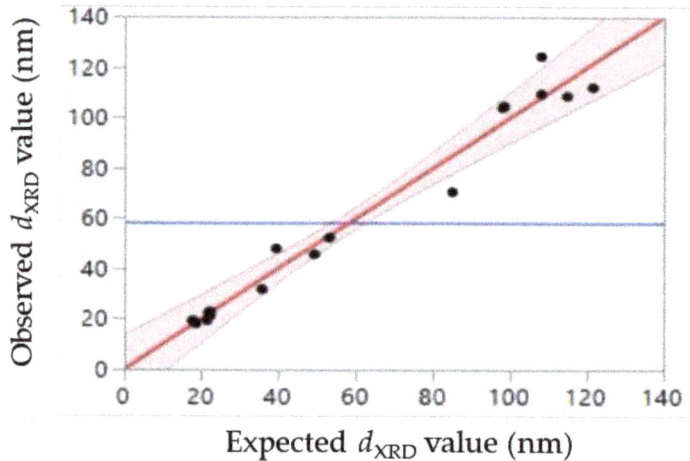

Figure 4. Observed d_{XRD} values in function of expected d_{XRD} by the built model. The red area and the blue line represent the confidence interval and the mean value of the d_{XRD}, respectively.

Table 3. Adjustment summary.

Term	Value
R^2	0.9715
Adjusted R^2	0.9559
Square root of the mean error	84.968
Mean of the response	585.578
Number of observations	18

Table 4. Analysis of variance report.

Source	Degree of Freedom	Sum of Squares	Mean Square	F Ratio	Prob. > F
Model	6	27,042.891	4507.15	624.301	<0.0001
Error	11	794.146	72.20	-	-
Corrected total	17	27,837.037	-	-	-

The total sum of squares is the sum of the squared differences between the response values and the sample mean. It represents the total variation in response values (27,837.037). The error sum of squares is the sum of squared differences between the fitted values and the actual values. It represents the variability that remains unexplained by the model (794.146). The model sum of squares is the difference between the total sum of squares and the error sum of squares; therefore, it represents the variability explained by the model [26]. In this case, the variability explained by the model is equal to 27,042.891, which is much higher than 794.146, which remained unexplained. The mean square is the sum of squares divided by the related degree of freedom.

The F ratio is a statistical test, the ratio between the model and the error mean squares. The Prob. > F is the *p*-value of the F test. The *p*-value is used to quantify the statistical significance of a result under a null hypothesis. *p*-value is a measure of the probability of obtaining an F ratio as large as what is observed. In other words, a very small *p*-value means that such an extreme observed outcome would be very unlikely under the null hypothesis [26]. Here, the Prob. > F is lower than 0.0001, which indicates that there is at least one significant effect in the model.

Figure 5 shows the importance of each parameter in the model (*p*-value and Log-Worth). The LogWorth is defined as $-\log_{10}(p\text{-value})$, this is a transformation adjusting *p*-values to provide an appropriate scale for graphing. Generally, a *p*-value lower than 0.01 corresponds to a presumption against the null hypothesis that is very high [26]. The reference blue line represents the $-\log_{10}(0.01)$, which is equal to two. A parameter that has a LogWorth value greater than two is therefore considered significant. Assuming these considerations, the most influencing effect on the crystallite size is the pH; a LogWorth equal to 8.812 and an exact *p*-value of 1.5×10^{-9} were obtained. To a lesser extent, the second parameter (pH * pH) also possesses a LogWorth value higher than two. On the basis of this analysis, we can conclude that the ZnO crystallite size depends on the pH, and that the function has a slight quadratic component (pH^2). Because of the significance of the model, it is possible to predict the d_{XRD} for chosen values of the three parameters, using the prediction profiler.

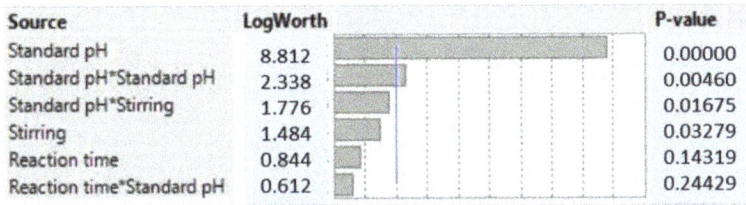

Figure 5. Effects summary. The blue line represents the LogWorth reference equal to two.

These results are consistent with the observations made during the experiments. Indeed, the clear solution transformed into a milky white suspension after the addition of the first drops of NaOH. Nucleation begins immediately, because of the insolubility of ZnO crystallites in the solution. Thus, stirring and reaction time do not have much impact. Whereas, the higher the pH, the higher the NaOH concentration, and, therefore, the higher the supersaturation of the solution, which leads to faster nucleation. When the pH is lower, the concentration of NaOH in the solution is lower as well, and the crystallites can grow slower and are therefore bigger.

2.2. Synthesis 2

ZnO NPs were also prepared by a second protocol, as described in Section 3.1.2, giving sample Z19. The obtained NPs were also characterized by PXRD and TEM analyses. The aim of using this second synthesis is to produce regular spherical NPs with a maximum size of 10 nm.

The collected diffraction peaks (Figure 6) are in very good agreement with the wurtzite reference (JCPDS = 36–1451). The d_{XRD} calculated by the Scherrer equation, using the (102) peak plane as well, is equal to 8 nm (Table 5).

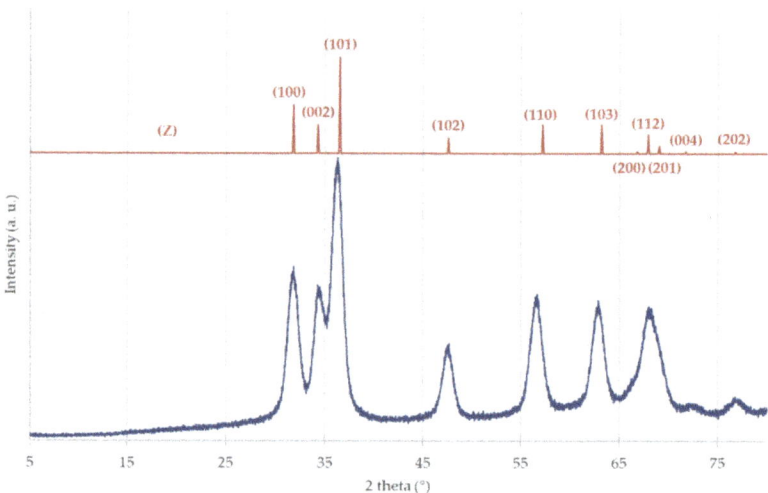

Figure 6. XRD diffraction pattern obtained for ZnO sample (Z19) prepared by synthesis 2 (blue line) as compared to a wurtzite reference diffraction spectrum (Z—red line).

Table 5. Comparison of the main reaction parameters of the two ZnO syntheses, the crystallite size (d_{XRD}) and the morphology observed by TEM.

Synthesis	pH	Basic Titrant	Temperature (°)	d_{XRD} (nm)	NPs Shape
1	8.00	NaOH	25	105	geometrical
1	10.25	NaOH	25	45	geometrical
1	12.50	NaOH	25	20	spherical
2	10.16	KOH	60	8	spherical

The Z19 sample was also analyzed by XPS, which confirmed the ZnO composition (Figure S2).

TEM analyses were performed both for the sample before any washing, and after the washing steps and redispersion in technical EtOH. The TEM images (Figure 7a) revealed that most of the NPs are spherical and well dispersed. Since the NPs are dispersed, the size measurement of individual NPs (d_{TEM}) was performed using ImageJ® software, in order to increase the precision. The obtained average size of NPs is 5 nm ($\sigma = 1$ nm), close to the d_{XRD} calculated by the Scherrer formula. On the TEM image after the washing steps (Figure 7b), NPs were found to be agglomerated into clusters of an average size of 113 nm ($\sigma = 70$). This result suggests that the centrifugation between each washing step may induce aggregation of the NPs.

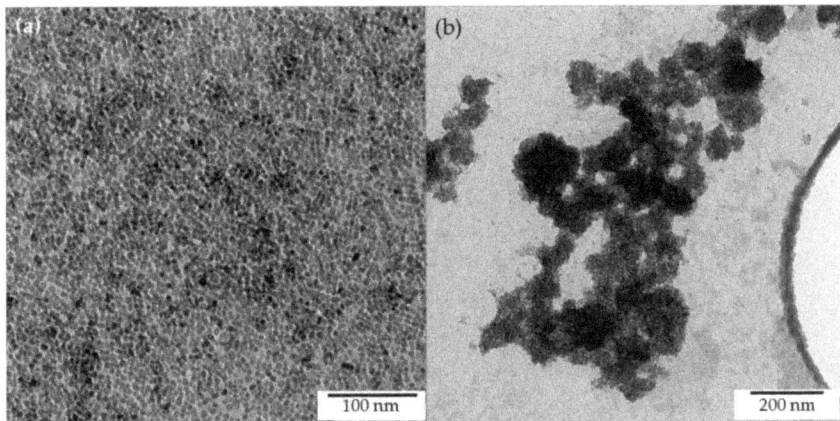

Figure 7. TEM images of ZnO sample (Z19) prepared by synthesis 2. (**a**) Sample analyzed before washing steps, NPs are well dispersed; (**b**) sample analyzed after washing steps, NPs are agglomerated in particle clusters.

In the article reporting the protocol that we adapted for this second synthesis, Shamhari et al. [10] emphasize that the utilization of absolute EtOH as a solvent is required to produce NPs with a uniform shape and size. Therefore, a variant of this synthesis, using technical EtOH as a solvent, was also performed, in order to study the influence of the EtOH grade during the reaction process.

As shown in Figure 8, NPs prepared using technical EtOH (Z20), and after washing steps, have the same morphology as those in the case of the synthesis performed in absolute EtOH. The average NPs size is 5 nm (σ = 2 nm), which is comparable to the first synthesis test using absolute EtOH (Z19). Therefore, the grade of the EtOH used as a solvent does not impact the morphology and size of the prepared NPs. As compared to NPs synthesized in absolute EtOH, analyzed before any washing and redispersion steps, the NPs are less well dispersed compared to when absolute EtOH is used as a solvent. This result seems to confirm the hypothesis that ZnO NP aggregation occurs during the centrifugation and washing steps.

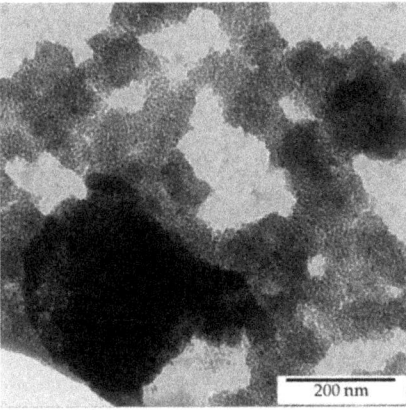

Figure 8. TEM image of ZnO sample (Z20) prepared by a variant of synthesis 2 using technical EtOH as a solvent, after washing steps.

2.3. Comparison of Both Synthesis Methods

In order to compare the two ZnO syntheses performed, Table 5 gathers the mean reaction parameters (pH, basic titrant, temperature (T)), the crystallite size formed (d_{XRD}), and their morphology, as observed on TEM images. As discussed in Section 2.1.3, pH is the most influencing parameter on the d_{XRD}, in the case of synthesis 1. As a comparison, the pH in synthesis 2 is also shown in Table 5. At the same pH value, different particle sizes were obtained using both methods. The reaction medium of synthesis 2 became iridescent after 2 h, while the reaction mediums of the others turned into a milky white solution right after the addition of the very first drops of the basic titrant. ZnO crystallization seems to be slower during synthesis 2, and, therefore, KOH is a basic titrant that is more appropriate for the synthesis of smaller ZnO NPs. The higher temperature also allows better solubility of zinc salt, and reduces the agglomeration.

2.4. Photocatalytic Activity

Among different semiconductors, ZnO is a successful and popular photocatalyst that has demonstrated high photosensitivity and chemical stability [27,28]. As developed in Sections 2.1 and 2.2, a large variety of ZnO nanostructures were obtained from the different syntheses. Because the influence of the morphology of ZnO nanostructures on their photocatalytic activity is not fully investigated in the literature, it seemed interesting to test some of the prepared ZnO samples in such applications. In this case, the tests were performed under UV-A light because of the high band gap value of ZnO (>3 eV).

The degradation percentage of p-nitrophenol (PNP, structure presented in Figure 9), D_{PNP}, is given in Equation (2) [29], where $[PNP]_t$ is the concentration in PNP at time t and $[PNP]_0$ is the initial concentration of PNP at time t = 0.

$$D_{PNP}(\%) = \left(1 - \frac{[PNP]_t}{[PNP]_0}\right) \times 100\% \qquad (2)$$

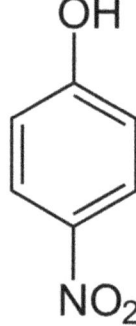

Figure 9. Structure of PNP molecule.

2.4.1. Experiments under Two UV-A Lamps

In order to study the influence of the morphology of the ZnO NPs on the photocatalytic activity, different ZnO samples, prepared with synthesis 1 and 2, exhibiting different nanoshapes, were selected to be tested (Figures 2 and 7). The experimental methods are detailed in Section 3.3. Samples Z2 and Z13 exhibit the same morphology (triangular aggregates), but their sizes are not the same (Z2:29 nm; Z13:20 nm) (Table 2). This is why they were both chosen. The others are either nanotubes (Z10), faceted (Z18), round aggregates (Z17), hexagonal aggregates (Z14), or very small spherical NPs (Z19).

First, a blank test was performed as follows: PNP was irradiated alone under UV-A light, in order to determine whether its degradation occurs after 7 h of light exposure. Figure S5 shows the PNP absorption spectra for PNP, as follows: not irradiated; irradiated with UV-A light during 7 h non-filtered; the same, but filtered with the same syringe filter as used for the catalysts; and filtered, followed by the addition of a drop of HNO_3 (1 mol·L^{-1}), respectively. As observed in Figure S5, PNP is not degraded upon 7 h of UV-A light exposure. The filter (polypropylene, 13 mm diameter, 0.2 m pore size, Whatman™) does not adsorb the PNP. Thus, this filter can be used in further tests.

The photocatalytic activity of the seven selected ZnO samples for PNP degradation was tested under two UV-A lamps. The degradation percentage of PNP (D_{PNP}) was calculated using Equation (2), and the results are gathered in Figure 10. Commercial Evonik P25 TiO_2 was also tested as a reference. Indeed, it is the most used commercial photocatalyst, and no commercial ZnO photocatalyst is available to date.

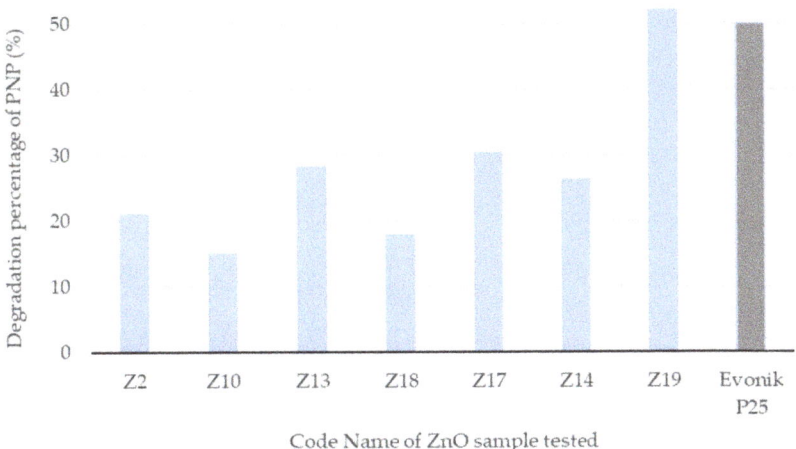

Figure 10. Degradation percentage of PNP (%) for the 7 ZnO samples tested (under two UV-A lamps) and for commercial Evonik P25 TiO_2 used as reference.

The best results were obtained with Z19 (D_{PNP} = 52%), which presented the smallest particle sizes. The second highest PNP degradation percentage (D_{PNP} = 30%) was connected to the Z17 sample, whose particle size (9 nm) was similar to that of Z19, but these were agglomerated into bigger particles (203 nm), as can be appreciated from Figure 2e. The descending D_{PNP} order for the other ZnO samples tested approximately followed the order of the NPs size, as follows: Z13 (20 nm) > Z14 (30 nm) > Z2 (29 nm) > Z18 (37 nm). The sample that provided the lowest PNP degradation rate was Z10. Indeed, as shown in Figure 2b, the morphology of the ZnO structure in this case is long and tubular. The best photocatalyst (Z19) has similar activity to commercial Evonik P25 (~50% after 7 h). It is important to be reminded that this commercial photocatalyst is made by a high-temperature aerosol process [22], while the ZnO samples presented here are made at low temperature, without a calcination step.

2.4.2. Specific Surface Area and Optical Properties

In order to obtain the values of the direct band gap energy ($E_{g;direct}$) and the specific surface area of the NPs, which are two important parameters of semiconductor photocatalysts, diffuse reflectance ultraviolet–visible spectroscopy (DRUVS) and 5 points Brunauer–Emmett–Teller (BET) surface area analyses were performed on the seven ZnO samples selected for the photocatalytic tests, and also on commercial Evonik P25.

From DRUVS, it is possible to determine the $E_{g;direct}$ by plotting $(F(R_\infty)h\nu)^2$ as a function of the photon energy, and by extrapolating the linear part of the curve to the intersection with the x-axis (Equations (5) and (6) from [29]). For commercial TiO_2, it is the indirect band gap that is calculated, as TiO_2 is an indirect semiconductor [30].

Figure 11 shows the plot of $(F(R_\infty)h\nu)^2$ as a function of the photon energy $h\nu$ for the seven ZnO samples tested. Table 6 gathers the results of the band gap energies found. All the ZnO samples possess a slightly different $E_{g;direct}$ value, which is similar to values found in the literature (~3.2 eV) [9,31]. There is no correlation between $E_{g;direct}$ and NP size in this case. The light used for the photocatalytic tests is UV-A light, and its spectrum extends from a wavelength of 400 nm to 315 nm, which corresponds to 3.10 eV to 3.93 eV. Therefore, the $E_{g;direct}$ of each sample is compatible with UV-A light activation, which is why all the samples were active photocatalysts, and were able to degrade PNP.

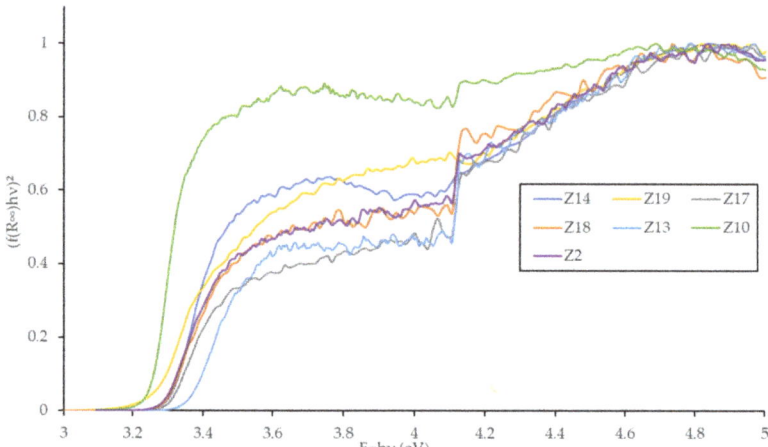

Figure 11. Determination dy DRUVS of the direct band gap energy for the seven ZnO samples tested in photocatalysis.

Table 6. Band gap energy ($E_{g,direct}$), PNP degradation efficiency (D_{PNP}), surface area from 5 points BET (S_{BET}) and corresponding calculated value (S_{th}) for the seven ZnO samples. S_{th} was not calculated for Z10 due to its tubular morphology not allowing the NPs size to be measured from TEM images.

Code Name	S_{BET} (m²·g⁻¹)	S_{th} (m²·g⁻¹)	D_{PNP} (%)	D_{PNP}/S_{BET}	$E_{g,direct}$ (eV)
Z2	5	20	21	4.2	3.28
Z10	4	*	15	3.8	3.25
Z13	23	56	28	1.2	3.36
Z18	4	22	18	4.5	3.28
Z17	6	23	30	5.0	3.30
Z14	3	10	26	8.7	3.29
Z19	30	129	52	1.7	3.26
Evonik P25	55	77	50	0.9	3.05 **

* = not determined, indeed for Z10 samples the tubular morphology is not compatible with the model. ** For commercial Evonik P25 TiO₂, it is the indirect band gap that is calculated.

From the 5 points nitrogen physisorption measurements, it is possible to determine the value of BET surface area (m²·g⁻¹). Table 6 gathers the results provided by the nitrogen adsorption 5 points BET (S_{BET}), as compared to the theoretical model surface (S_{th}) calculated by Equation (3), for the seven ZnO samples tested in photocatalysis, and for Evonik P25.

$$S_{th} = \frac{3}{\rho * r} \quad (3)$$

where S_{th} (m²·g⁻¹) is the theoretical model surface area; ρ (g·m⁻³) is the density, a value of 5.61 g·cm⁻³ was taken for ZnO [32], and a value of 3.89 g·cm⁻³ for TiO₂ [22]; and r is the radius (m) of the crystallite. Here, it is chosen to take r as the radius determined by PXRD (i.e., $d_{XRD}/2$).

As shown in Table 6, the highest surface area measured by BET is for the sample Z19, which was expected, according to its high PNP degradation result. The S_{BET} and S_{th} values are consistent with each other, even if all the S_{BET} values measured are much lower than expected. This is probably due to the agglomeration of NPs in larger aggregates in the dried powders, which results in a decrease in the specific surface area. Indeed, the specific surface area is inversely proportional to the size of NPs, and agglomerated particles decrease the apparent size of NPs. The calculation of the S_{th} also has some of the following limitations: all the NPs are considered spherical, which is not always the case (Z18 for example), and all the NPs are considered monodispersed. Nevertheless, it can give an idea of the importance of agglomeration in the samples, and its impact on the specific surface area measurements. It can be concluded that the photocatalytic activity is mainly governed by the available surface of NPs. Commercial Evonik P25 presented the highest specific surface area (~55 $m^2 \cdot g^{-1}$).

If the activity is normalized by the specific surface area (Table 6, fifth column), the order of photoefficiency is different, and all the ZnO samples are better than the commercial P25 catalyst. Especially, the Z14 sample, composed of hexagon, presents a high normalized photoefficiency that is ten times higher than P25. These results suggest that if the surface area of the ZnO samples could be increased, an activity much higher than the one of Evonik would be obtained.

Some ways to increase the specific surface area would be, for example, to use some surfactant during the synthesis, which would create additional porosity inside the ZnO materials, due to a templating effect. A thermal treatment of the synthesized samples could also create some microporosity if unreacted reagents were still present. However, the temperature of such a treatment must be optimized in order to avoid sintering of the particles, which would increase the particle sizes and decrease the specific surface area. Another way to increase the specific surface area would be to synthesize smaller nanoparticles below 5 nm, but these would be more difficult to recover.

2.4.3. Experiments under Four UV-A Lamps with the Best ZnO Photocatalyst

The ZnO NPs prepared by synthesis 2 (Z19), which possess an average NPs size of 8.3 nm, were the best photocatalysts among all the ZnO samples tested, due to their higher specific surface area. The photocatalytic activity of Z19 was, therefore, also tested under four UV-A lamps, in order to enhance the PNP degradation. After 7 h under four UV-A lamps, Z19 exhibited a DPNP equal to 80%, which is a very competitive result.

2.4.4. Recycling Experiments with the Best ZnO Photocatalyst

To assess the stability of the ZnO material, recycling experiments were conducted with the Z19 sample, knowing that ZnO suffers from photocorrosion [13]. In Table 7, the results of three consecutive photoactivity tests on PNP degradation are presented, with the mean value and the standard deviation of these numbers. The degradation activity remained constant during the three experiments, showing that the material kept its integrity for 21 h. Moreover, an XRD measurement was taken on the Z19 sample, after the three experiments. The pattern (Figure S6) was the same as in Figure 6, showing that the material maintained its crystallinity during the recycling experiments.

Table 7. Recycling photocatalytic experiments on PNP degradation with Z19 sample.

Sample	1st Photocatalytic Experiment (Figure 10) (%)	2nd Photocatalytic Experiment (%)	3rd Photocatalytic Experiment (%)	Mean PNP Degradation Value (%)	Standard Deviation
Z19	52	47	49	48	3

Concerning the evolution of the samples while working with real samples of polluted water, carbon contamination of the surface of ZnO materials can occur during degradation, depending on the concentration of pollutants to eliminate. This contamination can be removed by regeneration of the catalyst with UV exposition, or a thermal treatment if the amount is very high.

3. Materials and Methods

All the chemical reagents and solvents (purity and their supplier) used are gathered in alphabetic order hereafter: Ammonium hydroxide (25% in water, extrapure, Acros Organics, Fisher Scientific, Hampton, NH, USA), Ethanol (99%, EURO DENATURATED Techni Solv®, VWR Chemicals, Radnor, PA, USA), Ethanol absolut (AnalaR NORMAPUR® Reag. Ph. Eur.,VWR Chemicals, Radnor, PA, USA), Methanol (Technical, VWR Chemicals, Radnor, PA, USA), Nitric acid (68%, VWR Chemicals, Radnor, PA, USA), p-Nitrophenol (99%, Acros Organics, Fisher Scientific, Hampton, NH, USA), Potassium hydroxide (>85% for laboratory use, Ph. Eur., Chem-Lab NV, Zedelgem, Belgium), Sodium Hydroxide (99%, AnalaR NORMAPUR® Reag. Ph. Eur., VWR Chemicals, Radnor, PA, USA), Zinc acetate dihydrate (99.5%, for analysis, E. Merck, Fort Kennerworth, NJ, USA).

3.1. ZnO Synthesis

Zinc oxide NPs were synthesized by means of different methods in order to obtain different morphologies. The parameters influencing the ZnO crystallite sizes have been studied.

3.1.1. Sol-Gel Synthesis 1

Zinc oxide (ZnO) NPs were first synthesized based on a sol-gel method reported by Alias et al. [9]. In a 250 mL Erlenmeyer flask, 20 mmol of zinc acetate dihydrate ($Zn(CH_3COO)_2 \cdot 2H_2O$) was dissolved into 100 mL of EtOH at room temperature by magnetic stirring. In order to study the influence of the pH, the agitation and the reaction time on the ZnO crystallite size, an experimental design (Table 1) was performed assisted by JMP Pro 2015 software (JMP 12, SAS, Cary, NC, USA, 2015). The pH of the solution was followed (Electrode pH, BlueLine, SI ANALYTICS®, Servilab) during the titration by sodium hydroxide solution (NaOH 1 mol·L^{-1}). The clear solution transformed into a milky white suspension immediately after the addition of the first drops of NaOH. After the predefined reaction time, under or without stirring, the product was extracted by centrifugation (ALC® centrifuge PK 120, 6000 rpm, 10 min, ALC international S.r.l., Cologno Monzese, Italy) and washed twice with deionized water and twice with EtOH. The product was dried at 120 °C overnight and ground with a mortar and pestle.

3.1.2. Sol-Gel Synthesis 2

ZnO NPs were also prepared following the solvo-thermal procedure reported by Shamhari et al. [10]. In a 250 mL Erlenmeyer flask, 6.7 mmol of $Zn(CH_3COO)_2 \cdot 2H_2O$ was dissolved in 63 mL of absolute EtOH and heated to 60 °C using a liquid oil bath with constant magnetic stirring. In a second flask, 13.2 mmol of potassium hydroxide (KOH) was dissolved in 33 mL of absolute ethanol under the same conditions as zinc acetate dihydrate. When both dissolutions were complete, KOH solution was added drop by drop into the zinc acetate dihydrate solution at 60 °C with vigorous stirring. The reaction mixture was stirred for 3 h at 60 °C. The solution turned iridescent and then a white precipitate formed.

The product was separated by centrifugation (ALC® centrifuge PK 120, 9000 rpm, 30 min) and washed twice with deionized water and twice with ethanol. The obtained product was dried at 120 °C overnight and ground with a mortar and pestle.

3.2. Material Characterizations

X-ray diffraction patterns were recorded with a D8 Advance (Bruker, Billerica, MA, USA) diffractometer, a Cu-Kα (0.15409 Å) anode and a LynxEye detector. A few milligrams of each sample were deposited on an epoxy sample holder, previously covered with a very thin layer of commercial moisturizer (Nivea®). Data were collected in the 2θ range from 5° to 80°, with a step of 0.15° and a time/step of 0.15 s at room temperature. The obtained 2D diffractograms were analyzed using DIFFRAC.EVA software (Bruker, Billerica, MA, USA); they were azimuthally integrated using the Fit2D software and calibrated with a LaB6 standard diffractogram.

The Scherrer formula (Equation (1)) was used to estimate the nanoparticle crystallite size, d_{XRD} (nm) [8].

Nitrogen sorption measurements were performed on a Micromeritics ASAP 2020 instrument (Bruker) in order to determine the BET specific surface area (S_{BET}). About a hundred milligrams of solid sample were degassed at 120 °C for 6 h prior to analysis.

Transmission electron microscopy was performed on an LEO 922 OMEGA energy filter transmission electron microscope operating at 120 kV. Sample preparation consisted of dispersing a few milligrams of each sample into an appropriate solvent, using sonication (VWR ultrasound cleaner, power level 9, 30 min). Then, a few drops of the supernatant were placed on a holed carbon film deposited on a copper grid (CF-1.2/1.3-2 Cu-50, C-flat™, Protochips, Morrisville, NC, USA). The grid was then carefully deposited on a filter paper and dried overnight under vacuum at room temperature. About one hundred particles were measured and an average value was estimated (d_{TEM}).

DRUVS analyses were performed on a UV 3600 Plus UV–VIS–NIR spectrophotometer from Shimadzu Kyoto Japan. Measurements were performed in diffuse reflectance mode. The spectra range of analysis was from 200 to 600 nm. The baseline was realized using Spectralon as a reference. The solid sample preparation consisted of filling the 3 mm diameter microsampling cup using the appropriate funnel (praying mantis™ sampling kit, Harrick). The sample was then flattened with a microscope glass slide and then introduced into the praying mantis™. The spectra were transformed using the Kubelka–Munk function [22,33] to produce a signal, normalized for comparison between samples, enabling the band gaps ($E_{g,direct}$) to be calculated. The details of this treatment method were widely described elsewhere [23,29].

XPS analyses were performed with a FISONS SSI-X-probe 100/206 spectrometer (Surface Science Instruments), equipped with an electron beam (8 keV) for surface charge neutralization. A few milligrams of each sample were deposited on a double-sided adhesive support, clung onto a brass cup, and then introduced into a Macor® carousel topped by a nickel grid in order to avoid charge effects. The analyses were then performed, without further sample preparation at room temperature with an analysis chamber pressure of 10^{-6} Pa.

The main peaks analyzed in the different samples were C 1s, N 1s, O 1s, and Zn 2p. Data treatment was executed with the CasaXPS software (Casa Software Ltd., Teignmouth, UK) using a Gaussian/Lorentzian (85/15) decomposition treatment and a Shirley-type baseline subtraction. All XPS spectra were calibrated using the C–(C,H) component of the C 1s peak localized at 284.8 eV.

3.3. Photocatalytic Experiments

In a quartz sealed round-bottom flask (Figure 12), 10 mg of each ZnO NPs were suspended in 10 mL of a PNP aqueous solution (1×10^{-4} mol·L^{-1}). The solutions were stirred and irradiated with two UV-A lamps (TL Mini Blacklight Blue, TL 8W BLB 1FM/10 × 25CC, λ_{max} = 370 nm, Philips, Amsterdam, The Nederlands). The best sample was also irradiated with four lamps. Aluminum foil was used to cover the outer wall of the reactor and prevent any side reactions with ambient light. A 2 mL sample was taken after 7 h of exposure to UV-A. ZnO catalysts were filtered off with a syringe filter (polypropylene, 13 mm diameter, 0.2 µm pore size, Whatman™, Maidstone, UK). The concentration of PNP was measured by UV/Vis spectroscopy (1700 UV/visible spectrophotometer, Shimadzu) at 317 nm, which correspond to the acidic form of PNP. Previously, a calibration curve was made for the PNP acidic form. Previously, a control experiment was performed to assess if PNP does not undergo direct photolysis under this ultraviolet A (UVA) illumination.

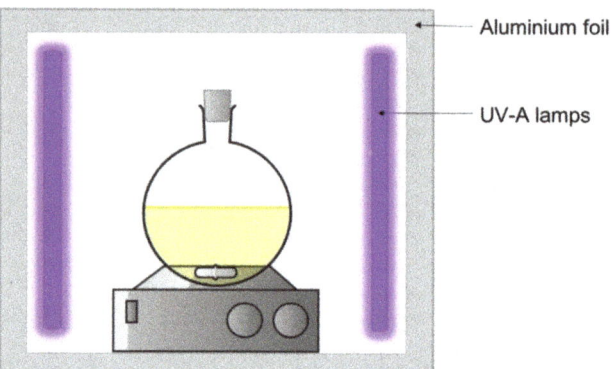

Figure 12. Schematic representation of the photocatalytic set used for the irradiation experiments.

It has been reported that if the degradation of PNP is not complete, specific peaks of the intermediate molecules appear in the range 200–500 nm of the UV/visible spectrum [34,35]. Due to the absence of these peaks during the measurements, complete mineralization can be considered here. Moreover, total mineralization of PNP during homologous photocatalytic tests using a similar installation has been shown in a previous work [36].

3.4. Recycling Photocatalytic Experiments

To evaluate the stability of the photoactivity of ZnO materials, photocatalytic recycling experiments were carried out with the Z19 sample. The same protocol as explained in Section 3.3 was performed with this catalyst. After the first photocatalytic experiment, the sample was recovered by centrifugation (9000 rpm for 30 min) followed by drying at 120 °C overnight. A second and third cycle of photocatalytic tests as described above were applied to the re-used catalysts, with washing and drying steps between each photocatalytic experiment. So, the Z19 sample was used for three successive photocatalytic experiments, for a total of 21 h of illumination. A mean PNP degradation rate of the three experiments was calculated. An XRD measurement was performed after the three photocatalytic experiments in order to check the integrity of the crystalline material.

4. Conclusions

In this work, ZnO nanoparticles were synthesized by a sol-gel process, in order to obtain different morphologies in green conditions. The goals were to study both the influence of the synthesis parameters on the resulting morphology, and the influence of these different morphologies on the photocatalytic activity, for the degradation of an organic pollutant in water.

Two different experimental protocols were tested to produce ZnO. All the samples provided the same ZnO crystalline phase (wurtzite), according to PXRD patterns. In order to study the influence of the synthesis parameters on the resulting morphology, an experimental plan was designed with JMP® software, to study the influence of three parameters (pH, stirring (or not), and reaction time) in the first synthesis method, which was a sol-gel process performed at room temperature and using NaOH as a basic titrant. Many different sizes and morphologies were obtained, according to TEM images. A statistical study performed with JMP® software revealed that pH is the most influencing parameter in the crystallite size response for the ZnO prepared with this synthesis method.

The second synthesis process (sol-gel at 60 °C, using KOH as a basic titrant) provided ZnO nanoparticles with the most regular shape, the smallest size (8 nm from XRD), and with the highest dispersion.

In the second part of this work, the ZnO nanoparticles were tested in the degradation of a model water pollutant, namely, *p*-nitrophenol, by advanced oxidation processes, under UV-A light. The ZnO photocatalysts exhibited notable degradation after only 7 h. Nanoparticles with different shapes and sizes were tested. The obtained results indicated that the photoactivity of ZnO is mainly governed by the specific surface area. Optimization of the nanoparticle size was therefore also critical for the water decontamination application.

Moreover, comparison with a commercial TiO_2 photocatalyst (Evonik P25) showed that the best ZnO produced with this green process can reach similar photoactivity without a calcination step.

By normalization of the photodegradation by the specific surface area, the ZnO samples showed higher activities than the commercial photocatalyst, up to ten times. This suggests that if the specific surface area of the ZnO can be further increased, the photoactivity will reach a higher level than the commercial photocatalyst. It is important to be reminded that the ZnO materials were obtained with a green synthesis process, without an energy-consuming calcination step.

Supplementary Materials: The following are available online at https://www.mdpi.com/article/10.3390/catal11101182/s1, Figure S1: XRD diffraction patterns obtained for a ZnO sample (Z1) before and after washing; Figure S2: XPS spectra of Z19 ZnO sample: (a) Zn 2*p* region, (b) O 1*s* region, (c) N 1*s* region and (d) C 1*s* region; Figure S3: Observed dXRD values in function of expected dXRD by the built model. The red area and the blue line represent, respectively, the confidence interval and the mean value of the dXRD; Figure S4: Effects summary. The blue line represents the LogWorth reference equal to two; Figure S5: Evolution of the absorbance of PNP between 250 and 500 nm for pure PNP (blue line), after 7 h of UV-A irradiation: not filtered (red line), filtered (yellow line) and with a drop of HNO3 (green line), in order to evaluate the effects of these parameters on the PNP absorbance before performing the photocatalytic experiments; Figure S6: XRD pattern of Z19 sample after three successive photocatalytic experiments in PNP degradation. Table S1: Analysis of variance report.

Author Contributions: Conceptualization, methodology, J.G.M., L.L. and S.H.; investigation, analysis, J.G.M., L.L., T.H., N.B., S.D.K., R.H.M.M. and S.H.; writing—original draft preparation, J.G.M., L.L. and S.H.; supervision, funding acquisition and project administration, B.E., C.-A.F. and S.H. All the authors corrected the paper before submission and during the revision process. All authors have read and agreed to the published version of the manuscript.

Funding: This research was funded by Innoviris Brussels, through the Bridge project—COLORES.

Data Availability Statement: The raw/processed data required to reproduce these findings cannot be shared at this time as the data also forms part of an ongoing study.

Acknowledgments: S.H. and N.B. thank the Belgian National Funds for Scientific Research (F.R.S.-FNRS) for their Research Director position and Fund for Research Training in Industry and Agriculture (FRIA) grant, respectively. The authors thanks Jean-Francois Statsyns and François Devred for BET measurements and DRUVS measurements, respectively. The authors also thank Tom Leyssens for the JMP simulations.

Conflicts of Interest: The authors declare no conflict of interest.

References

1. Khan, M.A.; Ghouri, A.M. Environmental Pollution: Its effects on life and its remedies. *J. Arts Sci. Commer.* **2011**, *2*, 276–285.
2. Qian, L.; Wang, S.; Xu, D.; Guo, Y.; Tang, X.; Wang, L. Treatment of municipal sewage sludge in supercritical water: A review. *Water Res.* **2016**, *89*, 118–131. [CrossRef] [PubMed]
3. Macova, M.; Toze, S.; Hodgers, L.; Mueller, J.F.; Bartkow, M.; Escher, B.I. Bioanalytical tools for the evaluation of organic micropollutants during sewage treatment, water recycling and drinking water generation. *Water Res.* **2011**, *45*, 4238–4247. [CrossRef] [PubMed]
4. Turolla, A.; Fumagalli, M.; Bestetti, M.; Antonelli, M. Electrophotocatalytic decolorization of an azo dye on TiO_2 self-organized nanotubes in a laboratory scale reactor. *Desalination* **2012**, *285*, 377–382. [CrossRef]
5. Fox, K.R. Water Treatment and Equipment Decontamination Techniques. *J. Contemp. Water Res. Educ.* **2009**, *129*, 18–21. [CrossRef]
6. Oturan, M.A.; Aaron, J.J. Advanced oxidation processes in water/wastewater treatment: Principles and applications. A review. *Crit. Rev. Environ. Sci. Technol.* **2014**, *44*, 2577–2641. [CrossRef]
7. Verma, S.; Younis, S.A.; Kim, K.H.; Dong, F. Anisotropic ZnO nanostructures and their nanocomposites as an advanced platform for photocatalytic remediation. *J. Hazard. Mater.* **2021**, *415*, 125651. [CrossRef]
8. Mahy, J.G.; Lambert, S.D.; Tilkin, R.G.; Poelman, D.; Wolfs, C.; Devred, F.; Gaigneaux, E.M.; Douven, S. Ambient temperature ZrO_2-doped TiO_2 crystalline photocatalysts: Highly efficient powders and films for water depollution. *Mater. Today Energy* **2019**, *13*, 312–322. [CrossRef]
9. Alias, S.S.; Ismail, A.B.; Mohamad, A.A. Effect of pH on ZnO nanoparticle properties synthesized by sol-gel centrifugation. *J. Alloys Compd.* **2010**, *499*, 231–237. [CrossRef]
10. Shamhari, N.M.; Wee, B.S.; Chin, S.F.; Kok, K.Y. Synthesis and characterization of zinc oxide nanoparticles with small particle size distribution. *Acta Chim. Slov.* **2018**, *65*, 578–585. [CrossRef]
11. Khanizadeh, B.; Khosravi, M.; Behnajady, M.A.; Shamel, A.; Vahid, B. Mg and La Co-doped ZnO nanoparticles prepared by sol–gel method: Synthesis, characterization and photocatalytic activity. *Period. Polytech. Chem. Eng.* **2020**, *64*, 61–74. [CrossRef]
12. Ong, C.B.; Ng, L.Y.; Mohammad, A.W. A review of ZnO nanoparticles as solar photocatalysts: Synthesis, mechanisms and applications. *Renew. Sustain. Energy Rev.* **2018**, *81*, 536–551. [CrossRef]
13. Pirhashemi, M.; Habibi-Yangjeh, A.; Rahim Pouran, S. Review on the criteria anticipated for the fabrication of highly efficient ZnO-based visible-light-driven photocatalysts. *J. Ind. Eng. Chem.* **2018**, *62*, 1–25. [CrossRef]
14. Talam, S.; Karumuri, S.R.; Gunnam, N. Synthesis, Characterization, and Spectroscopic Properties of ZnO Nanoparticles. *ISRN Nanotechnol.* **2012**, *2012*, 372505. [CrossRef]
15. Hasnidawani, J.N.; Azlina, H.N.; Norita, H.; Bonnia, N.N.; Ratim, S.; Ali, E.S. Synthesis of ZnO Nanostructures Using Sol-Gel Method. *Procedia Chem.* **2016**, *19*, 211–216. [CrossRef]
16. Nagornov, I.A.; Mokrushin, A.S.; Simonenko, E.P.; Simonenko, N.P.; Gorobtsov, P.Y.; Sevastyanov, V.G.; Kuznetsov, N.T. Zinc oxide obtained by the solvothermal method with high sensitivity and selectivity to nitrogen dioxide. *Ceram. Int.* **2020**, *46*, 7756–7766. [CrossRef]
17. Parashar, M.; Shukla, V.K.; Singh, R. Metal oxides nanoparticles via sol–gel method: A review on synthesis, characterization and applications. *J. Mater. Sci. Mater. Electron.* **2020**, *31*, 3729–3749. [CrossRef]
18. Brinker, C.J.; Scherer, G.W. *Sol-Gel Science: The Physics and Chemistry of Sol-Gel Processing*; Academic Press: London, UK, 2013.
19. Léonard, G.L.M.; Pàez, C.A.; Ramírez, A.E.; Mahy, J.G.; Heinrichs, B. Interactions between Zn^{2+} or ZnO with TiO_2 to produce an efficient photocatalytic, superhydrophilic and aesthetic glass. *J. Photochem. Photobiol. A Chem.* **2018**, *350*, 32–43. [CrossRef]
20. Ba-Abbad, M.M.; Kadhum, A.A.H.; Bakar Mohamad, A.; Takriff, M.S.; Sopian, K. The effect of process parameters on the size of ZnO nanoparticles synthesized via the sol-gel technique. *J. Alloys Compd.* **2013**, *550*, 63–70. [CrossRef]
21. Preethi, S.; Anitha, A.; Arulmozhi, M. A comparative analysis of the properties of zinc oxide (ZnO) nanoparticles synthesized by hydrothermal and sol-gel methods. *Indian J. Sci. Technol.* **2016**, *9*, 1–6. [CrossRef]
22. Malengreaux, C.M.; Douven, S.; Poelman, D.; Heinrichs, B.; Bartlett, J.R. An ambient temperature aqueous sol–gel processing of efficient nanocrystalline doped TiO_2-based photocatalysts for the degradation of organic pollutants. *J. Sol-Gel Sci. Technol.* **2014**, *71*, 557–570. [CrossRef]
23. Mahy, J.G.; Cerfontaine, V.; Poelman, D.; Devred, F.; Gaigneaux, E.M.; Heinrichs, B.; Lambert, S.D. Highly efficient low-temperature N-doped TiO_2 catalysts for visible light photocatalytic applications. *Materials* **2018**, *11*, 584. [CrossRef]
24. Hargreaves, J.S.J. Some considerations related to the use of the Scherrer equation in powder X-ray diffraction as applied to heterogeneous catalysts. *Catal. Struct. React.* **2016**, *2*, 33–37. [CrossRef]
25. Langfors, J.I.; Wilson, J.C. Seherrer after Sixty Years: A Survey and Some New Results in the Determination of Crystallite Size. *J. Appl. Crystallogr.* **1978**, *11*, 102–113. [CrossRef]
26. Salvatore, D.; Reagle, D. *JMP Statistics and Graphics Guide*; SAS Institute Inc.: Cary, NC, USA, 2008; ISBN 0071395687.
27. Reza, M.; Khaki, D.; Saleh, M.; Aziz, A.; Raman, A.; Mohd, W.; Wan, A. Application of doped photocatalysts for organic pollutant degradation—A review. *J. Environ. Manag.* **2017**, *198*, 78–94.
28. Delsouz Khaki, M.R.; Shafeeyan, M.S.; Raman, A.A.A.; Daud, W.M.A.W. Evaluating the efficiency of nano-sized Cu doped TiO_2/ZnO photocatalyst under visible light irradiation. *J. Mol. Liq.* **2018**, *258*, 354–365. [CrossRef]

29. Mahy, J.G.; Lambert, S.D.; Léonard, G.L.M.; Zubiaur, A.; Olu, P.Y.; Mahmoud, A.; Boschini, F.; Heinrichs, B. Towards a large scale aqueous sol-gel synthesis of doped TiO2: Study of various metallic dopings for the photocatalytic degradation of p-nitrophenol. *J. Photochem. Photobiol. A Chem.* **2016**, *329*, 189–202. [CrossRef]
30. López, R.; Gómez, R. Band-gap energy estimation from diffuse reflectance measurements on sol-gel and commercial TiO_2: A comparative study. *J. Sol-Gel Sci. Technol.* **2012**, *61*, 1–7. [CrossRef]
31. Benhebal, H.; Chaib, M.; Salmon, T.; Geens, J.; Leonard, A.; Lambert, S.D.; Crine, M.; Heinrichs, B. Photocatalytic degradation of phenol and benzoic acid using zinc oxide powders prepared by the sol-gel process. *Alex. Eng. J.* **2013**, *52*, 517–523. [CrossRef]
32. Fauduet, H. *Mécanique des Fluides et des Solides Appliquée à la Chimie*; Lavoisier: Paris, France, 2011; ISBN 274301315X.
33. Kubelka, P. New contributions to the optics of intensely light-scattering materials. *J. Opt. Soc. Am.* **1948**, *38*, 448–457. [CrossRef]
34. Paola, A.D.; Augugliaro, V.; Palmisano, L.; Pantaleo, G.; Savinov, E. Heterogeneous photocatalytic degradation of nitrophenols. *J. Photochem. Photobiol. A Chem.* **2003**, *155*, 207–214. [CrossRef]
35. Augugliaro, V.; Palmisano, L.; Schiavello, M.; Sclafani, A.; Marchese, L.; Martra, G.; Miano, F. Photocatalytic degradation of nitrophenols in aqueous titanium dioxide dispersion. *Appl. Catal.* **1991**, *69*, 323–340. [CrossRef]
36. Malengreaux, C.M.; Léonard, G.M.-L.; Pirard, S.L.; Cimieri, I.; Lambert, S.D.; Bartlett, J.R.; Heinrichs, B. How to modify the photocatalytic activity of TiO2 thin films through their roughness by using additives. A relation between kinetics, morphology and synthesis. *Chem. Eng. J.* **2014**, *243*, 537–548. [CrossRef]

Article

Tuneable Functionalization of Glass Fibre Membranes with ZnO/SnO$_2$ Heterostructures for Photocatalytic Water Treatment: Effect of SnO$_2$ Coverage Rate on the Photocatalytic Degradation of Organics

Vincent Rogé [1,*], Joffrey Didierjean [1], Jonathan Crêpellière [1], Didier Arl [1], Marc Michel [1], Ioana Fechete [2,3], Aziz Dinia [4] and Damien Lenoble [1]

1. Materials Research and Technology (MRT) Department, Luxembourg Institute of Science and Technology, 41 rue du Brill, L-4422 Belvaux, Luxembourg; joffrey.didierjean@list.lu (J.D.); jonathan.crepelliere@list.lu (J.C.); didier.arl@list.lu (D.A.); marc.michel@list.lu (M.M.); damien.lenoble@list.lu (D.L.)
2. LASMIS, Département Physique, Mécanique, Matériaux et Nanotechnologie (P2MN), Université de Technologie de Troyes (UTT)-Antenne de Nogent, 26 rue Lavoisier, 52800 Nogent, France; ioana.fechete@utt.fr
3. Institut de Chimie et Procédés Pour l'Energie, l'Environnement et la Santé (ICPEES), UMR 7515, 25 rue Becquerel, CEDEX 2, 67087 Strasbourg, France
4. Institut de Physique et Chimie des Matériaux de Strasbourg (IPCMS), UMR 7504, 23 rue du Loess, CEDEX 2, 67034 Strasbourg, France; aziz.dinia@ipcms.unistra.fr
* Correspondence: vincent.roge@list.lu; Tel.: +352-2758-88-45-55

Received: 2 June 2020; Accepted: 30 June 2020; Published: 2 July 2020

Abstract: The construction of a ZnO/SnO$_2$ heterostructure is considered in the literature as an efficient strategy to improve photocatalytic properties of ZnO due to an electron/hole delocalisation process. This study is dedicated to an investigation of the photocatalytic performance of ZnO/SnO$_2$ heterostructures directly synthesized in macroporous glass fibres membranes. Hydrothermal ZnO nanorods have been functionalized with SnO$_2$ using an atomic layer deposition (ALD) process. The coverage rate of SnO$_2$ on ZnO nanorods was precisely tailored by controlling the number of ALD cycles. We highlight here the tight control of the photocatalytic properties of the ZnO/SnO$_2$ structure according to the coverage rate of SnO$_2$ on the ZnO nanorods. We show that the highest degradation of methylene blue is obtained when a 40% coverage rate of SnO$_2$ is reached. Interestingly, we also demonstrate that a higher coverage rate leads to a full passivation of the photocatalyst. In addition, we highlight that 40% coverage rate of SnO$_2$ onto ZnO is sufficient for getting a protective layer, leading to a more stable photocatalyst in reuse.

Keywords: photocatalysis; ZnO; SnO$_2$; atomic layer deposition

---

1. Introduction

Photocatalytic water treatment has been intensively described in the last two decades. The amount of polluted water is constantly increasing, making the maintenance of reserves of clean drinkable water more and more challenging [1]. Current water depollution technologies (filtration membranes, reverse osmosis, adsorption, coagulation, deep UV with H$_2$O$_2$, etc.) have high operating costs and consume a lot of energy [2–5]. Consequently, the development of green and energy-efficient depollution technologies is attracting much attention. Photoactive materials working under sunlight are part of them. Among the different photocatalysts studied in the literature, semiconductors like ZnO or TiO$_2$ appear to be promising candidates as they are abundant, safe, thermally stable and display high photocatalytic properties [6–9]. ZnO has a direct band gap of 3.2–3.3 eV at room temperature (≈380 nm) and an exciton binding energy of 60 meV, making it photoactive in the ultraviolet (UV) range. Nevertheless, the large-scale use of photocatalysis for water treatment is limited due to the fast recombination of photogenerated charge carriers in those materials.

In order to improve photocatalytic properties of ZnO, different strategies have been proposed, for example, by using doping elements to improve the photoresponse range [10,11] or developing heterostructures (heterojunctions) with other semiconductors [12,13]. Based on their band alignment, heterostructures can be classified into three types (Figure 1): type I (symmetric), type II (staggered) and type III (broken). Type I heterostructures are often found in light emitting diode (LED) systems, as they promote the recombination of photogenerated electrons/holes [14]. Type II heterostructures are particularly interesting for photocatalytic applications, since they allow the respective delocalisation of photogenerated charge carriers. Holes are driven in the valence band maximum (VBM) of one semiconductor and electrons in the conduction band minimum (CBM) of the second one [15]. Consequently, photogenerated electrons/holes' lifetimes are increased. Type III heterostructures can be applied in tunnelling field-effect transistors [16].

ZnO-based type II heterostructures can be produced using different metal oxides or metal sulphides, like TiO_2 [17], CdS [18], CdSe [19], or SnO_2 [20]. Among those, the ZnO/SnO_2 heterojunction is highly attractive for photocatalytic applications, as SnO_2 is a thermally and chemically stable material, insoluble in water, and has a band gap of 3.6 eV (≈345 nm). This is higher than that of ZnO, and thus, it is almost transparent in the 3.2–3.6 eV range. In addition, ZnO and SnO_2 have different Fermi energy levels [21] and they both possess valence bands potentials around 3.0 V/ENH and 3.8 V/ENH, respectively, i.e., higher than the H_2O/OH^- redox couple (2.8 V/ENH).

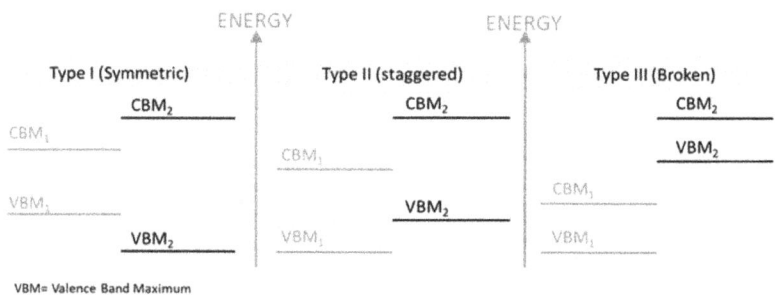

Figure 1. Schematic representation of the three different possible types of heterostructures.

According to the ZnO/SnO_2 heterostructure band alignment (Figure 2), photogenerated holes in the space charge area are delocalised in the valence band of ZnO, and electrons drift in the conduction band of SnO_2.

The synthesis of Janus-like nanoparticles, with both ZnO and SnO_2 exposed to the solution to be cleaned, is one of the most described structures in the literature [22–24]. The main advantage of such heterostructures is that holes and electrons are available for both the oxidation and the reduction of water in the form of $OH^·$ and $O_2^{·-}$ radicals, respectively. It has already been shown that $OH^·$ radicals are the most efficient ones for water treatment [25], as they are strong oxidisers (2.8 V/ENH) able to oxidise the C–C bonds of organic molecules [26–28]. $O_2^{·-}$ radicals however, follow an indirect pathway through H_2O_2 and then $OH^·$. Therefore, the recombination rate of those radicals is higher than that of $OH^·$ ones, and thus their photocatalytic degradation performance is usually reduced. ZnO/SnO_2 heterostructures are mostly found in the literature in the form of nanoparticles [29,30], nanorods [31] or fibres [13,32]. Various synthesis methods have been reported, such as liquid phase processes (i.e., sol-gel or hydrothermal growth) [33,34], electrospinning [35] or gas phase techniques [36]. However, one of the main drawbacks of this configuration is that it requires some post-treatment filtering process. As a matter of fact, a direct contact between (photo)catalytic nanoparticles and fauna and/or flora can be extremely harmful [37]. To circumvent this problem, some recent developments proposed to have the photocatalyst directly supported on a substrate [38]. Membranes are already widely used in the water treatment; coupling their physical separation properties with the photocatalytic activity of photocatalysts appears to be a promising strategy for the development of safe-by-design supported photocatalysts. This is such a strategy being pursued in the work reported here.

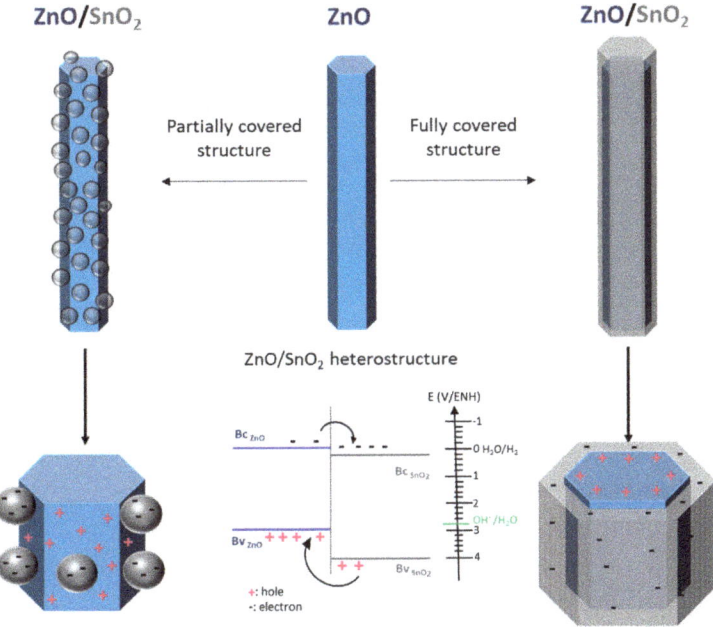

Figure 2. Schematic representation of the two possible ZnO/SnO$_2$ morphologies envisaged in this work. On the left, one can see that ZnO is partially covered by SnO$_2$ nanoparticles. In the scheme on the right, the ZnO is fully covered with a SnO$_2$ thin film.

In this publication, we propose to study the photocatalytic properties of ZnO/SnO$_2$ structures by adjusting the coverage rate of SnO$_2$ particles grown on ZnO nanowires. Therefore, we investigate the synthesis and characterisation of a ZnO/SnO$_2$ heterostructure based on ZnO nanorods/SnO$_2$ nanoparticles supported on glass fibres membranes. The functionalisation of glass fibres by ZnO nanorods has been performed using a liquid phase hydrothermal process and SnO$_2$ nanoparticles have been deposited using a gas phase Atomic Layer Deposition (ALD) process.

As presented in Figure 2, two different strategies can be foreseen to develop the desired ZnO/SnO$_2$ heterostructure. The first one consists in a core/shell type structure [39] obtained by a full coverage of ZnO (i.e., ZnO nanorods), with a continuous SnO$_2$ thin film (Figure 2, right). The advantage of this strategy is that the ZnO will be completely protected by the insoluble and stable SnO$_2$ film. Indeed, one of the drawbacks of ZnO is its known instability in water when the pH drops below 6 or increases above 8, unlike SnO$_2$, known to be stable and insoluble over a larger pH range. However, with the ZnO being completely covered by the SnO$_2$ thin film, photogenerated holes may be trapped inside the nanowire and not be available anymore on the surface for water oxidation (OH$^-$ formation). In this case, only reducing species will be active for the photocatalytic degradation of contaminants via O$_2$$^{·-}$ radical-induced reactions. The second strategy aims at partially covering ZnO nanowires with SnO$_2$

particles (Figure 2, left). This structure is close to a Janus one. By doing so, we expect to have a part of the ZnO surface available for photocatalytic reactions. It is yet unclear if the heterostructure formed between ZnO and SnO_2 will be efficient enough to balance the loss of ZnO exposed surface area.

As we showed in a previous work that the stability of ZnO in water could be strongly improved when protected with SnO_2, even at coverage rates below 100% [40]; the stability of the ZnO/SnO_2 over multiple reuse tests will also be studied.

2. Results and Discussion

In order to control the coverage rate of SnO_2 nanoparticles on ZnO nanorods, we used an ALD process in gas phase, with a chlorinated tin precursor. ALD deposition processes are typically used to grow conformal thin films on complex substrates. However, in some particular cases, they can be used for the synthesis of particles [41,42]. This is often attributed to the use of halogenated precursors (mainly chlorinated ones) that attack the film during the growth process, leading to particle structures [40]. This can be observed on Figure 3, where Scanning Electron Microscopy (SEM) images highlight that the growth of ZnO nanorods on each fibre of the membrane seems to be homogeneous, even on the deepest fibres. In addition, the SEM pictures point out that at a rather low number of SnO_2 ALD cycles (~500), small particles around 10 nm in size are observed on the ZnO nanorods. At 1000 cycles, the particles are slightly bigger, around 15 nm, and their density is much higher. After 1500 cycles, this effect is more pronounced, with particles around 25 nm in diameter. At 2000 cycles, ZnO nanorods seems to be almost completely covered with aggregated SnO_2 particles. Over 2500 ALD cycles, ZnO nanorods are completely covered with a granular SnO_2 film.

Figure 3. SEM images of a glass fibre membrane, ZnO nanorods grown in the glass fibre membrane, ZnO/SnO_2 after 500, 1000, 1500, 2000, 2500 and 3500 SnO_2 ALD cycles.

SnO_2 coverage rates have been estimated from the obtained SEM pictures using an image processing software (ImageJ software, thresholding process). A contrast is observed between SnO_2 particles and the ZnO underneath. Therefore, the image analysis is based on the roundness particles edge detection (SnO_2 particles) versus the background correction (here ZnO). Results are presented in

Figure 4. It highlights that about 8% SnO$_2$ coverage is achieved for 500 ALD cycles. After 1000 cycles, around 40% of the surface of nanorods is covered. A slower deposition rate is observed after 1500 cycles, with around 70% coverage. Above 2000 ALD cycles, the coverage rate is close to 100%, with some porosity due to the structure of the SnO$_2$ film.

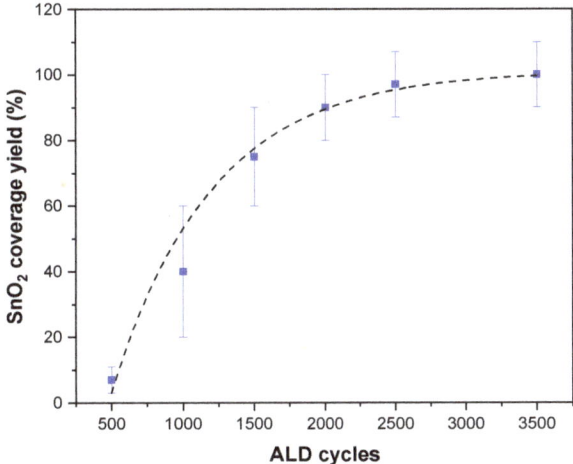

Figure 4. SnO$_2$ coverage rate as a function of the number of ALD cycles used (estimated from SEM pictures).

An Energy Dispersive X-ray (EDX) analysis (Figure 5a) of the synthesized ZnO/SnO$_2$ structure covered at 70% with SnO$_2$ nanoparticles reveals the presence of oxygen (Kα = 0.52 keV), Zinc (Lα = 1.01 keV, Kα1 = 8.63 keV and Kα2 = 9.53 keV), silicon (Kα = 1.74 keV) and tin (Lα = 3.44 keV and Lβ = 3.46 keV). The EDX spectrum is in accordance with the corresponding SEM picture, as we observe an intense peak of Zn related to the ZnO being the major component of the ZnO/SnO$_2$ structure. Peaks related to Tin are weak compared to the Zn one. This is related to the relatively small overall quantity of SnO$_2$ deposited on the ZnO nanowires. Indeed, the inherent volume of interaction of EDX probing down to few micrometres leads to a higher contribution of ZnO as well as the detection of silicon due to the glass fibre membrane used as support.

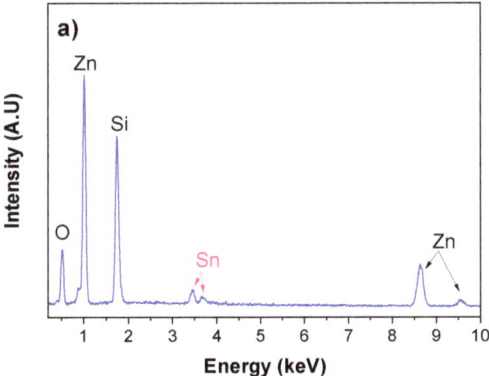

Figure 5. *Cont.*

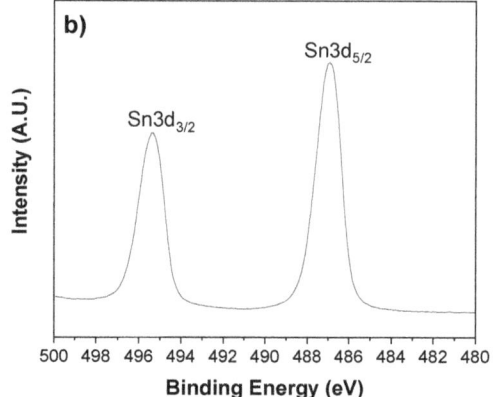

Figure 5. (**a**) EDX analysis of ZnO/SnO$_2$ (70% SnO$_2$ surface coverage). (**b**) High resolution XPS spectrum of the Sn3d peak.

Besides the first chemical screening performed by EDX analysis, an elemental composition of the developed membrane has been determined by X-ray Photoelectron Spectroscopy (XPS) analysis, with a specific focus on the oxidation state of Sn. Figure 5b corresponds to a high-resolution analysis of the Sn3d peak. In this figure, one can see the position of the Sn3d$_{3/2}$ peak at 495.4 eV and the Sn3d$_{5/2}$ peak at 486.7 eV, distinctive of a Sn^{4+} oxidised state of Sn in SnO$_2$ [13,43]. In addition, the coupled spin orbit splitting between the Sn3d$_{3/2}$ peak and the Sn3d$_{5/2}$ peak is exactly 8.5 eV, featuring the Sn–O bonding. The sharp shape of both peaks confirms one chemical bonding contribution: Sn–O. This further confirms that particles are composed of SnO$_2$.

Crystalline structures of functionalized photocatalytic membranes have been investigated by X-Ray Diffraction (XRD). Resulting diffractograms are presented in Figure 6 With a 70% SnO$_2$ surface coverage rate, the hexagonal wurtzite structure of ZnO is detected. The three main diffraction planes of the ZnO wurtzite structure, at 31.75°, 34.45° and 36.25°, corresponding to the (100), (002) and (101) diffraction planes, respectively, are intense and sharp. ZnO is well crystallised, which is a critical feature for efficient water treatment by photocatalysis [44]. In this sample, no SnO$_2$ crystalline structure can be identified. This may be due to the excessively small amount of material deposited, to the fact that the SnO$_2$ could have grown in an amorphous state or to very small crystallite sizes. The XRD diffractogram recorded for the sample completely covered with a SnO$_2$ film exhibits the same ZnO hexagonal wurtzite structure, but weak diffraction peaks characteristic of the tetragonal cassiterite structure of SnO$_2$ are also visible. They correspond to the (100), (101) and (211) diffraction planes at 26.54°, 33.89° and 51.78° respectively. The presence of those peaks confirms that the SnO$_2$ is crystalline. Nevertheless, the relative amount deposited compared to ZnO is too low to see some intense and well-defined peaks. Considering the configuration of the ZnO/SnO$_2$ heterostructure grown in a glass fibre membrane (with circular fibres), the probing of the surface with a grazing angle XRD analysis in very challenging to set up and not fully representative of the global structure.

Crystalline ZnO has a direct optical and electronic band gap of approximately 3.2 eV. It absorbs UV light and show photoluminescent properties with a near band edge (NBE) emission around 380 nm (3.2 eV), corresponding to the excitonic radiative recombination. This emission band is usually sharp and intense for highly crystalline ZnO materials. A second band is often observed in the visible region, centred in the green zone around 530 nm (2.33 eV), corresponding to deep level emission (DLE), due to defects in the ZnO matrix [45,46].

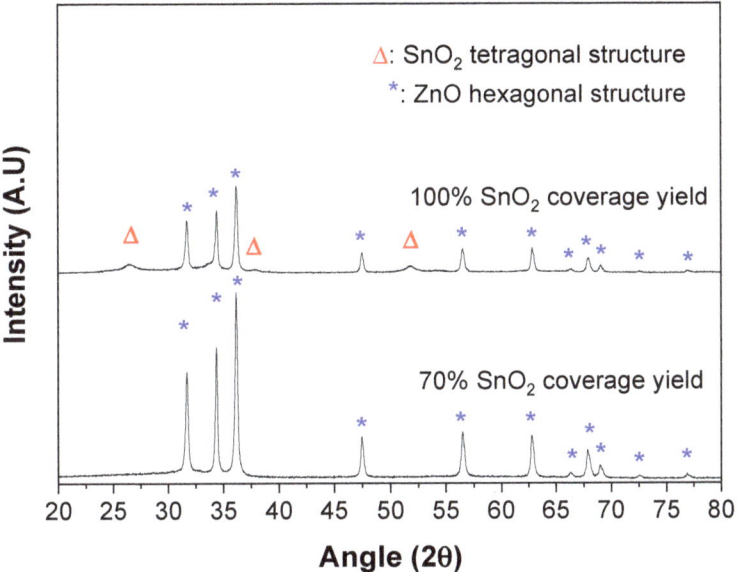

Figure 6. XRD diffractograms of the ZnO/SnO$_2$ heterostructure synthesized in glass fibre membranes after different SnO$_2$ coverage rate.

The optical properties of the synthesized ZnO nanorods and ZnO/SnO$_2$ structures after 70% and 100% coverage rate are presented in Figure 7. An intense and sharp peak is observed at 384 nm for ZnO nanorods. The emission peak related to defects in this case is relatively weak. This suggests that ZnO nanorods are highly crystalline with low defects, further reinforcing the conclusion drawn from the XRD analysis.

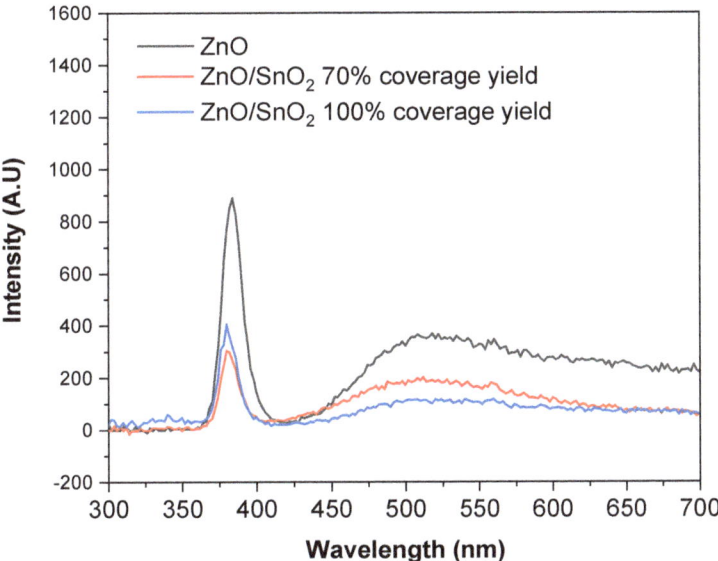

Figure 7. Photoluminescence spectra of ZnO nanorods, ZnO/SnO$_2$ with 70% coverage rate and ZnO/SnO$_2$ with 100% coverage rate.

When ZnO nanorods are covered by SnO$_2$, partially or totally (70% or 100%), the NBE emission is lowered in intensity, but still present. This lowering of intensity can be assigned to the presence of the heterostructure between ZnO and SnO$_2$, which stabilises photogenerated carriers, and thus, limits the radiative recombination and consequently the NBE emission intensity. Interestingly, the NBE emission intensity is in the same order of magnitude for the ZnO covered at 70% by SnO$_2$ and for the ZnO fully covered by SnO$_2$.

This reveals two important features when considering photocatalytic applications. Firstly, SnO$_2$ remains transparent to UV light, so that the ZnO underneath can still be excited, generating electron/hole carriers, even when fully covered. Secondly, a 70% coverage rate of SnO$_2$ is enough to prevent charge carriers recombination. The presence of the broad peak in the visible range can be attributed to remaining defects in the ZnO or SnO$_2$, particularly oxygen vacancies. The annealing treatment of ZnO, as well as the growth temperature of SnO$_2$, were limited to 300 °C because of the substrate instability above this temperature. Thus, it is likely that all oxygen vacancies were not completely eliminated [47,48].

The photocatalytic degradation properties of ZnO/SnO$_2$ structures have been investigated with standard methylene blue (MB). MB is a well-known and widely used chemical probe for the simple investigation of photocatalytic efficiencies of metal oxide photocatalysts. Five samples have been tested: ZnO nanorods without SnO$_2$, ZnO/SnO$_2$ with 8% coverage rate, ZnO/SnO$_2$ with 40% coverage rate, ZnO/SnO$_2$ with 70% coverage rate and ZnO/SnO$_2$ with 100% coverage rate. Results are presented in Figure 8.

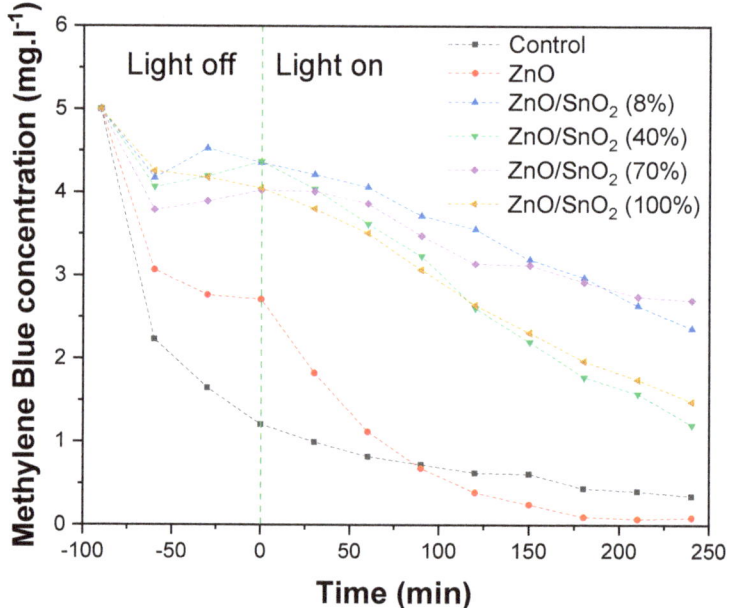

Figure 8. Photocatalytic degradation of Methylene blue under UV light (365 nm, 8 W) over ZnO or ZnO/SnO$_2$ photocatalysts. Percentages in bracket indicate the coverage rate of SnO$_2$ nanoparticles around ZnO.

In a first step, membranes were exposed to the solution in the dark for 90 min in order to stabilise the adsorption/desorption of MB on membranes. This process is crucial for a reliable determination of the photocatalytic degradation kinetics. If not taken into account, this may induce errors due to the uncertainty of distinguishing between adsorption and degradation phenomena. The control membrane (glass fibre only without any photocatalyst) showed high adsorption properties of the organic methylene Blue in the dark, as the measured concentration in solution decreased drastically from 5 mg·L^{-1} down to 1 mg·L^{-1}. The membrane functionalized with ZnO nanorods also showed good adsorption properties toward MB, but less than glass fibres, as the concentration dropped from 5 mg·L^{-1} down to 2.7 mg·L^{-1}. ZnO/SnO$_2$ heterostructure-based membranes exhibited lower affinity for MB, as less than 1 mg·L^{-1} was adsorbed (5 mg·L^{-1} down to more than 4 mg·L^{-1}), independently of the coverage rate. UV light irradiation on the control membrane (which corresponds to t = 0) had no visible effect on its behaviour toward MB. The picture in Figure 9 confirms that the MB is just adsorbed on the glass fibres, and not degraded. Indeed, the control membrane on the left-hand side of Figure 9 is completely blue after the photocatalytic test, which is not the case of the ZnO/SnO$_2$ photocatalytic membrane on the right-hand side of the picture. The ZnO functionalised membrane showed a peculiar behaviour compared to the control after exposure to UV light. A decrease in the MB concentration following pseudo first order kinetics can be observed. In less than 200 min, the solution has been completely decoloured. Surprisingly, all ZnO/SnO$_2$ membranes revealed very low photocatalytic properties, independently from the SnO$_2$ coverage rate. The adsorption process in the dark revealed a poor affinity with all surfaces, which could explain the slow degradation of MB. Another hypothesis could be that impurities trapped in the film or at the surface may act as scavengers for photogenerated carriers (e$^-$/h$^+$) or photogenerated OH· radicals. Considering the growth process of SnO$_2$ with the chlorinated precursor SnCl$_4$, chlorine could be present in/on the ZnO/SnO$_2$ structure and strongly affect the resulting photocatalytic properties [49]. In order to investigate the role of surface defects or residual chlorine, a model structure has been prepared by growing a ZnO/SnO$_2$ thin film on a silicon wafer, covered with 2–3 nm native oxide. Depth profiling of the sample was performed by Secondary Ions Mass Spectrometry (SIMS) analysis.

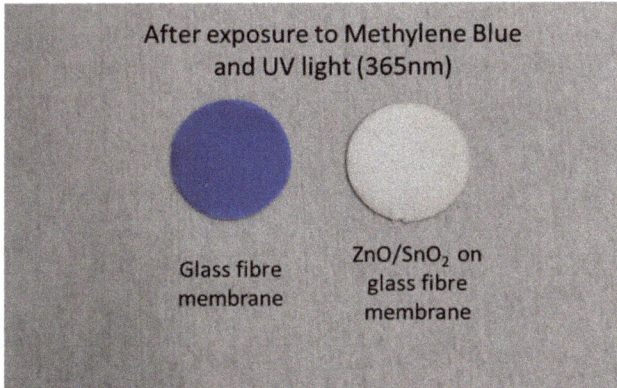

Figure 9. Picture of a glass fibre membrane after exposure to methylene blue (left) and glass fibre membrane functionalised with ZnO nanorods/SnO$_2$ (70% coverage rate) (right).

Figure 10 presents the uncalibrated concentration vs. depth of five elements tracked during the SIMS analysis: silicon, tin, zinc, chlorine and oxygen. The stack SnO$_2$/ZnO is featured by a high contribution of tin in the first 400 s of pulverisation followed by the zinc contribution from 500 to 1200 s. The oxygen contribution remains stable along the depth profile (pulverisation time from 0 s to 1200 s). After 1200 s of pulverisation, the zinc and oxygen contributions disappear in favour of silicon corresponding to the substrate contribution. Interestingly, a high contribution of chlorine is detected in the SnO$_2$ film. When the ZnO film is formed, the contribution of chlorine is lowered drastically. However, some chlorine is still visible in the ZnO film. This clearly evidences the fact that Cl is trapped within the SnO$_2$ films during its growth and that it slightly diffuses into the ZnO underneath. No chlorine is detected in the substrate level.

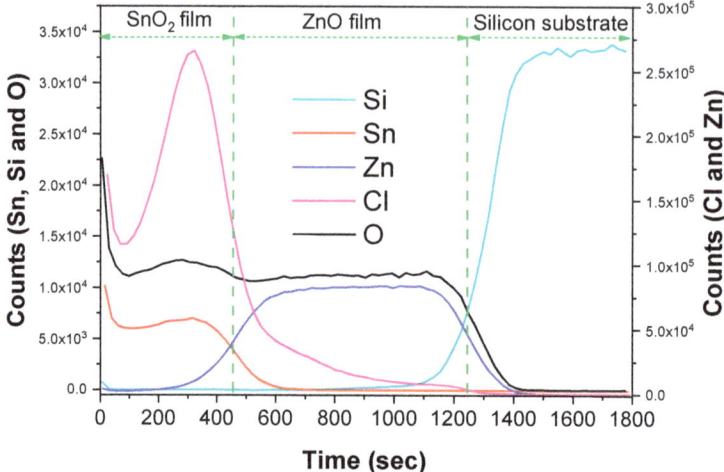

Figure 10. Depth profile SIMS analysis of SnO$_2$ grown by ALD on ZnO (a mirror-polished silicon wafer was used as substrate).

In order to remove defects like oxygen vacancies or chlorine, as-prepared ZnO/SnO$_2$ membranes have been cleaned under a UV/ozone atmosphere (254 nm, 20 W) for 30 min. Compared to plasma or thermal post-treatments, the dry UV/ozone (also called UVO, for ultra-violet ozone) post-treatment has been favoured for its ability to generate, at room temperature, clean and well-oxidised metal oxide structures with very low impact on their morphologies [50]. Also, this technique is known for being able to effectively remove chlorine defects from SnO$_2$ structures [51]. Photocatalytic degradation tests of MB have been performed again. They are presented on Figure 11.

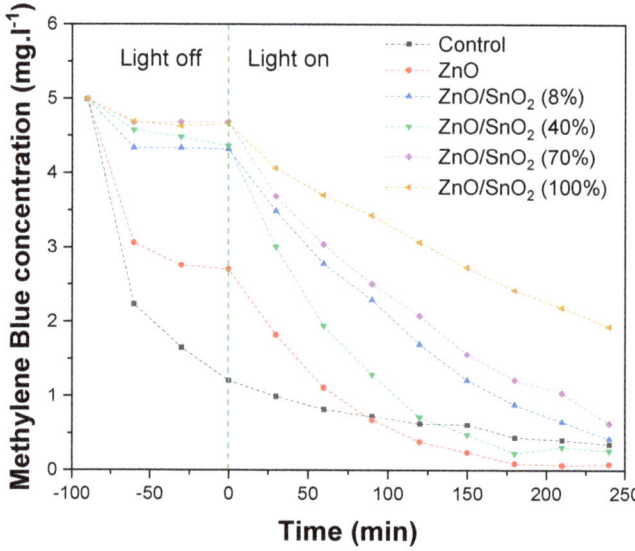

Figure 11. Photocatalytic degradation of methylene blue under UV light (365 nm, 8 W) over ZnO or ZnO/SnO$_2$ photocatalysts after UV/ozone treatment. Percentages in brackets indicate the coverage rate of SnO$_2$ nanoparticles around ZnO.

After cleaning, the affinity with the surface in the dark is not enhanced, but the photocatalytic degradation properties under UV light have been greatly improved for ZnO/SnO$_2$ heterostructures with 8%, 40% and 70% coverage rates. Among them, the heterostructure with 40% coverage rate is the quickest at cleaning the solution of MB according to the steep slope (starting from t = 0). Concerning the ZnO/SnO$_2$ heterostructure with 100% coverage rate, the cleaning had absolutely no impact on its poor photocatalytic degradation properties. The disappearance of MB remains slow compared to other synthesized heterostructures.

The photocatalytic degradation of MB over ZnO and ZnO/SnO$_2$ photocatalysts seems to follow pseudo-first order degradation kinetics, as usually observed when considering the photocatalytic degradation of pollutants in water [52,53]. Consequently, from the results obtained on Figure 8 (before cleaning) and Figure 11 (after cleaning), we determined the first order degradation rate constant k (in min^{-1}) of the photocatalysts by using the following equation:

$$\frac{C_0}{C} = e^{kt} \tag{1}$$

where C_0 is the initial concentration after 90 min in dark (mg·L^{-1}), C the concentration (mg·L^{-1}) at the time t (min). Calculated k, extracted from a plot of ln(C_0/C) versus t, are reported on Figure 12.

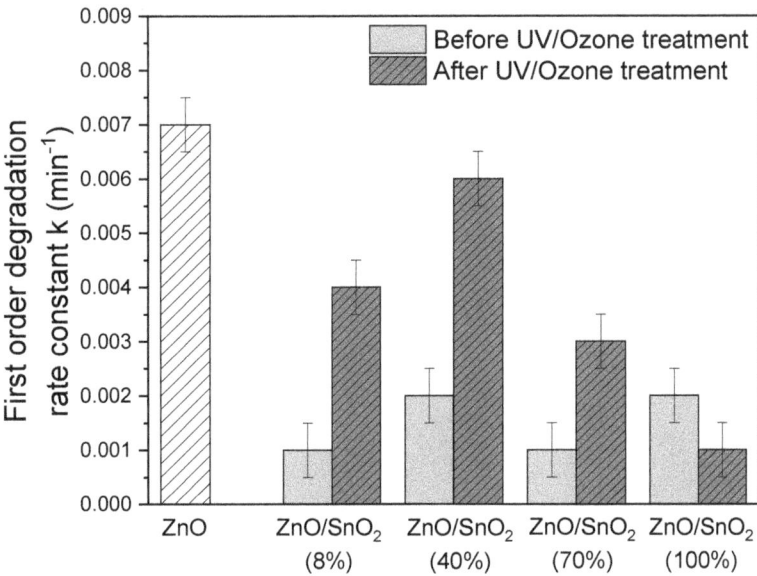

Figure 12. First order degradation rate constant k (min^{-1}) of ZnO and ZnO/SnO$_2$ photocatalysts before and after UV/Ozone cleaning.

The first order degradation rate constant of ZnO nanorods is found to be 7×10^{-3} min^{-1}. When covered by SnO$_2$ (even partially) without any cleaning, the rate drops to 1×10^{-3} – 2×10^{-3} min^{-1} for all samples. Chlorine defects seems to inhibit completely the photocatalytic activity of the material. However, after cleaning, degradation rates increased from 1×10^{-3} min^{-1} up to 4×10^{-3} min^{-1} for the ZnO/SnO$_2$ heterostructure with 8% SnO$_2$ coverage rate, 2×10^{-3} min^{-1} up to 6×10^{-3} min^{-1} for 40% SnO$_2$ coverage rate, 1×10^{-3} min^{-1} up to 3×10^{-3} min^{-1} for 70% SnO$_2$ coverage rate. With 100% SnO$_2$ coverage rate, however, no significant improvement is observed. Those results highlight an optimum coverage rate of ZnO by SnO$_2$ of around 40%, with a maximum rate value obtained of 6×10^{-3} min^{-1}. As discussed above in Figure 2, a trade-off exists between the ZnO surface availability for the photocatalytic degradation and the number of SnO_2 nanoparticles available at the surface for the heterostructure-based stabilisation of charge carriers. In the case of a very low SnO_2 nanoparticle coverage (8%), the loss of specific surface area of ZnO free for the photodegradation is more impactful than the presence of the heterostructure. For a coverage of 70% and beyond, the heterostructure delocalises the photogenerated holes in the core of the nanorod and inhibits the photocatalytic performance of the material. Around 40% coverage, the loss of specific surface area is compensated by the effect of the heterostructure on the surface. The ZnO/SnO_2 heterostructure developed at 40% coverage rate shows a photocatalytic efficiency close that of ZnO alone. Although this is not a strong improvement, it is still interesting as the SnO_2 acts as a protective coating, preventing the dissolution of ZnO in water, even when not completely covering the surface. We demonstrated this tendency in a previous work [40], where SnO_2 protected ZnO inside mesoporous anodic aluminium oxide membranes.

In the present case, we highlight the same protective behaviour of the ZnO/SnO_2 heterostructure through reusability photocatalytic tests and SEM pictures. In Figure 13, the reusability of both ZnO (a) and ZnO/SnO_2 (40%) (b) membranes after five photocatalytic degradation tests is presented. For both systems, the photocatalytic performance is slightly improved after several reuse tests. The reason for this improvement is not yet known, but it is likely that after exposure to UV for several hours, membranes surfaces get cleaner (degradation/removal of adsorbed surface carbon) and thus more reactive toward MB. Those reusability tests demonstrate the excellent performance of ZnO and ZnO/SnO_2 membranes for water depollution over time.

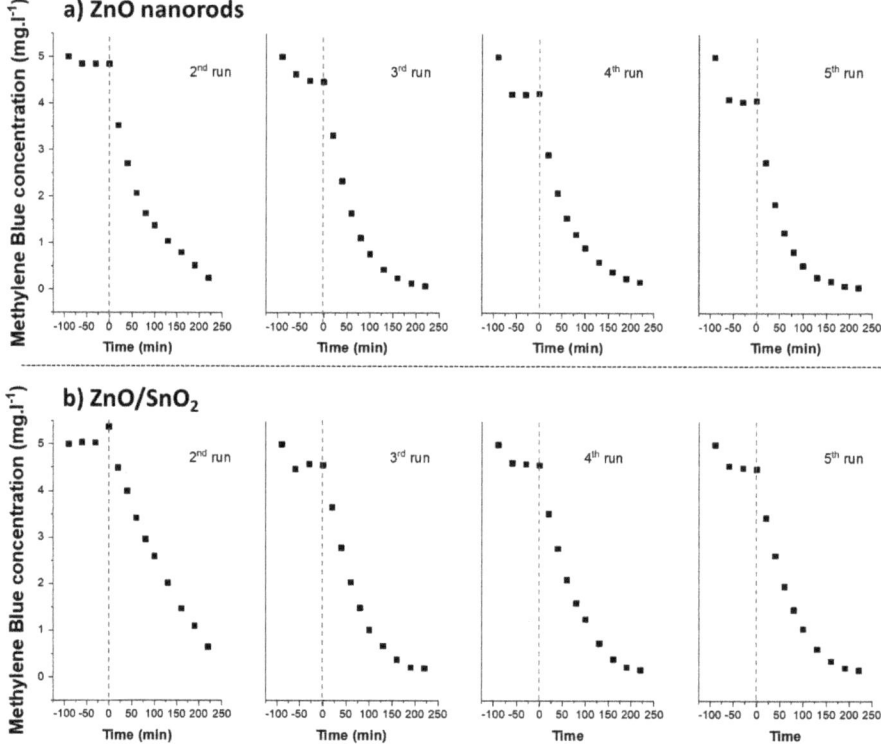

Figure 13. Reusability of ZnO (**a**) and ZnO/SnO_2 (40%) (**b**) functionalized membranes. Five photocatalytic tests have been performed.

If the excellent stability of the two membranes has been determined over five successive tests, the surface state of ZnO and ZnO/SnO$_2$ after those reusability tests is really different. Figure 14 presents SEM images for both ZnO and ZnO/SnO$_2$ functionalized membranes after the five photocatalytic degradation tests. On Figure 14a, we can clearly see that the ZnO nanorod structure has been damaged. On some glass fibres, the ZnO nanorods have completely disappeared. The remaining ZnO nanorods are shorter in length and diameter than before photocatalysis. In addition, an organic matrix, most likely some remaining traces of methylene blue, is present on their surface. This indicates an incomplete degradation of the pollutant. This hypothesis is confirmed by the presence of an intense carbon peak on the EDX spectrum (no carbon was detected before the photocatalytic degradation tests). In the case of the ZnO/SnO$_2$ heterostructure with 40% SnO$_2$ coverage rate (Figure 14b), the morphology of the photocatalyst is the same after five runs as before. The ZnO nanorods are undamaged and SnO$_2$ nanoparticles at the surface are still visible. Moreover, the surface seems to be clean, without any organic traces. The EDX analysis of the photocatalyst confirms that no carbon is detected on the surface. In addition, the peak related to the tin element attests to the presence of SnO$_2$ nanoparticles on the surface of the photocatalyst. Those results clearly highlight that over several degradation cycles, the ZnO photocatalytic properties will inexorably be lowered. Conversely, the ZnO/SnO$_2$ membrane appears to be more stable as a function of the degradation time. We demonstrate here the huge potential of the heterostructure, both as an efficient photoactive material and a stable heterojunction, due to the protective role of the SnO$_2$ around the ZnO.

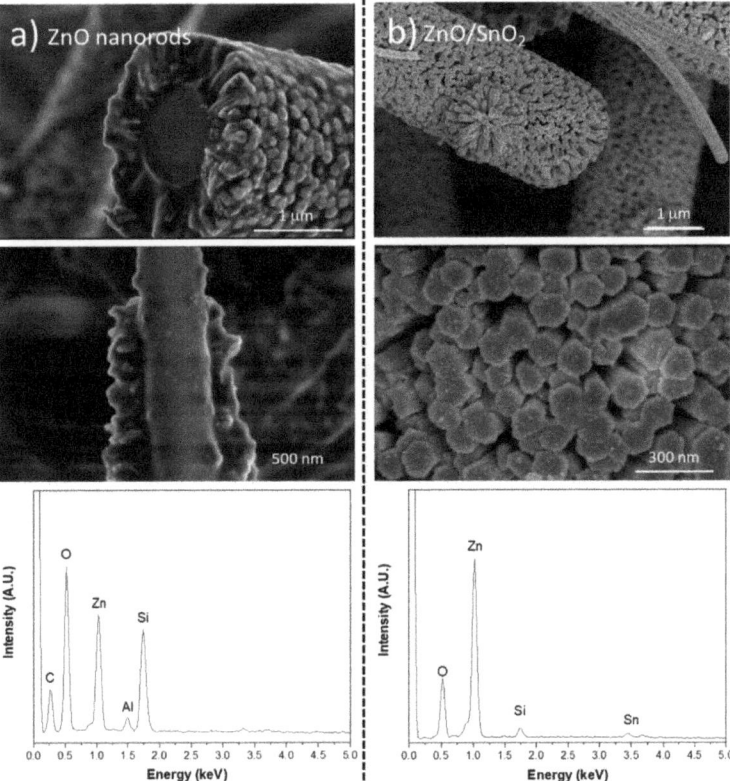

Figure 14. SEM images and EDX analysis of (**a**) ZnO nanorods and (**b**) ZnO/SnO$_2$ (40%) heterostructure, after five photocatalytic degradation tests.

3. Experimental Section

3.1. Materials and Experimental Processes

All chemicals were purchased from Sigma Aldrich (St. Louis, Missouri, United States), and used as received. Glass fibres membranes (APFB, 1 µm pore, 25 mm diameter without binder) were provided by Merck Millipore (Darmstadt, Germany). ZnO nanorods were synthesized in liquid phase by using zinc acetate (99.999%) and 98% anhydrous hydrazine in ultrapure water. Typically, a 25 mM zinc acetate solution was prepared with ultra-pure water in a flask equipped with a reflux system. Then, 25 mM of anhydrous hydrazine was added under vigorous steering (400 rpm). After homogenisation, the glass fibre membrane was dipped in the solution using a home-made holder. The reaction temperature was set to 80 °C for 2 h, under stirring and reflux. At the end of the reaction, the membrane was cleaned in ultra-pure water, dried, and annealed at 300 °C under an air atmosphere. The annealing process allows for the elimination of defects in the ZnO and enhances the crystallinity of ZnO rods, leading to a photocatalytic performant material. The SnO_2 growth on ZnO nanorods has been achieved by a gas phase ALD process, in a TFS200 instrument (BENEQ®, Espoo, Finland). Tin (IV) chloride ($SnCl_4$) precursor was used as the tin source and water as the oxidant. Precursors were stored at room temperature and low pressure in canisters. All precursors were introduced into the reaction chamber without any carrier gas. The reaction was performed between 1 to 5 mbar with nitrogen as carrier and purging gas. The chamber temperature was set at 300 °C during the reaction. A repeated number of $SnCl_4$/purge/H_2O/purge cycles allowed the control of the coverage rate of SnO_2 around ZnO nanorods. A typical ALD cycle can be described as follows (based on preliminary studies not shown here): 300 ms pulse of $SnCl_4$/2 s purge with nitrogen/300 ms pulse of H_2O/2 s purge with nitrogen. The number of cycles has been fixed between 500 and 3500 in order to investigate different SnO_2 particle densities, defined as coverage rate.

3.2. Characterisation Techniques

High-resolution Scanning Electron Microscopy (SEM) images were obtained on a Helios Nanolab 650 microscope (FEI, Eindhoven, The Netherlands), at an acceleration voltage of 2 kV and a current of 25 mA. Energy Dispersive X-ray (EDX) analyses were performed with a 50 mm² Xmax spectrometer (Oxford Instruments, Abington, UK), connected to a Helios Nanolab SEM. The working acceleration voltage was set to 10 kV for a current of 50 mA. XPS analyses were fulfilled with an Axis Ultra DLD X-ray spectrometer from Kratos Analytical (Manchester, UK), working with an Al Kα X-ray source (λ = 0.8343 nm, hν = 1486,6 eV) at 150 W. The crystallographic structures of ZnO/SnO_2 photocatalysts were studied by X-ray diffraction (XRD) in a Brüker D8 (Billerica, MA, USA) Discover diffractometer, with a Cu Kα X-ray source (λ = 0.1542 nm) in θ-2θ mode. The photoluminescent properties of ZnO/SnO_2 were determined with an Infinite M1000 pro spectrometer (TECAN, Männedorf, Switzerland), at an excitation wavelength of 280 nm and a detection range from 300 nm to 700 nm. The secondary ion mass spectrometry (SIMS) technique has been used to determine the different elements present in the photocatalysts. Experiments were performed in a SC Ultra system (CAMECA, Gennevilliers, France), with Cs ions accelerated at an energy of 1 keV. To perform this analysis, the ZnO/SnO_2 structure (20 nm thick SnO_2 film on 50 nm thick ZnO) has been grown on (100) one side polished silicon wafers covered with a 2–3 nm native oxide layer.

3.3. Photocatalytic Degradation of Methylene Blue

The photocatalytic characterisation of ZnO/SnO_2 heterostructures on methylene blue (MB) was carried out in 6-well plates (from Greiner, Kremsmünster, Austria, 35 mm well diameter, 2 mm height, 15 mL maximum volume), with 5 mL of a solution of MB at 5 mg·L^{-1}, under homogeneous stirring. The samples to be analysed were placed in separated wells. Like in many publications dealing with photocatalytic materials' performances, methylene blue has been used as a chemical probe to determine the degradation kinetic induced by different ZnO/SnO_2 photocatalysts [54–56]. MB concentration during the photocatalytic degradation process was determined from optical absorption measurements at 666 nm using the TECAN Infinite M1000UV-visible spectrometer. Photocatalysts were irradiated with a tubular UV lamp (from Hitachi, Chiyoda City, Tokyo, Japan, F8T5, 8 W) working at 365 nm, with a measured power density of 2.28 mW·cm^{-2}.

4. Conclusions

ZnO/SnO$_2$ heterostructures have been synthesised in macroporous glass fibres membranes using a hydrothermal process to grow ZnO nanorods along with a gas phase ALD process for the SnO$_2$ growth. We show that by adjusting the number of ALD cycles, it is possible to synthesize SnO$_2$ particles with a fine control of the coverage rate. Those functionalised membranes have been tested for the photocatalytic degradation of methylene blue under UV light. It has been shown that an optimum coverage rate of approximately 40% led to the most efficient photocatalytic activity against MB. Indeed, it appeared that the exposure of the ZnO surface to the solution to be cleaned is an important parameter for efficient photocatalysts, and that higher coverage rates inhibit the ZnO/SnO$_2$ structures' activity. We also point out that the ZnO/SnO$_2$ heterostructure with 40% coverage rate was highly stable in water after many reuse tests, whereas ZnO nanorods alone were damaged.

Author Contributions: Conceptualization, V.R., D.A., M.M., I.F., D.L.; methodology, V.R., J.D., J.C.; validation, I.F., A.D., D.L.; formal analysis, V.R., J.D., J.C.; writing—original draft preparation, V.R., J.D., J.C., D.A., M.M., I.F., A.D., D.L.; supervision, I.F., A.D., D.L.; project administration, M.M., D.A., D.L.; funding acquisition, D.L. All authors have read and agreed to the published version of the manuscript.

Funding: This research was funded by the "Fond National de la Recherche Luxembourgeoise" (FNR) on the NaneauII project (2015, project number C10/SR/799842).

Conflicts of Interest: The authors declare no conflict of interest.

References

1. Boretti, A.; Rosa, L. Reassessing the projections of the World Water Development Report. *NPJ Clean Water* **2019**, *2*, 1–6. [CrossRef]
2. Sharma, S.; Bhattacharya, A. Drinking water contamination and treatment techniques. *Appl. Water Sci.* **2016**, *7*, 1043–1067. [CrossRef]
3. Macedonio, F.; Drioli, E. Membrane Engineering for Green Process Engineering. *Engineering* **2017**, *3*, 290–298. [CrossRef]
4. Wünsch, R.; Plattner, J.; Cayon, D.; Eugster, F.; Gebhardt, J.; Wülser, R.; Von Gunten, U.; Wintgens, T. Surface water treatment by UV/H2O2 with subsequent soil aquifer treatment: Impact on micropollutants, dissolved organic matter and biological activity. *Environ. Sci. Water Res. Technol.* **2019**, *5*, 1709–1722. [CrossRef]
5. Epimakhov, V.N.; Oleinik, M.S.; Moskvin, L.N. Reverse-Osmosis Filtration Based Water Treatment and Special Water Purification for Nuclear Power Systems. *At. Energy* **2004**, *96*, 234–240. [CrossRef]
6. Fu, D.; Han, G.; Meng, C. Size-controlled synthesis and photocatalytic degradation properties of nano-sized ZnO nanorods. *Mater. Lett.* **2012**, *72*, 53–56. [CrossRef]
7. Kahng, S.; Yoo, H.; Kim, J.H. Recent advances in earth-abundant photocatalyst materials for solar H2 production. *Adv. Powder Technol.* **2020**, *31*, 11–28. [CrossRef]
8. Ishchenko, O.M.; Rogé, V.; Lamblin, G.; Lenoble, D. TiO$_2$- and ZnO-Based Materials for Photocatalysis: Material Properties, Device Architecture and Emerging Concepts. In *Semiconductor Photocatalysis—Materials, Mechanisms and Applications*; IntechOpen: London, UK, 2016. [CrossRef]
9. Rogé, V.; Guignard, C.; Lamblin, G.; Laporte, F.; Fechete, I.; Garin, F.; Dinia, A.; Lenoble, D. Photocatalytic degradation behavior of multiple xenobiotics using MOCVD synthesized ZnO nanowires. *Catal. Today* **2018**, *306*, 215–222. [CrossRef]
10. Chang, C.-J.; Lin, C.-Y.; Hsu, M.-H. Enhanced photocatalytic activity of Ce-doped ZnO nanorods under UV and visible light. *J. Taiwan Inst. Chem. Eng.* **2014**, *45*, 1954–1963. [CrossRef]
11. Lavand, A.B.; Malghe, Y.S. Synthesis, Characterization, and Visible Light Photocatalytic Activity of Nanosized Carbon Doped Zinc Oxide. *Int. J. Photochem.* **2015**, *2015*, 1–9. [CrossRef]

12. Belver, C.; Hinojosa-Reyes, L.; Bedia, J.; Tobajas, M.; Alvarez, M.A.; Rodríguez-González, V.; Rodriguez, J.J. Ag-Coated Heterostructures of ZnO-TiO$_2$/Delaminated Montmorillonite as Solar Photocatalysts. *Materials* **2017**, *10*, 960. [CrossRef]
13. Zhang, Z.; Shao, C.; Li, X.; Zhang, L.; Xue, H.; Wang, C.; Liu, Y. Electrospun Nanofibers of ZnO–SnO$_2$ Heterojunction with High Photocatalytic Activity. *J. Phys. Chem. C* **2010**, *114*, 7920–7925. [CrossRef]
14. Wang, S.; Tian, H.; Ren, C.; Yu, J.; Sun, M. Electronic and optical properties of heterostructures based on transition metal dichalcogenides and graphene-like zinc oxide. *Sci. Rep.* **2018**, *8*, 12009. [CrossRef]
15. Hu, W.; Yang, J. Two-dimensional van der Waals heterojunctions for functional materials and devices. *J. Mater. Chem. C* **2017**, *5*, 12289–12297. [CrossRef]
16. Koswatta, S.O.; Koester, S.J.; Haensch, W. On the Possibility of Obtaining MOSFET-Like Performance and Sub-60-mV/dec Swing in 1-D Broken-Gap Tunnel Transistors. *IEEE Trans. Electron Devices* **2010**, *57*, 3222–3230. [CrossRef]
17. Lei, J.F.; Li, L.B.; Shen, X.H.; Du, K.; Ni, J.; Liu, C.J.; Li, W.S. Fabrication of Ordered ZnO/TiO$_2$ Heterostructures via a Templating Technique. *Langmuir* **2013**, *29*, 13975–13981. [CrossRef] [PubMed]
18. Ding, M.; Yao, N.; Wang, C.; Huang, J.; Shao, M.; Zhang, S.; Li, P.; Deng, X.; Xu, X. ZnO@CdS Core-Shell Heterostructures: Fabrication, Enhanced Photocatalytic, and Photoelectrochemical Performance. *Nanoscale Res. Lett.* **2016**, *11*, 205. [CrossRef]
19. Verma, S.; Ghosh, H.N. Carrier relaxation dynamics in type-II ZnO/CdSe quantum dot heterostructures. *Phys. Chem. Chem. Phys.* **2017**, *19*, 24896–24902. [CrossRef]
20. Upadhaya, D.; Talinungsang; Kumar, P.; Purkayastha, D.D. Tuning the wettability and photocatalytic efficiency of heterostructure ZnO-SnO$_2$ composite films with annealing temperature. *Mater. Sci. Semicond. Process.* **2019**, *95*, 28–34. [CrossRef]
21. Wang, L.; Li, J.; Wang, Y.; Yu, K.; Tang, X.; Zhang, Y.; Wang, S.; Wei, C. Construction of 1D SnO$_2$-coated ZnO nanowire heterojunction for their improved n-butylamine sensing performances. *Sci. Rep.* **2016**, *6*, 35079. [CrossRef]
22. Uddin, T.; Nicolas, Y.; Olivier, C.; Toupance, T.; Servant, L.; Müller, M.M.; Kleebe, H.-J.; Ziegler, J.; Jaegermann, W. Nanostructured SnO$_2$–ZnO Heterojunction Photocatalysts Showing Enhanced Photocatalytic Activity for the Degradation of Organic Dyes. *Inorg. Chem.* **2012**, *51*, 7764–7773. [CrossRef] [PubMed]
23. Hamrouni, A.; Moussa, N.; Parrino, F.; Di Paola, A.; Houas, A.; Palmisano, L. Sol–gel synthesis and photocatalytic activity of ZnO–SnO$_2$ nanocomposites. *J. Mol. Catal. A Chem.* **2014**, *390*, 133–141. [CrossRef]
24. Verma, N.; Yadav, S.; Mari, B.; Mittal, A.; Jindal, J. Synthesis and Charcterization of Coupled ZnO/SnO$_2$ Photocatalysts and Their Activity towards Degradation of Cibacron Red Dye. *Trans. Indian Ceram. Soc.* **2018**, *77*, 1–7. [CrossRef]
25. Nosaka, Y.; Nosaka, A. Understanding Hydroxyl Radical (•OH) Generation Processes in Photocatalysis. *ACS Energy Lett.* **2016**, *1*, 356–359. [CrossRef]
26. Hu, P.; Long, M. Cobalt-catalyzed sulfate radical-based advanced oxidation: A review on heterogeneous catalysts and applications. *Appl. Catal. B Environ.* **2016**, *181*, 103–117. [CrossRef]
27. Wang, N.; Zheng, T.; Zhang, G.; Wang, P. A review on Fenton-like processes for organic wastewater treatment. *J. Environ. Chem. Eng.* **2016**, *4*, 762–787. [CrossRef]
28. Guerra-Rodríguez, S.; Rodríguez, E.; Singh, D.; Rodríguez-Chueca, J. Assessment of Sulfate Radical-Based Advanced Oxidation Processes for Water and Wastewater Treatment: A Review. *Water* **2018**, *10*, 1828. [CrossRef]
29. Yu, B.; Zeng, J.; Gong, L.; Zhang, M.; Zhang, L.; Chen, X. Investigation of the photocatalytic degradation of organochlorine pesticides on a nano-TiO$_2$ coated film. *Talanta* **2007**, *72*, 1667–1674. [CrossRef]
30. Zheng, L.; Zheng, Y.; Chen, C.; Zhan, Y.; Lin, X.; Zheng, Q.; Wei, K. Facile One-Pot Synthesis of ZnO/SnO$_2$ Heterojunction Photocatalysts with Excellent Photocatalytic Activity and Photostability. *ChemPlusChem* **2012**, *77*, 217–223. [CrossRef]
31. Huang, X.; Shang, L.; Chen, S.; Xia, J.; Qi, X.; Wang, X.; Zhang, T.; Meng, X. Type-II ZnO nanorod–SnO$_2$ nanoparticle heterostructures: Characterization of structural, optical and photocatalytic properties. *Nanoscale* **2013**, *5*, 3828–3833. [CrossRef]
32. Zhu, L.; Hong, M.; Ho, G.W. Hierarchical Assembly of SnO$_2$/ZnO Nanostructures for Enhanced Photocatalytic Performance. *Sci. Rep.* **2015**, *5*, 11609. [CrossRef] [PubMed]

33. Wang, Z.; Gao, S.; Fei, T.; Liu, S.; Zhang, T. Construction of ZnO/SnO2 Heterostructure on Reduced Graphene Oxide for Enhanced Nitrogen Dioxide Sensitive Performances at Room Temperature. *ACS Sens.* **2019**, *4*, 2048–2057. [CrossRef]
34. Talinungsang; Upadhaya, D.; Kumar, P.; Purkayastha, D.D. Superhydrophilicity of photocatalytic ZnO/SnO$_2$ heterostructure for self-cleaning applications. *J. Sol-Gel Sci. Technol.* **2019**, *92*, 575–584. [CrossRef]
35. Lu, Z.; Zhou, Q.; Wang, C.; Wei, Z.; Xu, L.; Gui, Y. Electrospun ZnO-SnO$_2$ Composite Nanofibers and Enhanced Sensing Properties to SF6 Decomposition Byproduct H2S. *Front. Chem.* **2018**, *6*, 540. [CrossRef]
36. Çetinörgü, E.; Goldsmith, S.; Boxman, R.L. Optical properties of transparent ZnO–SnO$_2$ thin films deposited by filtered vacuum arc. *J. Phys. D Appl. Phys.* **2006**, *39*, 1878–1884. [CrossRef]
37. Al-Kandari, H.; Younes, N.; Al-Jamal, O.; Zakaria, Z.Z.; Najjar, H.; Alserr, F.; Pintus, G.; Al-Asmakh, M.A.; Abdullah, A.M.; Nasrallah, G.K. Ecotoxicological Assessment of Thermally- and Hydrogen-Reduced Graphene Oxide/TiO$_2$ Photocatalytic Nanocomposites Using the Zebrafish Embryo Model. *Nanomaterials* **2019**, *9*, 488. [CrossRef] [PubMed]
38. Athanasekou, C.P.; Moustakas, N.; Morales-Torres, S.; Pastrana-Martínez, L.M.; Figueiredo, J.L.; Faria, J.L.; Silva, A.M.; Rodríguez, J.M.D.; Romanos, G.E.; Falaras, P. Ceramic photocatalytic membranes for water filtration under UV and visible light. *Appl. Catal. B Environ.* **2015**, *178*, 12–19. [CrossRef]
39. Zheng, L.; Zheng, Y.; Chen, C.; Zhan, Y.; Lin, X.; Zheng, Q.; Wei, K.; Zhu, J. Network Structured SnO$_2$/ZnO Heterojunction Nanocatalyst with High Photocatalytic Activity. *Inorg. Chem.* **2009**, *48*, 1819–1825. [CrossRef]
40. Rogé, V.; Georgantzopoulou, A.; Lenoble, D.; Mehennaoui, K.; Fechete, I.; Garin, F.; Dinia, A.; Gutleb, A.C. Tailoring the optical properties of ZnO nano-layers and their effect on in vitro biocompatibility. *RSC Adv.* **2015**, *5*, 97635–97647. [CrossRef]
41. Wack, S.; Popa, P.L.; Adjeroud, N.; Guillot, J.; Pistillo, B.R.; Leturcq, R. Large-Scale Deposition and Growth Mechanism of Silver Nanoparticles by Plasma-Enhanced Atomic Layer Deposition. *J. Phys. Chem. C* **2019**, *123*, 27196–27206. [CrossRef]
42. Shi, J.; Sun, C.; Starr, M.; Wang, X. Growth of Titanium Dioxide Nanorods in 3D-Confined Spaces. *Nano Lett.* **2011**, *11*, 624–631. [CrossRef] [PubMed]
43. Li, Z.; Wang, R.; Xue, J.; Xing, X.; Yu, C.; Huang, T.; Chu, J.; Wang, K.-L.; Dong, C.; Wei, Z.; et al. Core–Shell ZnO@SnO$_2$ Nanoparticles for Efficient Inorganic Perovskite Solar Cells. *J. Am. Chem. Soc.* **2019**, *141*, 17610–17616. [CrossRef] [PubMed]
44. Rogé, V.; Lamblin, G.; Fechete, I.; Dinia, A.; Garin, F.; Bahlawane, N.; Lenoble, D. Improvement of the photocatalytic degradation property of atomic layer deposited ZnO thin films: The interplay between film properties and functional performances. *J. Mater. Chem. A* **2015**, *3*, 11453–11461. [CrossRef]
45. Willander, M.; Nur, O.; Sadaf, J.R.; Qadir, M.I.; Zaman, S.; Zainelabdin, A.; Bano, N.; Hussain, I. Luminescence from Zinc Oxide Nanostructures and Polymers and their Hybrid Devices. *Materials* **2010**, *3*, 2643–2667. [CrossRef]
46. Ahn, C.H.; Kim, Y.Y.; Kim, N.C.; Mohanta, S.K.; Cho, H. A comparative analysis of deep level emission in ZnO layers deposited by various methods. *J. Appl. Phys.* **2009**, *105*, 13502. [CrossRef]
47. Matsushima, Y.; Maeda, K.; Suzuki, T. Nature of dark-brown SnO$_2$ films prepared by a chemical vapor deposition method. *J. Ceram. Soc. Jpn.* **2008**, *116*, 989–993. [CrossRef]
48. Aljawfi, R.N.; Alam, M.J.; Rahman, F.; Ahmad, S.; Shahee, A.; Kumar, S. Impact of annealing on the structural and optical properties of ZnO nanoparticles and tracing the formation of clusters via DFT calculation. *Arab. J. Chem.* **2020**, *13*, 2207–2218. [CrossRef]
49. Gultekin, I.; Ince, N.H. Degradation of reactive azo dyes by UV/H2O2: Impact of radical scavengers. *J. Environ. Sci. Health Part A* **2004**, *39*, 1069–1081. [CrossRef]
50. Oluwabi, A.T.; Gaspar, D.; Katerski, A.; Mere, A.; Krunks, M.; Pereira, L.; Acik, I.O. Influence of Post-UV/Ozone Treatment of Ultrasonic-Sprayed Zirconium Oxide Dielectric Films for a Low-Temperature Oxide Thin Film Transistor. *Materials* **2019**, *13*, 6. [CrossRef]
51. Shi, S.; Li, J.; Bu, T.; Yang, S.; Xiao, J.; Peng, Y.; Li, W.; Zhong, J.; Ku, Z.; Cheng, Y.-B.; et al. Room-temperature synthesized SnO$_2$ electron transport layers for efficient perovskite solar cells. *RSC Adv.* **2019**, *9*, 9946–9950. [CrossRef]
52. Yi, Z.; Wang, J.; Jiang, T.; Tang, Q.; Cheng, Y. Photocatalytic degradation of sulfamethazine in aqueous solution using ZnO with different morphologies. *R. Soc. Open Sci.* **2018**, *5*, 171457. [CrossRef] [PubMed]

53. Zhang, Q.; Xu, M.; You, B.; Zhang, Q.; Yuan, H.; Ostrikov, K. (Ken) Oxygen Vacancy-Mediated ZnO Nanoparticle Photocatalyst for Degradation of Methylene Blue. *Appl. Sci.* **2018**, *8*, 353. [CrossRef]
54. Lin, J.; Luo, Z.; Liu, J.; Li, P. Photocatalytic degradation of methylene blue in aqueous solution by using ZnO-SnO$_2$ nanocomposites. *Mater. Sci. Semicond. Process.* **2018**, *87*, 24–31. [CrossRef]
55. Hou, C.; Hu, B.; Zhu, J. Photocatalytic Degradation of Methylene Blue over TiO$_2$ Pretreated with Varying Concentrations of NaOH. *Catalysts* **2018**, *8*, 575. [CrossRef]
56. Zhang, H.; Han, Y.; Yang, L.; Guo, X.; Wu, H.; Mao, N. Photocatalytic Activities of PET Filaments Deposited with N-Doped TiO$_2$ Nanoparticles Sensitized with Disperse Blue Dyes. *Catalysts* **2020**, *10*, 531. [CrossRef]

© 2020 by the authors. Licensee MDPI, Basel, Switzerland. This article is an open access article distributed under the terms and conditions of the Creative Commons Attribution (CC BY) license (http://creativecommons.org/licenses/by/4.0/).

Communication

Core/Shell Ag/SnO$_2$ Nanowires for Visible Light Photocatalysis

Anna Baranowska-Korczyc [1,2,*], Ewelina Mackiewicz [1], Katarzyna Ranoszek-Soliwoda [1], Jaroslaw Grobelny [1] and Grzegorz Celichowski [1,*]

[1] Department of Materials Technology and Chemistry, Faculty of Chemistry, The University of Lodz, Pomorska 163, 90-236 Lodz, Poland; e.a.mackiewicz@gmail.com (E.M.); katarzyna.soliwoda@chemia.uni.lodz.pl (K.R.-S.); jaroslaw.grobelny@chemia.uni.lodz.pl (J.G.)

[2] Department of Chemical Textiles Technologies, ŁUKASIEWICZ-Textile Research Institute, 5/15 Brzezinska Street, 92-103 Lodz, Poland

* Correspondence: anna.baranowska-korczyc@iw.lukasiewicz.gov.pl (A.B.-K.); grzegorz.celichowski@chemia.uni.lodz.pl (G.C.)

Abstract: This study presents core/shell Ag/SnO$_2$ nanowires (Ag/SnO$_2$NWs) as a new photocatalyst for the rapid degradation of organic compounds by the light from the visible range. AgNWs after coating with a SnO$_2$ shell change optical properties and, due to red shift of the absorbance maxima of the longitudinal and transverse surface plasmon resonance (SPR), modes can be excited by the light from the visible light region. Rhodamine B and malachite green were respectively selected as a model organic dye and toxic one that are present in the environment to study the photodegradation process with a novel one-dimensional metal/semiconductor Ag/SnO$_2$NWs photocatalyst. The degradation was investigated by studying time-dependent UV/Vis absorption of the dye solution, which showed a fast degradation process due to the presence of Ag/SnO$_2$NWs photocatalyst. The rhodamine B and malachite green degraded after 90 and 40 min, respectively, under irradiation at the wavelength of 450 nm. The efficient photocatalytic process is attributed to two phenomenon surface plasmon resonance effects of AgNWs, which allowed light absorption from the visible range, and charge separations on the Ag core and SnO$_2$ shell interface of the nanowires which prevents recombination of photogenerated electron-hole pairs. The presented properties of Ag/SnO$_2$NWs can be used for designing efficient and fast photodegradation systems to remove organic pollutants under solar light without applying any external sources of irradiation.

Keywords: AgNWs; SnO$_2$; silver nanowires; core-shell nanostructures; photocatalytic activity; visible-light photocatalysis

Citation: Baranowska-Korczyc, A.; Mackiewicz, E.; Ranoszek-Soliwoda, K.; Grobelny, J.; Celichowski, G. Core/Shell Ag/SnO$_2$ Nanowires for Visible Light Photocatalysis. *Catalysts* 2022, 12, 30. https://doi.org/10.3390/catal12010030

Academic Editors: Sophie Hermans and Julien Mahy

Received: 9 November 2021
Accepted: 24 December 2021
Published: 28 December 2021

Publisher's Note: MDPI stays neutral with regard to jurisdictional claims in published maps and institutional affiliations.

Copyright: © 2021 by the authors. Licensee MDPI, Basel, Switzerland. This article is an open access article distributed under the terms and conditions of the Creative Commons Attribution (CC BY) license (https://creativecommons.org/licenses/by/4.0/).

1. Introduction

In recent years, due to high environmental pollution and fast industrial development, considerable interest has been paid to designing efficient, rapid, and widely applicable photocatalytic systems that are based on semiconductor nanostructures. The unique physical and chemical properties of semiconductors such as TiO$_2$, ZnO, ZnS, and CdSe make them extensively studied materials as photocatalysts [1]. One of the most promising semiconducting metal oxides is SnO$_2$ because of its high chemical, thermal, and mechanical stability as well as its high performance of organic pollutant degradation. However, a key issue in applying semiconducting nanostructures for practical purposes is the impossibility of visible light utilization. The efficient application of SnO$_2$ is limited by a large band gap of 3.6–4.1 eV and a very quick recombination of photogenerated electrons and holes [2,3]. To overcome these limitations, various metal nanostructures have been applied to form a heterojunction with semiconductors to induce efficient carrier separation under visible light irradiation. Ag-SnO$_2$ nanocomposite that is synthesized using an electrochemically active biofilm was proposed as a visible light-driven photocatalyst for the degradation of methyl orange, methylene blue, 4-nitrophenol, and 2-chlorophenol [4]. TiO$_2$/Ag/SnO$_2$ ternary heterostructures that were obtained by a one-step reduction approach demonstrated a visible-light photocatalytic effect with high stability and reusability due to the Ag nanoparticle surface plasmon resonance (SPR) influence [5]. An Ag/SnO$_2$ composite was also fabricated by the one-pot hydrothermal method and revealed high efficiency towards the photodegradation of phenol under visible light irradiation [6]. At the same photocatalytic conditions, Ag/Ag$_2$O/SnO$_2$ nanoparticles removed malachite green [7]. Ag-doped SnO$_2$ nanoparticles that were modified with curcumin were found to be an efficient photocatalyst for the degradation of rhodamine B under visible light [8]. The composition of Ag and SnO$_2$ nanoparticles was applied for the photocatalytic removal of nitrogen oxide under solar light [9]. In addition, Ag-SnO$_2$ nanocomposites were presented not only as a photocatalytic agent but also their antibacterial and antioxidant properties were investigated [10].

Ag nanostructures of various dimensionalities were used to form Ag/SnO$_2$ composites for photocatalytic applications, but Ag nanowires (AgNWs), despite their many advantages, have not been applied previously for these purposes. AgNWs reveal high transmittance, excellent plasmonic properties, high electrical performance, mechanical flexibility, nanometric size, and one dimensional (1D) geometry [11]. Moreover, the facile separation process of the AgNWs by filtration or sedimentation allows simple processing and improves their applicability in comparison to other silver nanostructures. In the core/shell Ag/SnO$_2$NWs heterojunction, the advantageous properties of both the nanomaterials can be combined.

This study presents a new photocatalytic system that is based on AgNWs that are coated with an SnO$_2$ shell. The efficiency of the novel core/shell Ag/SnO$_2$NWs photocatalyst was studied based on the absorbance intensity decrease during the decomposition process of rhodamine B, used as a model dye and under light irradiation at the wavelength from the visible range. Rhodamine B decomposes within 50 and 90 min under 395 nm and 450 nm illumination, respectively. Additionally, malachite green decomposes completely within 40 min with the presence of a new Ag/SnO$_2$NWs catalyst whereas typical TiO$_2$ photocatalyst only slightly affects the dye under 450 nm irradiation due to the low absorption in the visible region. The proposed Ag/SnO$_2$NWs system combines 1D morphology and the excellent physico-chemical properties of silver nanowires such as photoabsorption from the visible light region as well as the photocatalytic ability of tin oxide. Moreover, the metal/metal oxide arrangement significantly enhances the semiconductor photocatalytic properties due to fast carrier separation and preventing the recombination of photogenerated electron-hole pairs.

2. Results and Discussion

Ag/SnO$_2$NWs that were applied to the study were prepared in two processes, the first one was polyol synthesis to obtain AgNWs and then hydrolysis of sodium stannate in the presence of the nanowires to form a SnO$_2$ shell on the nanowire surface (Figure 1a). The core/shell Ag/SnO$_2$ nanowires were composed of 14 nm ($\pm$2 nm) thick SnO$_2$ shell consisting of 7 nm ($\pm$2 nm) rutile-type crystals surrounding the metallic core.

Our previous report describes, in detail, both synthesis stages, stability studies, morphological, and structural analysis of the core/shell Ag/SnO$_2$NWs system [12]. Despite the fact that AgNWs are characterized by fast atmospheric corrosion, nine weeks are enough to decompose completely; the Ag/SnO$_2$NWs are stable for over four months at ambient conditions. The core/shell Ag/SnO$_2$NWs show significant stability in the highly complexing environment of KCN solution. They are resistant to harsh CN$^-$ ions at the concentration range of 0.01 to 0.0001 wt.%. The high stability allows further applying them as a catalyst into various environments for the different pollutant decompositions [12]. The core/shell arrangement allows the formation of the metal-semiconductor junction to prevent the recombination of the photogenerated electron-hole pairs and to enhance photocatalytic efficiency by photoabsorption from the visible region [13]. Moreover, the SnO$_2$ shell acts as a protective coating on silver nanostructures against the influence of different environmental conditions [12], which provides high stability of the system. The absorbance spectra of AgNWs shows two major bands that are centered around 349 nm and 376 nm, respectively, that are responsible for the longitudinal and transverse modes in the SPR of the nanowires (Figure 2a). After SnO_2 shell synthesis the above-described peaks were red-shifted to 360 nm and 420 nm for the longitudinal and transverse SPR modes of AgNWs, respectively (Figures 2a and S1). It was shown previously that the SPR of silver nanowires is sensitive to the applied different coating as a result of the surrounding media dielectric constant changes [14]. In this study, the redshift allowed tuning of the absorbance properties and shifting the maxima of main absorbance bands to the visible light region. The absorbance spectra, typical for SnO_2, showed a peak at about 193 nm [15]. To study, in detail, the optical properties of the system, the hydrolysis process of sodium stannate in the aqueous solution without the silver nanowires presence was carried out. In this reaction, the pure SnO_2 nanoparticles (SnO_2NPs) were prepared that were similar in morphology and structure to the SnO_2 shell (Figure 1b). The nanoparticles revealed a mean diameter of about 25 nm (STEM analysis) and their properties were described in our previous report [16]. The absorbance spectra of SnO_2NPs shows the band with the maximum at 195 nm (Figure S2) that was suitable for photocatalysis induced by UV light. The value of the band gap for semiconducting nanostructures that was obtained by our method was calculated based on the Tauc plot (Figure S2) to be about 4.17 eV. This value is in good agreement with the reports for tin oxide nanoparticles [2].

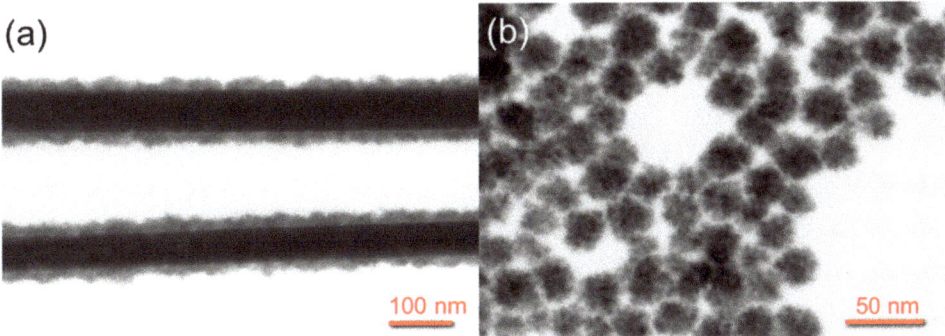

Figure 1. STEM images of (**a**) core/shell Ag/SnO_2NWs and (**b**) SnO_2NPs.

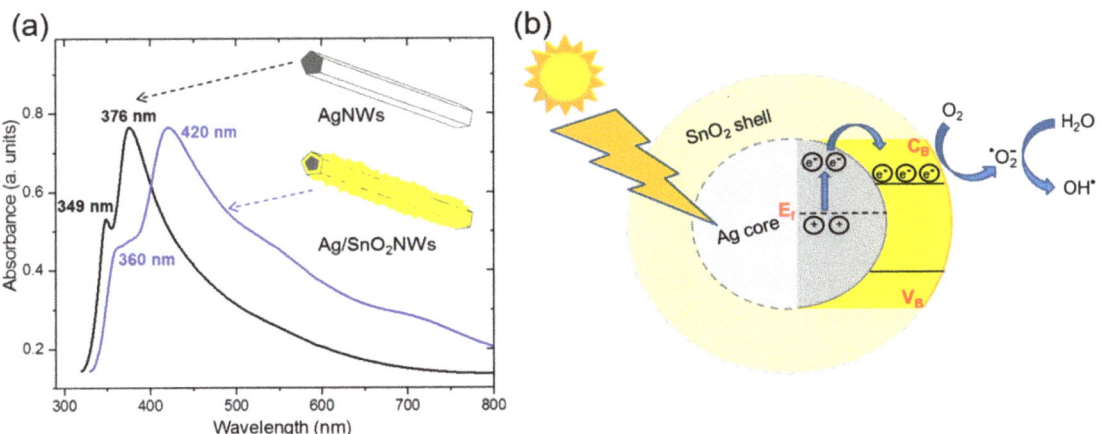

Figure 2. (**a**) The absorbance spectra of AgNWs and the core/shell Ag/SnO_2NWs. (**b**) Schematic illustration of the photocatalytic mechanism of Ag/SnO_2NWs under visible light irradiation.

The core/shell Ag/SnO$_2$NWs were used for the degradation of rhodamine B as a model organic dye that is characterized by an absorbance band with a maximum of 553 nm. Rhodamine B is widely applied for the photocatalytic model reaction study because it is broadly representative of organic compounds in its class and, since it strongly absorbs light, this allows facile monitoring of its degradation by UV/Vis spectroscopy [17]. The intensity of the band was studied under the irradiation of precisely defined sources, 395 nm and 450 nm LED lamps, corresponding electromagnetic radiation from the wavelength range of visible light. This approach can optimize the future photodegradation system for removing pollutants using solar irradiation instead of applying sophisticated and expensive UV-light sources (Figure S2). After coating AgNWs with an SnO$_2$ shell, the nanowires revealed effective photoabsorption at the visible light region due to the absorption band at 420 nm and partially by longitudinal mode at 360 nm (Figure 2a). The core/shell Ag/SnO$_2$NWs were dispersed homogeneously in an aqueous solution of rhodamine B. The shell formation prevented the nanowires from the aggregation process because SnO$_2$ is an efficient inorganic stabilizer of silver colloidal suspensions [16]. This phenomenon allows AgNWs introduction to the hydrophilic environment without aggregation and acting as a part of efficient photocatalyst as well as can also facilitate the redispersion process of Ag/SnO$_2$NWs after sedimentation. Figure S3a shows the separation process by sedimentation of the nanowires after 48 h. The nanowires were collected as sediment on the bottom of the vessel. This simple separation process of nanomaterials from the solution can be applied as a method of removing solvent with the decomposed pollutants after the photocatalysis process. A gentle mixing of the solution after sedimentation resulted in obtaining a homogenous mixture due to the SnO$_2$ coating on AgNWs, preventing the aggregation process (Figure S3b).

The intensity absorbance of rhodamine B was measured every 10 min after illumination by the selected light source. The samples were centrifuged for photocatalyst separation and further study of the optical properties of the supernatant with the dye. Figure 3a shows absorbance spectra of the supernatant with rhodamine B at the selected time points after 395 nm LED light irradiation. The absorbance intensity significantly decreased after 10 min of illumination and the dye was completely degraded in less than 1 h. It indicates the high efficiency of Ag/SnO$_2$NWs as a photocatalyst under the irradiation from the visible light region. To study the process in detail and the influence of different factors on the photodegradation process, rhodamine B degradation was studied under various conditions (Figure 3b). It is essential to study all the conditions and system components to avoid the possibility of spectral interferences by transformation intermediates which may absorb radiation at the wavelength of the dye's absorption maximum [18]. The ability of the degraded substance to inject electrons into the conduction band of a semiconductor should be primarily tested. Rhodamine B with the Ag/SnO$_2$NWs photocatalyst presence but not light-irradiated (dark experiment) did not show any degradation process (Figure S4). The dye sample without the Ag/SnO$_2$NWs but illuminated by 395 nm LED lamp revealed a slight decrease in the absorbance intensity (Figure S5) and proved that the photodegradation effect appears only in the presence of Ag/SnO$_2$NWs. Moreover, the contribution of SnO$_2$ in the photodegradation process in this range of illumination was verified. For these purposes, SnO$_2$ was synthesized using stannate precursor as in the case of the shell but without AgNWs presence. The process resulted in SnO$_2$ nanoparticles (SnO$_2$NPs) formation, described in our previous report [16]. To compare the influence of SnO$_2$ on the photocatalytic activity the Ag/SnO$_2$NWs, SnO$_2$NPs were added to the dye solution at the concentration of the whole core/shell complex (2 mg/mL) and at the concentration of the tin oxide in the shell (0.66 mg/mL) (Figure 3b). The weight percentage of the SnO$_2$ shell in Ag/SnO$_2$NWs complex was determined on the EDS studies and was about 33 wt.% (Figure S6). SnO$_2$ is known as an efficient photocatalyst under UV irradiation [19–21]. In our system, after irradiation by light from the visible region, both concentrations of SnO$_2$NPs showed a slight degradation rate of rhodamine B as a result of the negligible influence of this radiation on tin oxide nanostructures (Figures 3b and S7).

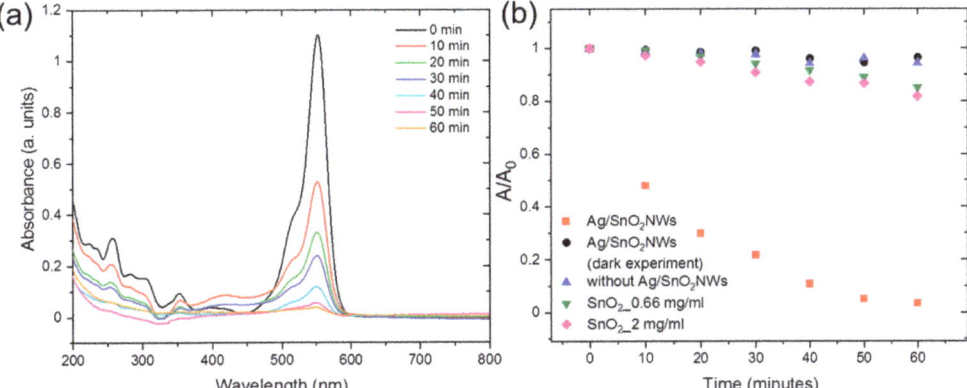

Figure 3. (**a**) The absorbance spectra of rhodamine B with Ag/SnO$_2$NWs photocatalyst presence under 395 nm light irradiation. (**b**) The degradation of rhodamine B with 2 mg/mL of Ag/SnO$_2$NWs catalyst (■), without photocatalyst (▲), with 0.66 mg/mL (▼), and 2 mg/mL (◆) of SnO$_2$NPs under 395 nm light irradiation and Ag/SnO$_2$NWs without irradiation (dark experiment, ●).

The rapid photodegradation of the dye under 395 nm illumination for about 60 min using Ag/SnO$_2$NWS is due to the formation of a metal/semiconductor heterostructure which prevents the fast recombination of photogenerated electron-hole pairs. The irradiation at the wavelength of 395 nm can excite only SPR of AgNWs in the core/shell composite and cause electron transfer from AgNWs to the conduction band (CB) of SnO$_2$ (Figure 2b). The carrier migration is a result of combining two materials with different work functions; Ag is characterized by a work function of 4.26 eV and SnO$_2$ with a work function of about 4.84 eV [22,23]. The photoinduced electrons can get sufficient energy to surmount the Schottky barrier on the Ag/SnO$_2$ interface despite the uniform energy levels of both components. The electrons are transferred to the semiconductor material to equilibrate the metal-metal oxide alignment and form a new Fermi energy level (Figure 2b). The transferred free electrons are trapped by dissolved oxygen molecules in the water and form a high oxidative species, such as superoxide radical anions (O$_2^{-\bullet}$) and hydroxyl radicals (HO$^\bullet$) [4]. In the core/shell heterostructure arrangement, the recombination of the photoinduced hole-electron pairs is inhibited mainly by forming a complex of free electrons from the CB with oxygen molecules. The trapped electrons can facilitate the formation of O$_2^{-\bullet}$ and HO$^\bullet$ reactive radicals and significantly enhance the rate of their formation, increase photocatalytic activity, and reduce organic substance degradation time. The absorbance intensity of rhodamine B decreased by half after 10 min of irradiation. Moreover, the strong confinement and anisotropic effect in the 1D core/shell metal/semiconductor structures can facilitate carrier separation and increase photocatalytic efficiency.

To the best of our knowledge, it is the first report presenting the photocatalytic properties of the 1D core/shell Ag/SnO$_2$NWs nano-system. It can broaden the range of AgNWs applications since SnO$_2$ revealed a high environmental stability [12]. The silver nanostructures tend to aggregate and dissolve in the aquatic environment, so the protective shells are also applied to increase stability, processability, and range of applications. The ability of the new catalyst to decompose organic compounds under solar irradiation combined with a high resistance even to harsh conditions can allow the designing of photodegradation systems for various environments and without any external irradiation sources. The advantage of the presented Ag/SnO$_2$NWs photocatalyst is its high stability under different conditions in comparison to other silver/wide band semiconductor composites. An example of that hybrid is Ag/ZnO heterostructure, which, although revealed high photocatalytic activity [24], both its components show a low stability. Silver nanostructures are affected by atmospheric corrosion due to the effective interaction of Ag+ ions with sulphides and the formation of a silver sulphide layer [25]. ZnO is easily degraded at the nanoscale in hydrophilic environments [26]. Another, more stable semiconductor of TiO$_2$ that was combined with various Ag nanostructures revealed photocatalytic activity that was characterized by a degradation time of 60 min [27] and more than 90 min [28], but under UV-light. Ag nanowires that were modified with an α-Fe$_2$O$_3$ show similar efficiency and time of the process; methylene blue was degraded for 30 min under visible light illumination, according to the authors, due to the synergetic effect of LSPR and the effective separation of photogenerated carriers between both materials [29]. A ternary TiO$_2$/Ag/SnO$_2$ system was applied for the photodegradation of methylene blue for more than 140 min for 40 mg of the photocatalyst of 3.12×10^{-5} methylene blue (100 mL) under visible light irradiation [5]. Ag/Ag$_2$O/SnO$_2$ nanocomposites removed malachite green (20 mg/L) after 120 min by using 30 mg of the photocatalyst [7]. The photocatalysis process under visible light with the addition of Ag-SnO$_2$ nanocomposites that were synthesized using electrochemically active biofilm was measured in hours, but was significantly efficient than for pure SnO$_2$ [4]. The sphere-like plasmonic Ag/SnO$_2$ photocatalyst revealed a phenol decomposition time of 50 min that was similar to our results but with different dimensionalities allowing the application of them to other purposes [6].

To study the photodegradation process with Ag/SnO$_2$NWs catalyst that was induced by the light from the visible region, an additional light source from this area at the wavelength of 450 nm was chosen. The removal efficiency of rhodamine B was similar for 395 nm and 450 nm light sources and was calculated to be about 87% and 88% for irradiation at 395 nm and 450 nm (Figure S8). However, the degradation time was not comparable; the process took about 50 and 90 min for irradiation at 395 nm and 450 nm, respectively. The efficiency of the degradation was also high, but the time that was needed to decompose the dye increased to about 90 min (Figure 4a). Both the SPR bands that were irradiated at 450 nm were not so effectively excited as for the 395 nm source. The AgNWs absorbed only the irradiation above 450 nm and their absorbance spectrum was only partially excited (Figure 2a). The process was still efficient and significantly higher than for pure SnO$_2$ irradiated under 450 nm (Figure 4b).

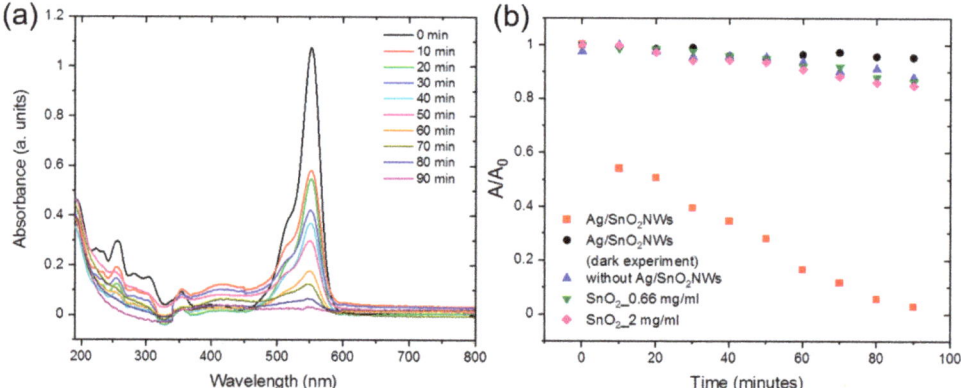

Figure 4. (a) The absorbance spectra of rhodamine B with Ag/SnO$_2$NWs photocatalyst presence under 450 nm light irradiation. (b) The degradation of rhodamine B with 2 mg/mL of Ag/SnO$_2$NWs catalyst (■), with no photocatalyst (▲), with 0.66 mg/mL (▼), and 2 mg/mL (♦) of SnO$_2$NPs under 450 nm light irradiation and Ag/SnO$_2$NWs without irradiation (dark experiment, ●).

Similar to the 395 nm excitation source (Figure 3b), the absorbance intensity of the dye under illumination at the wavelength of 450 nm without the photocatalyst did not change significantly, indicating the minimal influence of the irradiation on the optical properties of rhodamine B (Figures 4b and S9). The experiment with Ag/SnO$_2$NWs photocatalyst, without applying irradiation (dark experiment, Figure S4), showed only a minimal decrease in the absorbance intensity and demonstrated the stability of the dye and the essential influence of light illumination in the decomposition process. The irradiation at the wavelength of 450 nm on SnO_2NPs at both concentrations, 2 mg/mL as Ag/SnO_2NWs and 0.66 mg/mL corresponding to SnO_2 amount in the core/shell complex, showed a slight degradation rate of rhodamine B (Figure S10). It indicated that the selected irradiation source did not significantly affect the organic substances when using pure semiconductor as a photocatalyst, only the Ag/SnO_2 heterojunction can be considered as a source for the utilization of organic pollutants under visible light.

Ag/SnO_2NWs were also applied for the degradation of malachite green, which is present in the environment and its remediation is highly required. Malachite green is a dye that is commonly used in the textile and food industry. It should be removed after industrial processes due to the fact that it is highly toxic, especially for aquatic flora and fauna [30–32]. Figure 5a shows the rapid degradation of malachite green by Ag/SnO_2NWs under the light from the visible range (450 nm); malachite green was completely decomposed after 40 min. The absorbance intensity of the dye with Ag/SnO_2NWs photocatalyst presence but without any irradiation did not change significantly (Figures 5b and S11). The dark experiment proved the photodegradation mechanism for fast malachite green decomposition. Malachite green irradiation at the wavelength of 450 nm without photocatalyst revealed only a slight decrease in the absorbance intensity (Figures 5b and S12). The irradiation of SnO_2NPs at both concentrations, 2 mg/mL as Ag/SnO_2NWs and 0.66 mg/mL corresponding to SnO_2 amount in the core/shell complex also showed only a slight degradation rate of malachite green (Figures 5b and S13). Moreover, to compare Ag/SnO_2NWs photocatalyst efficiency at the visible range to a well-known commercial photocatalyst, malachite green was treated with TiO_2 under 450 nm irradiation. TiO_2 as a semiconductor that is characterized by absorption in the UV range, decomposed the dye only to a minimal extent (Figures 5b, S14 and S15) [33]. The core/shell Ag/SnO_2NWs system shows a high degradation rate of organic pollutants under visible light irradiation and can be used in practical photocatalysis reactions for efficient remediation.

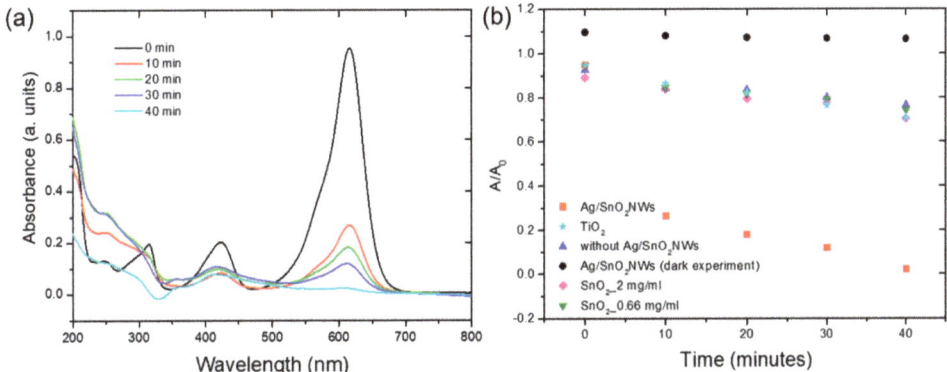

Figure 5. (a) The absorbance spectra of malachite green with Ag/SnO_2NWs photocatalyst presence under 450 nm light irradiation. (b) The degradation of malachite green with 2 mg/mL of Ag/SnO_2NWs catalyst (■), with no photocatalyst (▲), with 0.66 mg/mL of SnO_2NPs (▼), 2 mg/mL of SnO_2NPs (♦), and with 2 mg/mL TiO_2 catalyst (★) under 450 nm light irradiation and Ag/SnO_2NWs without irradiation (dark experiment, ●).

3. Experimental Section

3.1. Preparation of Core/Shell Ag/SnO_2NWs

In the first stage, AgNWs were synthesized by a polyol process and then covered by an SnO_2 shell as a result of the hydrolysis process of sodium stannate; both stages were described in detail in our previous studies [12,34]. In brief, 0.408 g of $AgNO_3$ solution (purity 99.9999%, Sigma-Aldrich, St Louis, MO, USA) in ethylene glycol (EG, POCH) was added (feed rate of 16 mL/h) to the mixture of 40 mL of EG, 2 g of polyvinyl pyrrolidone (PVP, molecular weight of 55 kDa, Sigma Aldrich) and 0.028 g of sodium chloride (NaCl, Chempur, Karlsruhe, Germany) was constantly heated to 170 °C, refluxed, and stirred at 570 rpm. Then, the solution was maintained at the same conditions for 1 h and cooled to room temperature. The mixture of AgNWs was diluted by acetone followed by dispersion in 60 mL of ethanol (anhydrous, POCH).

An SnO_2 shell on AgNWs was obtained in a one-step process by adding 5.051 g of an aqueous solution of sodium stannate trihydrate (0.25 wt.%. $Na_2SnO_3 \cdot 3H_2O$, Sigma-Aldrich, 95%) to the AgNWs mixture that was heated to 100 °C, refluxed, and stirred at 300 rpm. The solution was kept at the above-described conditions for 15 min and then cooled in a cold water-bath. The mixture of AgNWs was previously dispersed in water by adding 2.5 g of AgNWs that was obtained in polyol process (ethanol solution), to 92.7 g deionized water and 1 wt.% aqueous solution of sodium citrate ($Na_3C_6H_5O_7 \cdot 2H_2O$, purity 99.0%, Sigma Aldrich).

The core/shell Ag/SnO_2NWs were filtered (Merck Millipore (Burlington, MA, USA) type RTTR, Isopore membrane Filter, the pore size of 1.2 μm) to remove any by-products of the reactions and obtain high purity samples. To obtain pure SnO_2 that was relevant to the shell part of the complex for the control experiments, 5.70 g sodium stannate trihydrate (0.25 wt.% aqueous solution) was added to 40 g of boiling water and this sample was heated to 100 °C and stirred at 600 rpm for 15 min. As a result, the SnO_2 nanoparticles (SnO_2NPs) were synthesized. Their morphology and structure were described in our previous work [16].

3.2. Photocatalytic Activity Study

Photocatalytic activities of the core/shell Ag/SnO_2NWs were determined by the decomposition of rhodamine B (≥95%, Sigma Aldrich) as a model system. The aqueous solution of Ag/SnO_2NWs (2 mg/mL) and rhodamine B (5 mg/L, 10.4 μM) was irradiated by 395 nm and 450 nm LED lamps (100 W). The removal efficiency of rhodamine B by Ag/SnO_2NWs photocatalyst for both irradiation sources was calculated based on the initial (C_0) and final concentration (C) of the dye according to Equation (1).

$$H\,(\%) = \frac{C_0 - C}{C_0} \times 100 \tag{1}$$

As control samples, SnO_2NPs were also illuminated by 395 nm and 450 nm as well as 365 nm LED lamps at concentrations of 2 mg/mL (the same as Ag/SnO_2NWs system) and 0.66 mg/mL (as shell SnO_2 percentage wt.% in the whole complex). The weight percentage of the shell and core of the Ag/SnO_2NWs composite was determined based on EDS (Energy Dispersive X-ray Spectroscopy) studies using an FEI Nova NanoSEM 450 microscope that was equipped with EDAX Roentgen spectrometer (EDS) and an Octane Pro Silicon Drift Detector (SDD). The samples were previously collected on silicon wafers for EDS measurements.

Moreover, malachite green oxalate salt (7 mg/L, 15.1 μM, Sigma Aldrich) was degraded with the presence of Ag/SnO_2NWs (2 mg/mL) under irradiation of a 450 nm LED lamp (100 W). To compare the photocatalytic ability of Ag/SnO_2NWs to commercial photocatalyst, 2 mg/mL of TiO_2 (titanium (IV) oxide, anatase, nanopowder, <25 nm particle size, 99.7%, Sigma Aldrich) was added to malachite green oxalate salt solution (7 mg/L, 15.1 μM) and irradiated by 450 nm LED lamp (100 W). The degradation reactions of rhodamine B and malachite green were monitored by measuring the UV/Vis absorption spectra (UV5600 spectrophotometer, Biosens) of the sample solution taken out at regular intervals, every 10 min for 395 nm and 450 nm LED lamps or every 30 min for 365 nm LED lamp illumination. The sample solution was constantly stirred (400 rpm) and cooled during illumination. The LED lamps were fixed over the vessel that was filled with a sample solution. The absorbance spectra of each sample supernatant were recorded in the wavelength range of 200 to 800 nm after the centrifugation process (8000 rpm, 2 min) of a 2 mL sample after selected illumination time.

4. Conclusions

The study demonstrates Ag nanowires that are covered with SnO_2 shell as a new, efficient photocatalyst under the irradiation from the visible light range. The fast degradation process is the effect of the combination of advantages of both the components forming the core/shell Ag/SnO_2NWs hybrid as well as phenomena appearing on the material interfaces. The SPR absorbance spectrum of AgNWs after coating with SnO_2 shifts towards the visible region and facilitates excitation of the electrons in the nanowires by photons at this wavelength range. The photocatalytic activity of SnO_2 is enhanced significantly and achievable without UV irradiation. The excited electrons from the metal core are transferred to the metal oxide shell and captured by oxygen molecules and involved in the formation of reactive radicals that are essential for the degradation of organic compounds. This report utilizes rhodamine B as a model organic dye for studying the activity of novel 1D metal/semiconductor Ag/SnO_2NWs photocatalyst. The rhodamine B is degraded after 50 and 90 min under irradiation at the wavelength of 395 nm and 450 nm, respectively. Moreover, malachite green as an environmental organic pollutant is decomposed after 40 min by Ag/SnO_2NWs and only slightly degraded by the common catalyst of TiO_2 under 450 nm irradiation. High photocatalytic activity of the Ag/SnO_2NWs system is attributed to the core/shell metal/semiconductor arrangement which results in carrier separations and prevents the recombination of photogenerated electron-hole pairs. The facile processing of an Ag/SnO_2NWs hybrid by simple separation, such as filtration or sedimentation, is beneficial for photocatalytic applications.

Our findings indicate that the core/shell of Ag/SnO_2NWs represents a very promising material that is characterized by high environmental stability for the designing of future photocatalytic systems under solar irradiation for effective remediation processes of various environments.

Supplementary Materials: The following are available online at https://www.mdpi.com/article/10.3390/catal12010030/s1, Figure S1: Absorbance spectra of core/shell Ag/SnO_2NWs. Figure S2: The absorbance spectra of SnO_2NPs at the concentrations of 2 and 0.66 mg/mL and (inset) Tauc plot for SnO_2 energy gap value determination. Figure S3: The images of an aqueous solution of Ag/SnO_2NWS (a) left for 48h for sedimentation and (b) then gently mixed to redisperse the core/shell nanowires. Figure S4: The absorbance spectra of rhodamine B without any irradiation after centrifugation of Ag/SnO_2NWs photocatalyst (dark experiment). Figure S5: The absorbance spectra of rhodamine B that was irradiated under 359 nm without catalyst. Figure S6: The EDS spectrum of Ag/SnO_2NWs, inset: The weight percentage of O, Ag, and Sn in the hybrid, and STEM image of the sample area for EDS analysis. Figure S7: The absorbance spectra of rhodamine B and SnO_2NPs at a concentration of (a) 0.66 mg/mL and (b) 2 mg/mL under 395 nm light irradiation. Figure S8: The degradation of rhodamine B with Ag/SnO_2NWs photocatalysts presence under 395 nm and 450 nm irradiation. Figure S9: The absorbance spectra of rhodamine B that was irradiated under 450 nm. Figure S10: The absorbance spectra of rhodamine B with the presence of SnO_2NPs at a concentration of (a) 0.66 mg/mL and (b) 2 mg/mL under 450 nm light irradiation. Figure S11: The absorbance spectra of malachite green without any irradiation with the presence of Ag/SnO_2NWs photocatalyst (dark experiment). Figure S12: The absorbance spectra of malachite green that was irradiated under 450 nm without catalyst. Figure S13: The absorbance spectra of malachite green with the presence of SnO_2NPs at a concentration of (a) 0.66 mg/mL and (b) 2 mg/mL under 450 nm light irradiation. Figure S14: The absorbance spectra of malachite green with the presence of 2 mg/mL TiO_2 photocatalyst that was irradiated under 450 nm with. Figure S15: The absorbance spectra of TiO_2 (P25).

Author Contributions: Conceptualization, methodology, validation, and formal analysis, A.B.-K. and G.C.; investigation, A.B.-K., E.M. and K.R.-S.; resources, J.G. and G.C.; data curation, G.C.; writing—original draft preparation, A.B.-K.; writing—review and editing, A.B.-K. and G.C.; funding acquisition, G.C. All authors have read and agreed to the published version of the manuscript.

Funding: The research was financially supported by a grant from the National Science Centre, Poland (Opus 15 no. 2018/29/B/ST8/02016).

Data Availability Statement: The data presented in this study are available on request from the corresponding authors.

Conflicts of Interest: The authors declare no conflict of interest.

References

1. Xu, C.; Ravi Anusuyadevi, P.; Aymonier, C.; Luque, R.; Marre, S. Nanostructured materials for photocatalysis. *Chem. Soc. Rev.* **2019**, *48*, 3868–3902. [CrossRef]
2. Kamble, V.B.; Umarji, A.M. Defect induced optical bandgap narrowing in undoped SnO_2 nanocrystals. *AIP Adv.* **2013**, *3*, 082120. [CrossRef]
3. Al-Hada, N.M.; Kamari, H.M.; Baqer, A.A.; Shaari, A.H.; Saion, E. Thermal calcination-based production of SnO_2 nanopowder: An analysis of SnO_2 nanoparticle characteristics and antibacterial activities. *Nanomaterials* **2018**, *8*, 250. [CrossRef]
4. Ansari, S.A.; Khan, M.M.; Ansari, M.O.; Lee, J.; Cho, M.H. Visible light-driven photocatalytic and photoelectrochemical studies of Ag-SnO_2 nanocomposites synthesized using an electrochemically active biofilm. *RSC Adv.* **2014**, *4*, 26013–26021. [CrossRef]
5. Zhang, Z.; Ma, Y.; Bu, X.; Wu, Q.; Hang, Z.; Dong, Z.; Wu, X. Facile one-step synthesis of $TiO_2/Ag/SnO_2$ ternary heterostructures with enhanced visible light photocatalytic activity. *Sci. Rep.* **2018**, *8*, 10532. [CrossRef]
6. Saravanakumar, K.; Muthuraj, V. Fabrication of sphere like plasmonic Ag/SnO_2 photocatalyst for the degradation of phenol. *Optik* **2017**, *131*, 754–763. [CrossRef]
7. Tafik, A.; Paramarta, V.; Prakoso, S.P.; Saleh, R. Using Ag/Ag_2O/SnO_2 Nanocomposites to Remove Malachite Green by Photocatalytic Process. In *Journal of Physics: Conference Series*; IOP Publishing: Bristol, UK, 2017; Volume 820, p. 012017. [CrossRef]
8. Vignesh, K.; Hariharan, R.; Rajarajan, M.; Suganthi, A. Photocatalytic performance of Ag doped SnO_2 nanoparticles modified with curcumin. *Solid State Sci.* **2013**, *21*, 91–99. [CrossRef]
9. Bui, D.P.; Nguyen, M.T.; Tran, H.H.; You, S.J.; Wang, Y.F.; Van Viet, P. Green synthesis of Ag@SnO_2 nanocomposites for enhancing photocatalysis of nitrogen monoxide removal under solar light irradiation. *Catal. Commun.* **2020**, *136*, 105902. [CrossRef]
10. Sinha, T.; Ahmaruzzaman, M.; Adhikari, P.P.; Bora, R. Green and Environmentally Sustainable Fabrication of Ag-SnO_2 Nanocomposite and Its Multifunctional Efficacy As Photocatalyst and Antibacterial and Antioxidant Agent. *ACS Sustain. Chem. Eng.* **2017**, *5*, 4645–4655. [CrossRef]
11. Manning, H.G.; da Rocha, C.G.; Callaghan, C.O.; Ferreira, M.S.; Boland, J.J. The Electro-Optical Performance of Silver Nanowire Networks. *Sci. Rep.* **2019**, *9*, 11550. [CrossRef]
12. Baranowska-Korczyc, A.; Mackiewicz, E.; Ranoszek-Soliwoda, K.; Nejman, A.; Trasobares, S.; Grobelny, J.; Cieślak, M.; Celichowski, G. A SnO_2 shell for high environmental stability of Ag nanowires applied for thermal management. *RSC Adv.* **2021**, *11*, 4174–4185. [CrossRef]
13. Seong, S.; Park, I.S.; Jung, Y.C.; Lee, T.; Kim, S.Y.; Park, J.S.; Ko, J.H.; Ahn, J. Synthesis of Ag-ZnO core-shell nanoparticles with enhanced photocatalytic activity through atomic layer deposition. *Mater. Des.* **2019**, *177*, 107831. [CrossRef]
14. Ramasamy, P.; Seo, D.M.; Kim, S.H.; Kim, J. Effects of TiO_2 shells on optical and thermal properties of silver nanowires. *J. Mater. Chem.* **2012**, *22*, 11651–11657. [CrossRef]
15. Sarmah, S.; Kumar, A. Optical properties of SnO_2 nanoparticles. *Indian J. Phys.* **2010**, *84*, 1211–1221. [CrossRef]
16. Baranowska-Korczyc, A.; Mackiewicz, E.; Ranoszek-Soliwoda, K.; Grobelny, J.; Celichowski, G. Facile synthesis of SnO_2 shell followed by microwave treatment for high environmental stability of Ag nanoparticles. *RSC Adv.* **2020**, *10*, 38424–38436. [CrossRef]
17. Pingmuang, K.; Chen, J.; Kangwansupamonkon, W.; Wallace, G.G.; Phanichphant, S.; Nattestad, A. Composite Photocatalysts Containing $BiVO_4$ for Degradation of Cationic Dyes. *Sci. Rep.* **2017**, *7*, 8929. [CrossRef]
18. Barbero, N.; Vione, D. Why Dyes Should Not Be Used to Test the Photocatalytic Activity of Semiconductor Oxides. *Environ. Sci. Technol.* **2016**, *50*, 2130–2131. [CrossRef]
19. Kim, S.P.; Choi, M.Y.; Choi, H.C. Photocatalytic activity of SnO_2 nanoparticles in methylene blue degradation. *Mater. Res. Bull.* **2016**, *74*, 85–89. [CrossRef]
20. Li, Y.; Yang, Q.; Wang, Z.; Wang, G.; Zhang, B.; Zhang, Q.; Yang, D. Rapid fabrication of SnO_2 nanoparticle photocatalyst: Computational understanding and photocatalytic degradation of organic dye. *Inorg. Chem. Front.* **2018**, *5*, 3005–3014. [CrossRef]
21. Xing, L.; Dong, Y.; Wu, X. SnO_2 nanoparticle photocatalysts for enhanced photocatalytic activities. *Mater. Res. Express* **2018**, *5*, 085026. [CrossRef]
22. Ghosh, S.; Goudar, V.S.; Padmalekha, K.G.; Bhat, S.V.; Indi, S.S.; Vasan, H.N. ZnO/Ag nanohybrid: Synthesis, characterization, synergistic antibacterial activity and its mechanism. *RSC Adv.* **2012**, *2*, 930–940. [CrossRef]
23. Das, O.R.; Uddin, M.T.; Rahman, M.M.; Bhoumick, M.C. Highly active carbon supported Sn/SnO_2 photocatalysts for degrading organic dyes. *J. Phys. Conf. Ser.* **2018**, *1086*, 012011. [CrossRef]
24. Ziashahabi, A.; Prato, M.; Dang, Z.; Poursalehi, R.; Naseri, N. The effect of silver oxidation on the photocatalytic activity of Ag/ZnO hybrid plasmonic/metal-oxide nanostructures under visible light and in the dark. *Sci. Rep.* **2019**, *9*, 11839. [CrossRef] [PubMed]

25. Elechiguerra, J.L.; Larios-Lopez, L.; Liu, C.; Garcia-Gutierrez, D.; Camacho-Bragado, A.; Yacaman, M.J. Corrosion at the nanoscale: The case of silver nanowires and nanoparticles. *Chem. Mater.* **2005**, *17*, 6042–6052. [CrossRef]
26. Baranowska-Korczyc, A.; Kościński, M.; Coy, E.L.; Grześkowiak, B.F.; Jasiurkowska-Delaporte, M.; Peplińska, B.; Jurga, S. ZnS coating for enhanced environmental stability and improved properties of ZnO thin films. *RSC Adv.* **2018**, *8*, 24411–24421. [CrossRef]
27. Cheng, B.; Le, Y.; Yu, J. Preparation and enhanced photocatalytic activity of Ag@TiO$_2$ core-shell nanocomposite nanowires. *J. Hazard. Mater.* **2010**, *177*, 971–977. [CrossRef]
28. Wang, J.; Zhao, H.; Liu, X.; Li, X.; Xu, P.; Han, X. Formation of Ag nanoparticles on water-soluble anatase TiO$_2$ clusters and the activation of photocatalysis. *Catal. Commun.* **2009**, *10*, 1052–1056. [CrossRef]
29. Lei, R.; Ni, H.; Chen, R.; Gu, H.; Zhang, B.; Zhan, W. Ag nanowire-modified 1D α-Fe$_2$O$_3$ nanotube arrays for photocatalytic degradation of Methylene blue. *J. Nanopart. Res.* **2017**, *19*, 378. [CrossRef]
30. Dimova, S.; Zaharieva, K.; Ublekov, F.; Kyulavska, M.; Stambolova, I.; Blaskov, V.; Nihtianova, D.; Markov, P.; Penchev, H. Novel dye degradation photocatalyst nanocomposite powders based on polydiphenylacetylene-zinc oxide in polystyrene matrix. *Mater. Lett.* **2020**, *269*, 127683. [CrossRef]
31. Solís-Casados, D.A.; Martínez-Peña, J.; Hernández-López, S.; Escobar-Alarcón, L. Photocatalytic Degradation of the Malachite Green Dye with Simulated Solar Light Using TiO$_2$ Modified with Sn and Eu. *Top. Catal.* **2020**, *63*, 564–574. [CrossRef]
32. Tsvetkov, M.; Zaharieva, J.; Milanova, M. Ferrites, modified with silver nanoparticles, for photocatalytic degradation of malachite green in aqueous solutions. *Catal. Today* **2020**, *357*, 453–459. [CrossRef]
33. Flak, D.; Coy, E.; Nowaczyk, G.; Yate, L.; Jurga, S. Tuning the photodynamic efficiency of TiO2 nanotubes against HeLa cancer cells by Fe-doping. *RSC Adv* **2015**, *5*, 85139–85152. [CrossRef]
34. Giesz, P.; Mackiewicz, E.; Nejman, A.; Celichowski, G.; Cieślak, M. Investigation on functionalization of cotton and viscose fabrics with AgNWs. *Cellulose* **2017**, *24*, 409–422. [CrossRef]

Article

Morphology Regulation Mechanism and Enhancement of Photocatalytic Performance of BiOX (X = Cl, Br, I) via Mannitol-Assisted Synthesis

Patrycja Wilczewska [1], Aleksandra Bielicka-Giełdoń [1,*], Karol Szczodrowski [2], Anna Malankowska [1], Jacek Ryl [3], Karol Tabaka [1] and Ewa Maria Siedlecka [1]

1. Faculty of Chemistry, University of Gdansk, Wita Stwosza 63, 80-308 Gdansk, Poland; patrycja.wilczewska@phdstud.ug.edu.pl (P.W.); anna.malankowska@ug.edu.pl (A.M.); karol.tabaka@phdstud.ug.edu.pl (K.T.); ewa.siedlecka@ug.edu.pl (E.M.S.)
2. Faculty of Mathematics, Physics and Informatics, University of Gdansk, Wita Stwosza 57, 80-308 Gdansk, Poland; karol.szczodrowski@ug.edu.pl
3. Institute of Nanotechnology and Materials Engineering, Faculty of Applied Physics and Mathematics, Gdansk University of Technology, Narutowicza 11/12, 80-233 Gdansk, Poland; jacek.ryl@pg.edu.pl
* Correspondence: a.bielicka-gieldon@ug.edu.pl; Tel.: +48-585235226

Abstract: BiOX (X = Cl, Br, I) photocatalysts with dominant (110) facets were synthesized via a mannitol-assisted solvothermal method. This is the first report on the exposed (110) facets-, size-, and defects-controlled synthesis of BiOX achieved by solvothermal synthesis with mannitol. This polyol alcohol acted simultaneously as a solvent, capping agent, and/or soft template. The mannitol concentration on the new photocatalysts morphology and surface properties was investigated in detail. At the lowest concentration tested, mannitol acted as a structure-directing agent, causing unification of nanoparticles, while at higher concentrations, it functioned as a solvent and soft template. The effect of exposed (110) facet and surface defects ($Bi^{(3-x)+}$, Bi^{4+}, Bi^{5+}) of BiOX on the photocatalytic activity of nanomaterials under the UV–Vis irradiation were evaluated by oxidation of Rhodamine B (RhB) and 5-fluorouracil (5-FU), an anticancer drug, and by reduction of Cr(VI). Additionally, the influence of crucial factors on the formation of BiOX in the synthesis with mannitol was discussed extensively, and the mechanism of BiOX formation was proposed. These studies presented a new simple method for synthesizing BiOX without any additional surfactants or shape control agents with good photocatalytic activity. The study also provided a better understanding of the effects of solvothermal conditions on the BiOX crystal growth.

Keywords: BiOX; photocatalysis; mannitol; soft template; structure-directing agent

1. Introduction

Bismuth oxyhalides BiOX (X = Cl, Br, I) are among the most important groups of semiconductors and have drawn considerable attention for their potential applications as novel photocatalysts active in ultraviolet, visible, and UV–Vis light [1]. BiOX crystalizes in the tetragonal structure of the matlockite with slabs $[Bi_2O_2]^{2+}$ interleaved by slabs of halogen $[X]^-$. These layered structures affect highly anisotropic electrical, magnetic, and optical properties and make BiOX promising materials in wide industrial and environmental applications [1,2].

Inorganic materials' physical and chemical features, especially optical, magnetic, and electronic properties depending on size, dimensions, and morphology, are also strongly related to the morphology and dimensionality of the semiconductors in their photocatalytic activity [1]. Over the years, many efforts have been made to synthesized BiOX with different morphology such as nanobelts [3], nanowires [4], nanosheets [5], nanoplates [6], nanotubes [3], nanoparticles [7], nanoflowers [8], lamella structures [9], hollow structures [10], and hierarchical nanostructures [6,7].

Additionally, within described types of structure, attempts of facet engineering were made [8,11].

One of the common methods of obtaining various nanomaterials and BiOX itself is solvothermal synthesis. Fabrication of bismuth oxychloride (BiOCl) using ethanol (ETH) [12], ethylene glycol (EG) [13–15], diethylene glycol (DEG) [14,15], triethylene glycol (TEG) [14], glycerol (GLY) [16,17], N,N-dimethylformamide (DMF) [7,14], polyethylene glycol (PEG) [13,18], water [14,18,19], and mannitol (MAN) [15,19] in the temperature range from 140 °C to 160 °C and the time of 3–16 h was studied. Bismuth oxybromide has so far been obtained from water [20,21], ethanol [22–24], isobutanol (ISO) [21], ethylene glycol [20–22,24,25], glycerol [21,22,24], and mannitol [25] under condition of 20–160 °C and 8–15 h. Solvothermal synthesis of bismuth oxyiodine used EG [26,27], ETH [26,27], GLY [26], and water [26] as solvents, and the autoclave reactions were conducted in 160 °C for 12 h.

Additionally, a few studies have focused on interactions between bismuth and halogens precursor, surfactants, and the solvent mixtures' effect on these nanomaterials' morphology. BiOXs have been prepared by the solvothermal method with many surfactants—tetrabutylammonium halides [12], hexadecyltrimethylammonium halides [28], cetylpyridinium halide [25], polyvinyl pyrrolidone [29], polyvinyl alcohol [30], and sodium dodecyl sulfate [31]). Surfactants were used as templates during the synthesis reaction process. The physical and chemical properties of BiOCl were tunable by mixing PEG and water [18] or TEG and DMF [7] in different volume ratios. A similar investigation for BiOI with EG and water mixture was also conducted [26].

Although many different methods have been developed to modulate the morphology and crystallites size of BiOX (X = Cl, Br, I), expensive surfactants and various organic solvents were usually used for synthesis. Because the solvent plays a significant role in controlling bismuth oxyhalides' morphology, it is crucial to develop an additive-free synthesis method, which will be more convenient to fabricate these semiconductors. Moreover, the development of alternative ecological synthesis for nanomaterials may reduce the negative impact of widely used chemicals on human health and the environment. A control of semiconductor morphology is a promising approach to improve the photocatalytic activity of BiOX and should be investigated more thoroughly.

Mannitol (MAN) is a type of sugar alcohol used as a sweetener. Due to its excellent mechanical compressing properties and physical-chemical stability, it is used to produce pharmaceuticals. Furthermore, toxicity studies indicated that mannitol did not cause any considerable adverse effects. As an environmentally friendly and cheap bio-polyol, MAN can be used to prepare nanomaterial due to the presence of a long carbon chain and multi-hydroxyl groups. It was reported that the presence of mannitol during the synthesis favored the formation of uniform nanostructures [25], which is desirable in photocatalytic processes. However, information on mannitol as a solvent in the semiconductors' synthesis and their photoactivity is limited and requires further research.

This study described the effects of mannitol concentration (0.1 M, 0.5 M, and 1 M) on the (110) planes exposition and defects in the crystal lattice of BiOX material (where X = Cl, Br, I), enhancing its photoactivity in oxidation and reduction of micropollutants. Moreover, such prepared photocatalysts were used for the first time to cytostatic drug (5-fluorouracil (5-FU)) removal from water.

To the best of our knowledge, this is the first report on the exposed (110) facets-, size- and defects-controlled synthesis of BiOX (X = Cl, Br, I) achieved by solvothermal synthesis with mannitol as a solvent, capping agent, and/or soft template simultaneously. This research provides a new, simple strategy to fabricate BiOX without any additional surfactants or shape-controlled agents and enhanced photocatalytic activity. The obtained results indicated that BiOX prepared in mannitol solution could be used to remove a wide range of water micropollutants, including cytostatic drugs such as 5-fluorouracil.

2. Results and Discussion

2.1. Characterization

2.1.1. XRD Analysis

The purity and crystallinity of prepared BiOX photocatalysts were examined by X-ray powder diffraction (XRD). Figure 1 shows the XRD patterns of synthesized BiOX by the mannitol-assisted solvothermal method.

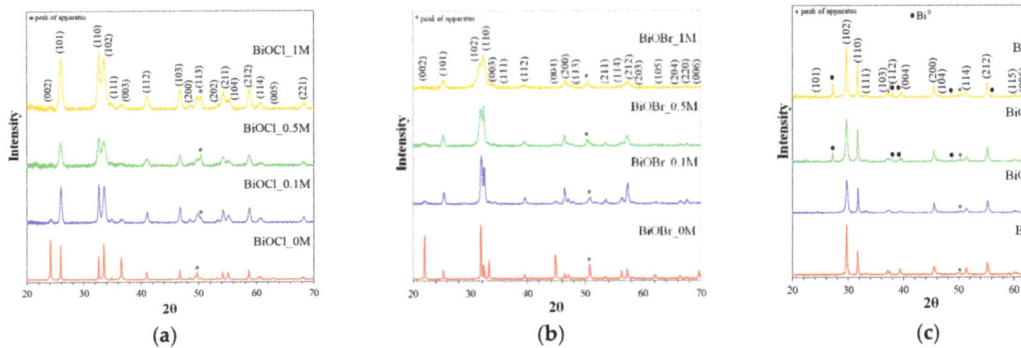

Figure 1. X-ray powder diffraction (XRD patterns) of (**a**) BiOCl, (**b**) BiOBr, and (**c**) BiOI prepared in synthesis with different mannitol concentrations.

As shown in Figure 1a, all diffraction peaks of BiOCl could be indexed to the tetragonal phase BiOCl (PDF 04-002-3608). No other diffraction peaks were detected, indicating the high purity of BiOCl. The signals of BiOCl_0M and BiOCl_0.1M diffractograms were intense, sharp, and narrow suggesting the high crystallinity of the samples. The higher concentration of the mannitol solution resulted in the formation of smaller crystallites, which can be observed as wider and shorted peaks in the BiOCl_0.5M and BiOCl_1.0M diffractograms [14,32]. The crystallite sizes obtained from the Scherrer equation and the half-maximum full width of the signal decreased from 95 nm (BiOCl_0M) to 32 nm (BiOCl_1M). In the absence of mannitol in the synthesis process, diffraction peaks of BiOCl_0M photocatalyst at 24.2°, 36.6°, 49.6°, and 63.2° have corresponded {001} family of planes. In contrast to distilled water, mannitol inhibited the formation of {001} planes due to the interaction of the hydroxyl groups of mannitol molecules and oxygen atoms in BiOCl (001) facets through hydrogen bonds. This phenomenon limited the growth along the [001] orientation of the BiOCl crystals, and the peaks of these family planes had lower intensity than the sample prepared in ultrapure water.

The XRD patterns of BiOBr are presented in Figure 1b. Signals were well-fitted to BiOBr (PDF 04-002-3609). The signals of BiOBr_0M were the sharpest and narrowest, similarly to the BiOCl series than the other BiOBr. The crystallite size of BiOBr was 90 nm and decreased after the introduction of the mannitol to the synthesis to 42 nm (BiOBr_0.1M). The rest of the BiOBr samples could not determine the crystallites' size from the most intense signal due to its poor separation. Bismuth oxybromide synthesized in ultrapure water (BiOBr_0M) was characterized by the {001} family of planes in the structure. In contrast, these planes' signals were absent in the series of BiOBr samples prepared in mannitol solutions. Furthermore, as previously described, the hydroxyl groups of alcohols are preferentially adsorbed on the (102) facets by the coordination of exposed Bi^{3+} ions, thus inhibiting growth along the (102) direction [33].

Our study observed these phenomena in preparation of a series of BiOBr (BiOBr_0.1M, BiOBr_0.5M, and BiOBr_1M), while this trend was not found for BiOCl. This fact could be related to the weak force interaction between mannitol and the smaller BiOCl particles. The diffraction intensity ratio (I_{110}/I_{102}) of the BiOX photocatalysts, listed in Table 1, increased with the increase of the mannitol concentration. The (110) and (102) facets were preferred for generation and accumulation of free e⁻ for the reduction reaction than other crystal facets [34,35]. Knowledge of the BiOX facets responsible for reduction is still limited in contrast to the oxidation processes connected with facets (001) and (010) [11,20].

Table 1. Physicochemical properties of BiOX (X = Cl, Br, I) samples synthesized in ultrapure water and mannitol solutions.

Photocatalyst	Particle Size	Morphology	Eg (eV)	MVB (eV)	MCB (eV)	I_{110}/I_{102}
BiOCl_0M	0.48–1.9 µm	microplate	3.3	3.51	0.21	0.65
BiOCl_0.1M	41.3–66.6 nm	nanoplate	3.2	3.46	0.26	1.03
BiOCl_0.5M	13.4–25.4 nm	nanoplate	3.05	3.38	0.33	1.14
BiOCl_1M	15–20 nm	nanoparticle	3.0	3.36	0.36	1.13
BiOBr_0M	1.45–5.12 µm	microplate	2.58	2.99	0.41	0.25
BiOBr_0.1M	40.8–81.0 nm	nanoplate	2.65	3.03	0.38	0.76
BiOBr_0.5M	58.5–72.7 nm	nanoplate	2.65	3.03	0.38	1.08
BiOBr_1M	530–734 nm	microstructure flower-like	2.56	2.98	0.42	1.40
BiOI_0M	4.13–5.65 µm	microplate	1.65	2.26	0.61	0.46
BiOI_0.1M	0.74–1.21 µm	microstructure rose-like	1.65	2.26	0.61	0.75
BiOI_0.5M	0.74–1.21 µm	microplate	1.58	2.23	0.65	0.72
BiOI_1M	183–361 nm	nanoplate	1.50	2.19	0.69	0.63

Figure 1c shows the XRD patterns of the BiOI samples prepared in highly purified water as a reference sample and mannitol solution. All samples could be indexed as BiOI (PDF 04-012-5693). The XRD results indicated that the series of BiOI photocatalysts synthesized by the solvothermal method were well crystallized and of high purity. The crystallite sizes of BiOI_0M, BiOI_0.1M, BiOI_0.5M, and BiOI_1M were 73.6 nm, 41.0 nm, 45.7 nm, and 52.6 nm, respectively. BiOI photocatalysts prepared in pure water and 0.1 M solution of mannitol (BiOI_0.1M) were characterized by almost identical XRD patterns. As previously described for the series of BiOCl and BiOBr, the samples prepared in a higher concentration of mannitol than 0.1 M the {001} planes were not formed. This phenomenon could be explained by the larger ionic radius of I⁻ than Br⁻ and Cl⁻ ionic radius and their various spatial packing of ions on the plane.

Furthermore, in the series of BiOI, the type of solvent used for the sample synthesis influenced the appearance of the additional Bi^0 phase. Increasing the concentration of mannitol from 0.1 M to 0.5 M and 1 M led to the formation of 12% and 17% of the Bi^0 phase in BiOI materials, respectively. It can be explained by the reducing properties of mannitol and its redox reaction with Bi^{3+} ions. It was previously reported that the organic compounds used in the synthesis, such as Na_2EDTA [36] and ethylene glycol [37], were capable of reducing metal ions at a high temperature of solvothermal synthesis. The appearance of Bi^0 in BiOI samples was probably related to a slower formation of bismuth oxyiodine nanosheets than other oxyhalides nanosheets and longer time of free Bi^{3+} ions exposition to the interaction with reductive mannitol. The slower formation of BiOI nanoplates due to the large radius of iodide ions was the reason for the favorable reduction of bismuth ions. Slower formation of BiOI was also observed after the introduction of Na_2EDTA, which could complex Bi^{3+} to the synthesis, and the rate of obtaining bismuth oxyiodide was decreased with increasing concentration of Na_2EDTA [38]. The obtained results revealed relevant coordination interactions between solvents and precursors in the formation of BiOX and were specific to each type of halide. Moreover, high crystallinity and purity, which are considered essential factors influencing nanomaterials' photocatalytic properties, were also solvent-dependent.

2.1.2. SEM Analysis

The morphologies and structures of selected BiOX (X = Cl, Br, I) photocatalysts examined by scanning electron microscopy are shown in Figure 2. The size and morphology of the samples are listed in Table 1.

Figure 2. SEM images of BiOX synthesized in (**a**–**c**) water; (**d**–**f**) 0.1 M mannitol solution; (**g**–**i**) 0.5 M mannitol solution; and (**j**–**l**) 1 M mannitol solution.

Figure 2a,d,g,j shows the series of BiOCl prepared in water and mannitol solutions in various concentrations. BiOCl_0M prepared in water was consisted of large amounts of nanosheets with a width of 0.48–1.9 µm and thickness of 50–70 nm. These nanoplates possessed rounded edges, and part of nanoplates resembled an irregular square. As shown in Figure 2d, BiOCl synthesized in the lowest concentration of mannitol contained more homogeneous and much smaller nanoplates with a width of 41.3–66.6 nm and thickness of 50–70 nm. The higher concentration of mannitol affected the BiOCl particle size. Thus, they were 13.4–25 nm and 10–15 nm for BiOCl_0.5M and BiOCl_1M, respectively. The changes in particle thickness were not observed. Similar results were obtained for BiOBr photocatalysts, except for BiOBr_1M synthesized in 1 M mannitol solution. Particles of BiOBr_1M (Figure 2k) were constructed into a flower-like microsphere of tight nanoplates with irregular edges. Our results are consistent with a previous report that mannitol can lead to the formation of hierarchical nanostructures [13]. It is worth mentioning that only the BiOBr_1M possessed a hierarchical structure, which indicated that the interaction between the solvent and precursors was more complex and required further research.

Furthermore, BiOI synthesized in water (Figure 2b) was composed of irregular square-like nanosheets, with a low tendency to agglomerate. The nanosheets size of the BiOI series was in the range of 1.51–3.26 μm, and its thicknesses were similar to BiOCl and BiOBr. In a 0.1 M mannitol solution, a nanoparticle of BiOI agglomerated into rose-like microstructures (0.74–1.21 μm) composed of dozen thicker and smaller nanoplates (213–357 nm) than those formed in water. The concentration of mannitol higher than 0.1 M prevented agglomerations of the forming nanoplates. This fact could be related to the mannitol solution's viscosity (Table 2), which increased with increasing mannitol concentration. Additionally, nanoplates of BiOI were smaller with the increasing concentration of mannitol. This trend was also observed during the preparation of the series of BiOCl samples.

Table 2. Physicochemical properties of the mannitol solutions (Temp: 20 °C).

Type of Solution	Density [g cm^{-3}]	Viscosity [cP]
deionized water	0.99823	1.005
0.1 M mannitol	1.00383	1.087
0.5 M mannitol	1.03534	1.269
1 M mannitol	1.06681	1.763

The inhibition of the growth of BiOX (X = Cl, Br, I) nanoparticles in the mannitol solutions clearly showed that the solvent with a long chain and polyhydroxy groups was both a soft template and structure-directing agent. The significant factors that influenced BiOX nanoparticles' formation were the size and interaction of halogen ions with the solvent and the concentration of the mannitol solution. Therefore, the products' morphology can be modulated by adjusting the appropriate concentration of mannitol as a solvent.

2.1.3. XPS Analysis

The surface chemical composition and the valance state of the BiOX (X = Cl, Br, I) were characterized using X-ray photoelectron spectroscopy (XPS). The results are shown in Table 3 and Figure 3.

Table 3. The binding energy of bismuth species observed character of atoms in a surface layer of prepared photocatalysts.

Sample	BE Bi 4f $_{7/2}$ (eV)				BE Bi 4f $_{5/2}$ (eV)			
	Bi$^{(3-x)+}$	Bi^{3+}	Bi^{4+}	Bi^{5+}	Bi$^{(3-x)+}$	Bi^{3+}	Bi^{4+}	Bi^{5+}
BiOCl_0M	158.8	160.1	-	-	163.7	165.4	166.6	-
BiOCl_0.1M	158.7	159.9	161.1	-	163.6	165.2	166.4	-
BiOCl_0.5M	158.4	159.8	161.1	-	163.4	165.0	166.3	-
BiOCl_1M	158.6	160.0	161.1	-	163.6	165.3	166.5	-
BiOBr_0M	158.0	160.1	-	-	163.3	165.4	-	-
BiOBr_0.1M	158.2	159.9	160.9	-	163.2	165.2	166.1	-
BiOBr_0.5M	158.3	159.8	161.0	-	163.1	165.1	166.3	-
BiOBr_1M	158.3	159.9	160.9	-	163.6	165.2	166.2	-
BiOI_0M	-	159.4	161.5	-	-	165.0	166.6	-
BiOI_0.1M	-	159.6	161.3	162.7	-	164.8	166.6	167.7
BiOI_0.5M	-	159.9	161.1	162.5	-	165.2	166.4	167.8
BiOI_1M	-	159.9	161.0	162.5	-	165.2	166.3	167.7

"-"—absence.

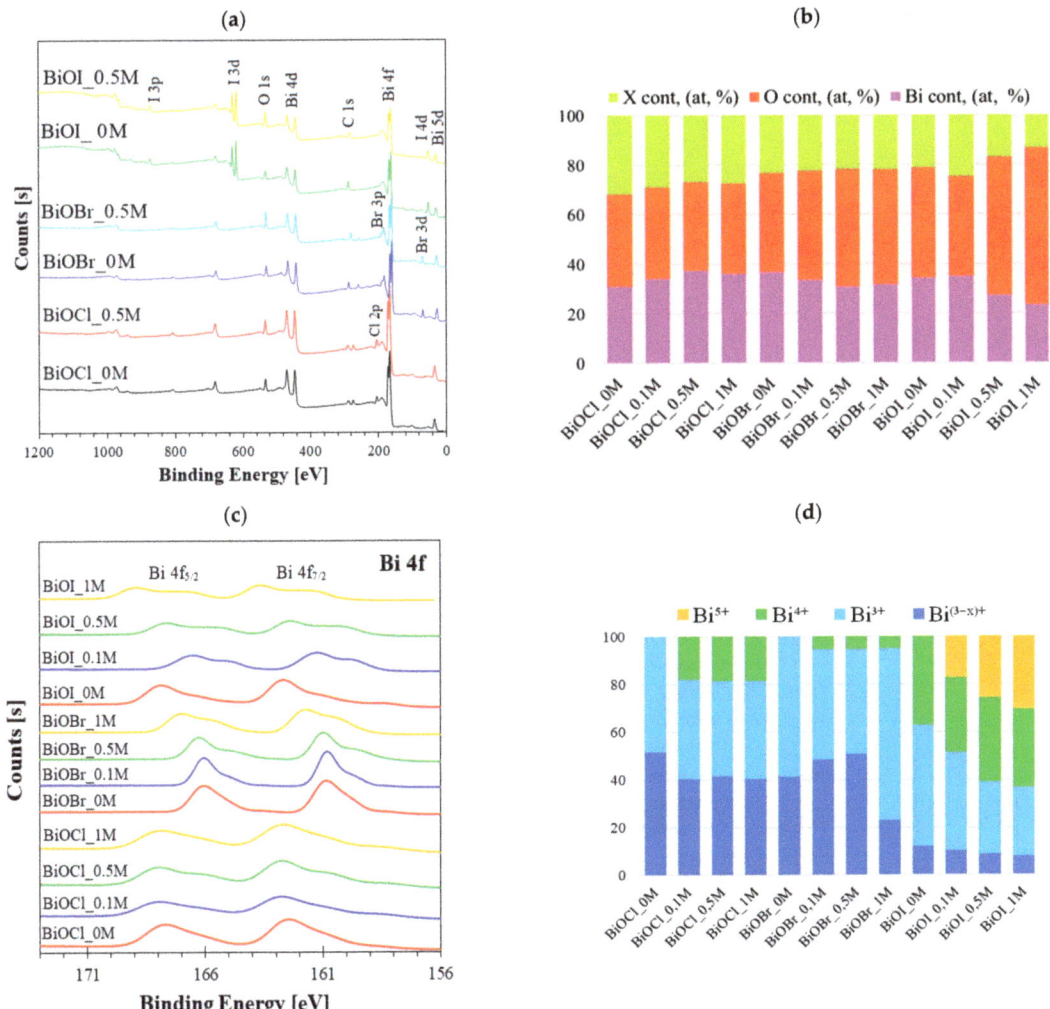

Figure 3. (**a**) Survey spectra of selected BiOX (X = Cl, Br, I); (**b**) surface elemental composition of BiOX (X = Cl, Br, I); (**c**) effect of mannitol on high-resolution spectra of Bi 4f in BiOX (X = Cl, Br, I); and (**d**) the concentration of mannitol resulting in the bismuth species ratio.

Interestingly, the Bi $4f_{5/2}$ and Bi $4f_{7/2}$ regions for BiOCl_0M and BiOBr_0M synthesized in water were fitted into the peaks attributed to Bi^{3+} and $Bi^{(3-x)+}$. This fact suggested the formation of oxygen vacancies in the samples. In the synthesis of BiOCl and BiOBr with mannitol solutions, additional new peaks appeared at the higher binding energies of 165.1 eV (±0.4 eV) and 161.1 eV (±0.4 eV). The peaks were attributed to the higher valance state of bismuth Bi^{4+}. BiOI prepared in the ultrapure water included Bi^{3+}, $Bi^{(3-x)+}$, and Bi^{4+}. The higher bismuth state appeared during the BiOI synthesis in a mannitol solution similar to BiOCl and BiOBr samples (Table 4). The higher concentration of mannitol in the synthesis of BiOI resulted in the coexistence of multiple bismuth species—Bi^{3+}, Bi^{4+}, and Bi^{5+}.

Table 4. The kinetic data for photocatalytic degradation of selected micropollutants over BiOX (X = Cl, Br, I).

Sample Label	Rh B		Cr^{6+}		5-FU	
	k_{app} [min^{-1}]	R^2	k_{app} [min^{-1}]	R^2	k_{app} [min^{-1}]	R^2
BiOCl_0M	0.152	0.9736	0.003	0.9627	0.076	0.9972
BiOCl_0.1M	0.045	0.9743	0.006	0.9703	0.008	0.9569
BiOCl_0.5M	0.084	0.9761	0.021	0.9715	0.022	0.9799
BiOCl_1M	0.070	0.9746	0.017	0.968	0.093	0.9989
BiOBr_0M	0.046	0.9667	0.001	0.9554	0.007	0.9909
BiOBr_0.1M	0.068	0.9822	0.010	0.9635	0.097	0.9521
BiOBr_0.5M	0.115	0.9977	0.022	0.9613	0.021	0.9902
BiOBr_1M	0.092	0.9986	0.025	0.9813	0.017	0.9796
BiOI_0M	0.008	0.9791	0.003	0.9782	inactive	
BiOI_0.1M	0.005	0.9387	0.004	0.9667	0.095	0.9886
BiOI_0.5M	0.021	0.9756	0.029	0.9938	inactive	
BiOI_1M	0.021	0.9700	0.012	0.9913	inactive	

In contrast to the XRD analysis, the XPS spectra did not show the characteristic peaks of metallic Bi^0 in the BiOI_0.5M and BiOI_1M samples. Therefore, it is supposed that Bi^0 was probably formed as the first step of the synthesis due to the high density and viscosity of mannitol solutions and its weak interaction with iodide anion. In the next step, BiOI was formed and covered the Bi^0 phase.

The percentages of Bi, O, X (Figure 3b) in the sample were similar to the theoretical amount in the pure phase of BiOX only in BiOCl_0M. For samples synthesized in mannitol solutions, a higher amount of oxygen than their stoichiometric amount in the BiOX crystal lattice was found. This trend was very clearly observed in the series of BiOI, where oxygen was 44.4% and 63.1% for BiOI_0M and BiOI_1M, respectively. A correlation between the oxygen amount and the number of Bi ions at a higher oxidation state in samples was also found (Figure 3b,d).

The existence of Bi^{4+} and Bi^{5+} in BiOX materials was surprising and rarely reported in the literature. Species such as Bi^{4+} and Bi^{5+} were presented in BiOBr prepared with shape controlling agent polyvinylpyrrolidone (PVP) [39] or in other bismuth-based materials (Bi_2WO_6) [40]. Bismuth at the higher oxidation state than Bi^{3+} was observed in $BiOCl_xI_{1-x}$ solid solution and was responsible for enhanced electron transfer, thus photocatalytic activity [41]. It worth mentioning that the previously reported Bi^{4+} and Bi^{5+} were generated in BiOX in a rather small amount (up to 18%). In the synthesis with surfactants, they were formed in 5.4–16.3% and 2.2–7.7% for Bi^{4+} and Bi^{5+}, respectively. In the presented study Bi^{4+} existed in 18.2–18.7%, 5.4%, and 31.4–37.3% for BiOCl, BiOBr, and BiOI, respectively. Additionally, Bi^{5+} was observed in a significant amount (17.2–30.6%) in BiOI prepared in mannitol solutions. The formation of these species was probably the result of the use of mannitol in the synthesis as Bi^{4+} was absent in BiOCl_0M and BiOBr_0M while Bi^{5+} was observed only in BiOI obtained via mannitol. Furthermore, the results indicated that coordination of Bi^{3+} ions by hydroxyl groups of mannitol played a crucial role in the formation of vacancies and ability to local electron transfer in BiOX, which could enhance the separation of photogenerated electron/hole charge pairs.

2.1.4. UV–Vis/DRS Analysis

The absorption of light by semiconductors is an important factor affecting their photocatalytic performance and is one of the key factors determining their high photocatalytic activity. The UV–Vis diffuse reflectance spectra (UV–Vis/DRS) of the series of BiOX samples are shown in Figure 4.

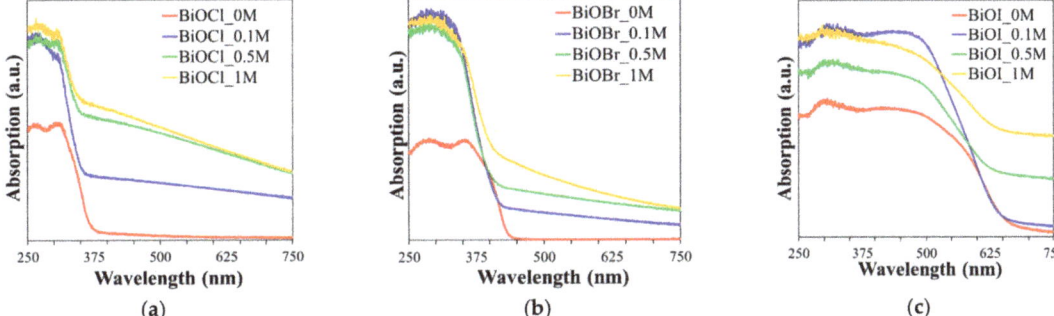

Figure 4. UV–Vis diffuse reflectance spectra of (**a**) BiOCl, (**b**) BiOBr, and (**c**) BiOI prepared with different concentrations of mannitol.

The BiOX samples of the mannitol series showed stronger absorption in the UV–Vis region compared to BiOX_0M. The increase in the concentration of the mannitol solution used for the photocatalyst preparation resulted in better absorption in the whole light range by sample due to the smaller particles and previously mentioned crystal lattice defects. With the higher mannitol concentration than 0.5 M, a slight enhancement of absorption was found. Moreover, all BiOX samples prepared in mannitol exhibited a blue-shift of absorption edges compared to BiOX obtained in water. The energy bandgap (Eg), position of maximum valance band (MVB), and minimum conduction band (MCB) were theoretically determined based on literature [18,42].

The BiOX samples of the mannitol series showed stronger absorption in the UV–Vis region compared to BiOX_0M. The increase in the concentration of the mannitol solution used for the photocatalyst preparation resulted in better absorption in the whole light range by sample. With the higher mannitol concentration than 0.5 M, a slight enhancement of absorption was found. Moreover, all BiOX samples prepared in mannitol exhibited a blue-shift of absorption edges compared to BiOX obtained in water. The energy bandgap (Eg), position of maximum valance band (MVB) and minimum conduction band (MCB) were theoretically determined based on literature [18,42] and presented in Table 3. Eg for series of BiOCl, BiOBr, and BiOI was estimated in the range of 3.0–3.3 eV, 2.56–2.65 eV, and 1.50–1.65 eV, respectively, which were consistent with the reported previously [14,23,43].

Additionally, it was observed that with increasing mannitol concentration, the bandgap of the BiOX (X = Cl, I) samples decreased, and also, the crystallite size decreased. The exception was the series of BiOBr samples, which showed a significant change in morphology (3D structure) with increasing mannitol concentration. Mannitol concentration influenced the position of the band edges Eg, as shown in Table 3. BiOX (X = Cl, Br, I) with the higher exposed facet (110) obtained in 0.5 M and 1 M mannitol characterized a lower minimum conduction band and a higher maximum valance band. The results confirmed that the appropriate synthesis conditions could offer the bang gap tuning and band-edge modification through morphology and size optimization of semiconductors.

2.1.5. PL Analysis

Photoluminescence (PL) study allowed to monitor the recombination rate of photoinduced charges pair electron-hole in the material. Figure 5 shows the PL spectra of the series of BiOCl, BiOBr, and BiOI samples excited at a wavelength of 315 nm. The emission spectra of each series had similar shapes, while the intensity of spectra decreased in the following order BiOCl > BiOBr > BiOI. This fact was related to the ability of the samples to separate e^-/h^+ charge pairs. High surface energy and reactive (001) and (110) facets found in BiOX_0M allowed for the formation of more catalytically active sits [44,45], which would explain the high intensity of the bismuth oxyhalides spectra.

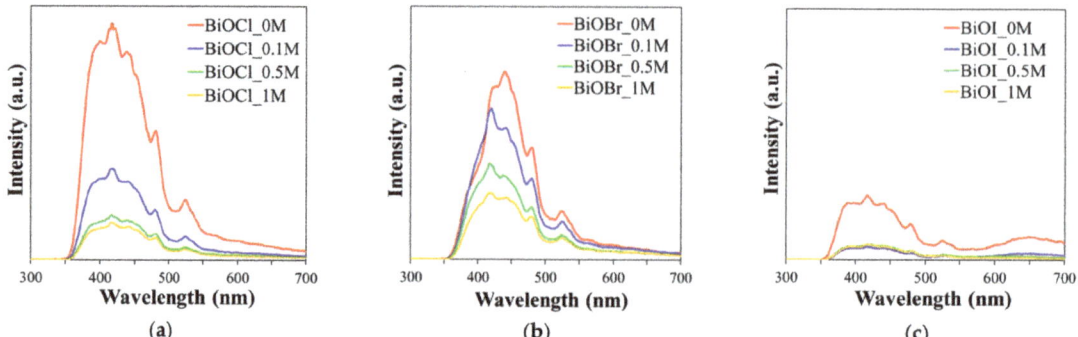

Figure 5. Photoluminescence (PL) spectra of (**a**) BiOCl, (**b**) BiOBr, and (**c**) BiOI synthesized in various concentrations of mannitol solution.

The intensity of PL spectra decreased after the introduction of mannitol to synthesize BiOX (X = Cl, Br, I) samples. Increasing the mannitol concentration decreased the PL intensity of the samples obtained in their solutions. This fact could indicate a better separation of photoexcited electron-hole pairs in these samples. The better separation of charge pairs in samples prepared in mannitol solutions was a result of (a) reduction in the size of the photocatalyst particles, (b) the appearance of highly reactive (110) facets [45], and (c) the existence of highly oxidized bismuth ions.

2.2. Photocatalytic Activity

Many factors, including morphology, crystallinity, or effective generation and separation of photoexcited charge pairs, play an essential role in influencing the efficiency of photodegradation of pollutants. Micropollutants such as Rhodamine B (RhB) cation dye, colorless cytostatic drug 5-FU, and heavy metal Cr(VI) in the form of anion form $Cr_2O_7^{2-}$ were selected for photodegradation study to examine the activity of BiOX photocatalysts. Direct and indirect photolysis of 5-FU and Cr(VI) was almost negligible, while RhB degraded by 16%, as was reported before [16].

The photocatalytic activities of the three series of BiOX (X = Cl or Br or I) in the photooxidation of RhB were conducted, and the results are shown in Figure 6. The pseudo-first-order rate constants k_{app} were calculated, and the results are presented in Table 4.

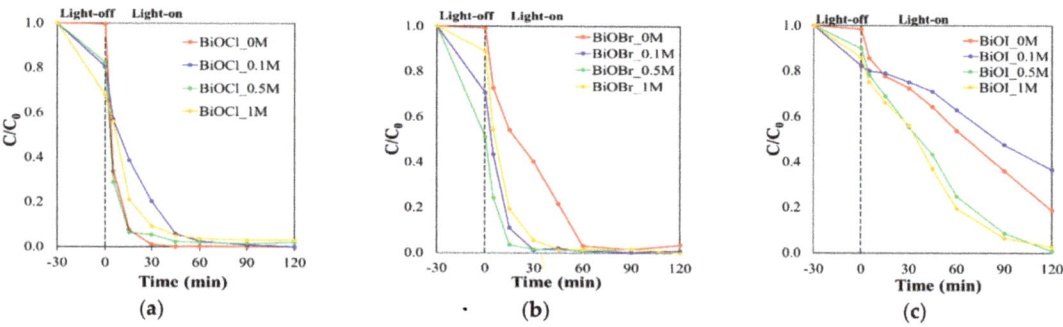

Figure 6. Photocatalytic degradation of RhB over (**a**) BiOCl, (**b**) BiOBr, and (**c**) BiOI under UV–Vis irradiation.

Regardless of the type of halogen in BiOX prepared in distilled water (BiOX_0M), the adsorption of RhB on the photocatalyst surface was not observed. Similar results for high crystallinity nanoplates BiOX were reported by other groups [6,15]. However, with the increase of mannitol concentration used in BiOX synthesis from 0.1 M to 1.0 M, the dye adsorption efficiency elevated. The trend for all prepared series was observed, but the largest increase in RhB absorption for the BiOBr samples was found. The adsorption of this dye reached the highest value of 49% for the BiOBr_0.5M photocatalyst. The adsorption of RhB was related instead to morphology changes than the $Bi^{(3-x)+}$, Bi^{4+}, and Bi^{5+} bismuth species in the crystal lattice.

The value of pseudo-first-order kinetic rate constant k_{app} of RhB photooxidation in the series of BiOCl was the highest for BiOCl_0M, while in the series of BiOBr and BiOI, the highest value of k_{app} was obtained for the samples prepared in 0.5 M mannitol solution.

In the literature, the photocatalyst activity in RhB oxidation was correlated with the structure of the prepared samples. Furthermore, density functional theory (DFT) computations showed a strong relationship with the importance of existence (001), (010), and (110) facets in BiOX and their photocatalytic activity [44,46].

The (001) plane with high thermodynamic stability has been previously reported as beneficial to photocatalytic performance, including RhB oxidation [46]. The remarkable photocatalytic degradation of RhB was observed over irradiation of the dye in the presence of BiOCl_0M, which possessed well-formed, exposed {001} planes. The samples prepared in mannitol solutions were characterized by a decrease in {001} planes with an increase in the mannitol concentration used in the synthesis. Therefore, the RhB removal rate was lower for BiOCl samples prepared in mannitol solutions.

Decomposition in the presence of BiOBr_0M with well-developed (002), (003), and (004) planes showed the lowest activity in series of BiOBr samples, which suggested that exposed facets were a minor factor in the photocatalytic oxidation of RhB in this series. Noticeably, the BiOX_0.5M (X = Br, I) with small uniform nanosheets and well exposed (110) crystal face exhibited the highest photocatalytic activity among the samples prepared in mannitol solutions. Therefore, the (110) crystal face with a higher electron density seems to be more critical to photocatalytic oxidation of RhB. The higher (110) active facet exposure found for BiOBr_0.5M allowed separate free electrons more effectively and enhanced the photocatalytic activity compared with other BiOBr samples. The previous research has indicated that the exposed (110) crystal plane facilitated the migration of oxidants holes and reduced the recombination of photogenerated electron-hole pairs in the BiOBr [35]. A similar trend was found for the series of BiOI.

Moreover, the analysis of the O/X ratio, where X = Cl, Br, I, and Bi described in the XPS results showed that the photodegradation efficiency of RhB gradually increased with the increase in the amount of oxygen in the semiconductors. That was probably the result of better separation of e^-/h^+ pairs and the effective mobility of holes due to defects in the samples.

The difference in the activity trend of the series of BiOCl and BiOBr/BiOI could be related to another phenomenon. RhB molecules could behave as photosensitizers, absorb light energy and convert into an excited state (RhB*). The excited state could inject electrons into the CB of the photocatalyst, trapped by dissolved O_2 in solution to generated oxidizing species such as superoxide radicals. The photocatalytic oxidation occurred between these active species and RhB^+. The following reactions to achieve the aim of degradation were possible (Equations (1)–(3)):

$$RhB + h\nu \rightarrow RhB^* \quad (1)$$

$$RhB^* + BiOX \rightarrow BiOX/\ e^- + RhB^{+\bullet} \quad (2)$$

$$RhB^{+\bullet} + O_2^{-\bullet} \rightarrow by\text{-}products \rightarrow H_2O + CO_2 \quad (3)$$

The synthesis of BiOX (X = Cl, Br, I) in mannitol solution favored surface creation and (110) facet exposition; thus, photoexcited electrons could have a lower recombination rate. Therefore, more electrons were trapped by the dissolved oxygen, and more $^\bullet O_2^-$ could be produced. The increase in superoxide radical's production is beneficial for RhB degradation. Therefore, this phenomenon can explain the higher activity toward RhB oxidation of the samples prepared in mannitol.

It is worth mentioning that the samples obtained by us in mannitol solution compare to the BiOX (X = Cl, Br, I) photocatalyst prepared via surfactants: hexadecyl(trimethyl)ammonium bromide (CTAB) [35,47,48], PVP [39], or sodium dodecyl sulfate (SDS) [39] reported in literature showed remarkable higher photocatalytic performance toward RhB oxidation.

To demonstrate the potential application of the presented BiOX materials in the oxidation of pharmaceuticals, the cytostatic drug 5-FU was used as a model compound. In contrast to RhB, 5-FU characterized the low partition coefficient $K_{o/w}$. The results of the 5-FU photocatalytic oxidation are shown in Figure 7.

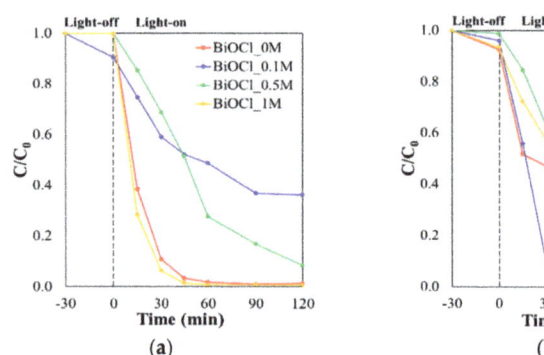

Figure 7. Photocatalytic degradation of 5-fluorouracil in the presence of (**a**) BiOCl, (**b**) BiOBr, and (**c**) BiOI under UV–Vis irradiation.

Previous studies have proved that mainly holes (h^+) were involved in the oxidation of 5-FU using BiOX photocatalysts, while superoxide radicals [17] were the minor oxidant of this drug. All prepared BiOX samples exhibited low 5-FU adsorption on their surface, and the highest values were 12%, 10%, and 10% for BiOI_0.1M, BiOCl_0.1M, and BiOI_0M, respectively. The pseudo-first-order rate constants k_{app} were calculated, and the results are listed in Table 5.

Table 5. Sample label of studied BiOX.

No.	Sample Label	BiOX Precursors	Solvent
\multicolumn{4}{c}{Bismuth oxychloride}			
1.	BiOCl_0M	2 mmol KCl, 2 mmol $Bi(NO_3)_3·5H_2O$	deionized water
2.	BiOCl_0.1M		0.1 M mannitol
3.	BiOCl_0.5M		0.5 M mannitol
4.	BiOCl_1M		1 M mannitol
\multicolumn{4}{c}{Bismuth oxybromide}			
5.	BiOBr_0M	2 mmol KBr, 2 mmol $Bi(NO_3)_3·5H_2O$	deionized water
6.	BiOBr_0.1M		0.1 M mannitol
7.	BiOBr_0.5M		0.5 M mannitol
8.	BiOBr_1M		1 M mannitol
\multicolumn{4}{c}{Bismuth oxyiodide}			
9.	BiOI_0M	2 mmol KI, 2 mmol $Bi(NO_3)_3·5H_2O$	deionized water
10.	BiOI_0.1M		0.1 M mannitol
11.	BiOI_0.5M		0.5 M mannitol
12.	BiOI_1M		1 M mannitol

The sample BiOCl_0M showed remarkable photocatalytic activity toward oxidation 5-FU (k_{app} = 0.076 min^{-1}). However, in the series of BiOCl, BiOCl_1M obtained in 1 M mannitol solution exhibited the highest degradability of 5-FU. After 45 min of irradiation, almost 100% of the drug was decomposed with the k_{app} of 0.093 min^{-1}. Preparation of BiOCl_0M in water resulted in the formation of many active sites, which allowed for the generation of charge carriers e$^-$/h$^+$, which was observed as a high intensity of PL spectra. The enhancement of BiOCl_1M activity could be attributed to both the diminishing and unifying of nanoplates and the creation of surface defects. These occurrences prolonged the life of photoexcited holes and increased the 5-FU removal rate. A similar result was reported for the 3D flower-like BiOCl obtained in glycerol [17]. The authors also explained the photocatalyst's high activity in the 5-FU decomposition by the long lifetime of holes. Due to the drug's low affinity to the BiOX surfaces, this factor was crucial for 5-FU decomposition with success. Mannitol solution used for the synthesis of BiOX regulated their morphologies and surface structures resulted in an enhanced degradation efficiency of selected micropollutants.

The BiOBr_0.1M sample prepared in 0.1 M mannitol solution exhibited the best photocatalytic activity toward 5-FU (k_{app} = 0.097 min^{-1}) in series of BiOBr. The efficiency removal of the drug, using BiOBr_0.1M, reached 91% after 30 min of UV–Vis irradiation, while for the remaining samples, it was in the range from 18% to 53%. The increase in activity of BiOBr_0.1M toward 5-FU degradation was probably related to the higher I_{110}/I_{102} ratio and the exposure of the (110) facet responsible for the reduction and generation of superoxide radicals. As it was discussed above, superoxide radicals could participate in the decomposition of 5-FU. Moreover, the contribution of surface defects in BiOBr_0.1M sample was 48.5% of Bi$^{(3-x)+}$ and 5.4% Bi^{4+}. Their existence in the photocatalyst slowed down the recombination of photogenerated charge pairs and enhanced the degradation rate of 5-FU. Compared to our previously reported Bi$_4$O$_5$Br$_2$ [16] and BiOCl$_{0.5}$Br$_{0.5}$ [17], the presented BiOBr prepared in a 0.1 M mannitol solution showed higher 5-FU photodegradation ability. The results indicated a crucial role of mannitol as a capping agent during BiOX synthesis.

The prepared series of BiOI exhibited much lower photocatalytic activity than BiOCl and BiOBr, which was connected to its narrower energy bandgap. The main factor influence activity toward 5-FU degradation was morphology, and surface defects prevented from the recombination of photogenerated holes and electrons in BiOI [47]. Only BiOI_0.1M with 3D rose-like microspheres decomposed 5-FU, and after 120 min of irradiation, 71% of the drug was removed at the k_{app} of 0.095 min^{-1}. BiOI synthesized in this work is characterized by the higher degradation rate of 5-FU than Bi$_4$O$_5$I$_2$ prepared via ionic liquids reported by our research group [16]. It is believed that Bi$_4$O$_5$I$_2$ has a stronger redox ability than BiOI [49,50]. However, the results implying the beneficial influence of mannitol on the redox potential of holes photogenerated under irradiated BiOI. The same trend was found in BiOCl and BiOBr materials.

The influence of mannitol concentration on the prepared samples' photoreduction ability was the next step of our investigations. Figure 8 shows Cr(VI) photoreduction activities under UV–Vis light on BiOX (X = Cl, Br, I).

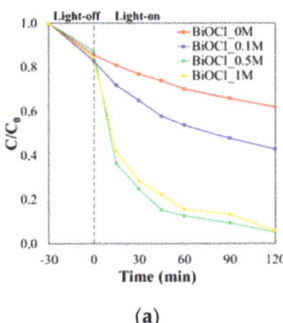
(a)

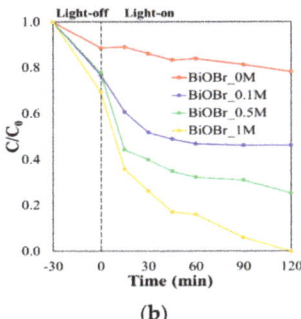
(b)

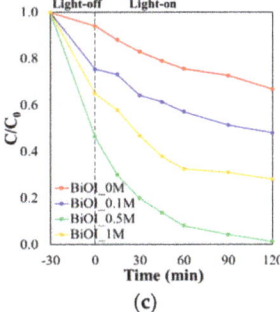
(c)

Figure 8. Photocatalytic reduction of hexavalent chromium over (**a**) BiOCl, (**b**) BiOBr, and (**c**) BiOI under UV–Vis irradiation.

The adsorption of Cr(VI) in the form of $Cr_2O_7^{2-}$ ions increased in the following order BiOCl < BiOBr < BiOI. Moreover, the adsorption efficiency of Cr(VI) ions increased with the elevation of mannitol concentration used for the BiOX synthesis. Defects such as Bi^{4+} and Bi^{5+} in the crystal lattice resulted in a higher positive charge of BiOX surfaces. Their number in the sample increased with the increase in the concentration of mannitol used for synthesis. The photocatalyst surface's positive charge was favorable for $Cr_2O_7^{2-}$ ions adsorption due to their electrostatic interaction.

In Table 5, the pseudo-first-order kinetic rate constant k_{app}, calculated based on the results, are presented. BiOCl_0.5M and BiOCl_1M gave much higher activity in Cr(VI) reduction than BiOCl_0M and BiOCl_0.1M. The efficiency of reduction of Cr(VI) reached 38%, 57%, 95% and 94% for BiOCl_0M, BiOCl_0.1M, BiOCl_0.5M and BiOCl_1M, respectively.

The series of BiOBr exhibited a higher degradation rate among the prepared BiOX, which increased from 0.001 min^{-1} (BiOBr_0M) to 0.025 min^{-1} (BiOBr_1M) with an increasing concentration of mannitol used in the synthesis. Among BiOBr samples, BiOBr_1M showed the highest rate and efficiency of Cr(VI) removal. Photoreduction of Cr(VI) in the form of $Cr_2O_7^{2-}$ increased with an increase in the concentration of mannitol used in the synthesis and resulted in 22%, 54%, 75%, and 100% degradation after 120 min in the presence of BiOBr_0M, BiOBr_0.1M, BiOBr_0.5M and BiOBr_1M, respectively.

The photocatalytic activity toward Cr(VI) among BiOI samples increased with increasing mannitol concentration used for synthesis up to 0.5M, and BiOI_0.5M exhibited the highest value of k_{app}. For 1 M mannitol concentration, the k_{app} decreased. The Cr(VI) in the solution was degraded in 32%, 52%, 99%, and 72% after 120 min of illumination over BiOI_0M, BiOI_0.1M, BiOI_0.5M, and BiOI_1M, respectively.

The results suggested that Cr(VI) reduction was facet dependent, and materials with dominated the (110) facet exhibited higher efficiency in its reduction. BiOX photocatalysts prepared in mannitol solutions were characterized by many exposed (110) crystal planes. The same trend was reported for BiOBr [51]. The number of exposed (110) crystalline planes in the sample correlated with Cr(VI) reduction activity. However, BiOX with many exposed (110) crystal planes was previously synthesized with PVP as a structure-directing agent [35]. BiOX with exposed (110) crystal planes was prepared in a less expensive and environmentally friendly mannitol solution in our study. The presence of mannitol in the synthesis affected the I_{110}/I_{102} ratio in the BiOX samples, which are summarized in Table 5. The exposure of (110) facets and value of the I_{110}/I_{102} ratio increased with the mannitol concentration in the series of BiOCl and BiOBr. In the series of BiOI, the activity of BiOI_0.5M and BiOI_1M in the Cr(VI) reduction process was inhibited.

Nevertheless, the XRD analysis showed that BiOI samples prepared in mannitol solutions possessed an additional metallic bismuth phase in a crystalline lattice, which could prolong photogenerated charge pairs' lifetime through separation acted as a sink for electrons and enhanced oxidation processes. However, in the photoreduction of $Cr_2O_7^{2-}$ ions, the excited electrons were the main species, and the presence of Bi^0 prevented reaction between electrons and Cr(VI) ions. The lower photocatalytic activity of BiOI_1M than BiOI_0.5M could be explained by the higher content of the metallic bismuth phase that participated in the photoreduction of chromium (VI).

Additionally, the enhanced photoreduction of Cr(VI) in the presence of irradiated BiOX could be attributed to the strong adsorption on the photocatalyst's surface, which was obtained by the use of mannitol for synthesis. The previously reported surfactants addition to the BiOX synthesis [35,52] and BiOX modifications [53] did not elevate the Cr(VI) adsorption on the photocatalyst surface to such an extent. Based on the results obtained, it could be concluded that the mannitol used in the synthesis as a template and capping agent played a crucial role in the exposure of (110) plane of BiOX (X = Cl, Br, I) and the creation of defects. These phenomena enhanced the adsorption and reduction of $Cr_2O_7^{2-}$ ions.

The obtained results indicated that the photocatalytic activity of BiOX (X = Cl, Br, I) was closely related to the presence of mannitol at a particular concentration, which was used in the synthesis process due to its dual roles in the morphology regulation and the surface defects creation.

2.3. Mechanism of BiOX Crystallites Formation

The processes for the formation of BiOX crystallites in mannitol solutions could be described by the following chemical equations (Equations (4)–(6)):

$$2Bi^{3+} + 2C_6H_8(OH)_6 \rightarrow Bi_2(C_6H_8O_6) + C_6H_8(OH_2)_6^{6+} \qquad (4)$$

$$Bi_2(C_6H_8O_6) + 2X^- + 2H_2O \rightarrow 2BiOX_{(s)} + C_6H_8O_6^{6-} + 4H^+ \qquad (5)$$

$$C_6H_8O_6^{6-} + C_6H_8(OH_2)_6^{6+} \rightarrow 2C_6H_8(OH)_6 \qquad (6)$$

The proposed formation mechanism is presented in Figure 9.

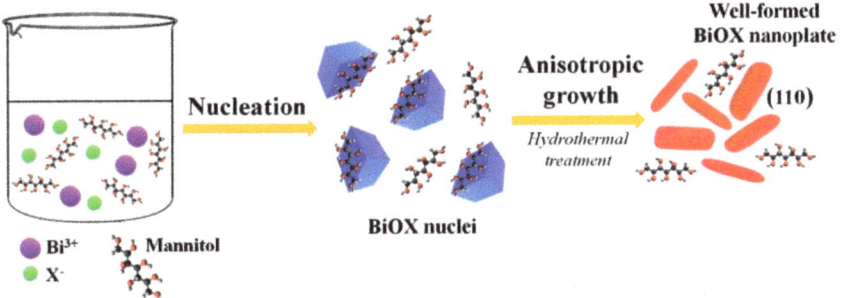

Figure 9. The illustration of the formation mechanism of BiOX mannitol-assisted hydrothermal treatment.

Mannitol with polyhydroxy structure mediated the nucleation and crystal growth of BiOX nanoparticles by creating the multi-dental bismuth ligand ($Bi_2(C_6H_8O_6)$) with strong interactions with Bi^{3+} ions. The viscosity and density of the 0.1 M mannitol solution were higher than that of H_2O, and these parameters increased with the elevation of mannitol concentration (Table 2). Therefore, the rate of halide ions diffusion was successively inhibited by elevation of mannitol concentration from 0.1 M to 1 M. Moreover, the halides' diffusion rate in the mannitol solution decreased with the increase of the anion radius (Cl^- < Br^- < I^-). The reaction rate between the two Bi^{3+} ions coordinated by one molecule of mannitol (bismuth alkoxide) with six ions of iodide in a 0.5 M concentration of mannitol was probably so low that reduction of Bi^{3+} by mannitol to metallic Bi^0 was observed. This phenomenon led to metallic bismuth formation during the preparation of BiOI_0.5M and BiOI_1M samples. The bismuth alkoxides could decompose during solvothermal treatment associated with high temperature and pressure, mainly to BiO^+. The forming metal-ligand complex and its slow decomposition at high concentration of mannitol, low availability of Bi^{3+}, and slow diffusion rate of halide ions, resulted in oxygen-rich BiOX Bi^{4+}/Bi^{5+} species. A higher amount of these species was reached in the BiOI samples obtained in the synthesis with me⁻ ions characterized by the largest radius of ion among halides.

Moreover, the mannitol concentration's increased viscosity significantly suppressed the nanocrystals' intrinsic anisotropic growth, as was confirmed by the SEM images. The results suggested that the nucleating speed could be controlled by the mannitol's density and viscosity, which is beneficial for nanosheets formation. Additionally, the higher concentration of mannitol prevents the aggregation of nanoparticles. The surface energy of the BiOBr nanosheets prepared in 1 M mannitol was too high; therefore, it aggregated into the hierarchical structure for the surface energy reduction. Moreover, in the solvothermal reaction, $C_6H_8O_6^{6-}$ was the by-product (Equation (5)) that could be easily adsorbed on the exposed Bi^{3+} (102) facets and favors formation {110} planes. Mannitol also limited the anisotropic crystal growth along the [001] direction due to repulsion of $C_6H_8O_6^{6-}$ and terminal oxygen of (001) facets.

In summary, mannitol acted as a structure-directing agent and a growth and shape control agent. The existence of electrostatic attraction and hydrogen bonds between mannitol molecules and the long carbon chain helped form a uniform nanosheet (BiOCl, BiOBr, and BiOI) or 3D flower-like heterostructure (BiOBr).

3. Materials and Methods

Bismuth nitrate, potassium chloride, potassium iodide, mannitol, sulphuric acid, acetone, and potassium dichromate were purchased from StanLab Sp. J. (STANLAB, Lublin, Poland). Potassium bromide was obtained from Alfa Aesar (Alfa Aesar, Karlsruhe, Germany). Ethyl alcohol, 1,5-diphenylcarbazide, and rhodamine B were purchased from POCh S.A. (POCH S.A. Gliwice, Poland). The drug 5-fluorouracil was obtained from Sigma-Aldrich (St. Louis, MO, USA). Orthophosphoric acid was provided by Chempur (Piekary Śląskie, Poland). All chemicals used in this study were commercially available analytical grade and were used without further purification.

3.1. Synthesis

The series of BiOX (X = Cl, Br, I) semiconductors were prepared by a solvothermal method. Briefly, 2 mmol of KX (X = Cl, Br, I) were dissolved in 20 mL of deionized water or mannitol solution, and 2 mmol of $Bi(NO_3)_3 \cdot 5H_2O$ were suspended by sonification for 15 min. The KX solution was then added dropwise to 20 mL of deionized water containing bismuth nitrate suspension under vigorous stirring. After 30 min continuously stirring, the mixture was transformed into a 50 mL Teflon-lined stainless-steel autoclave. Then the autoclave was sealed and heated at 160 °C for 16 h. After heat treatment, the autoclave was allowed to cool naturally to room temperature. The products were collected and washed with ethanol and deionized water thoroughly and dried at 80 °C. The series of BiOX was synthesized using various concentrations of mannitol, i.e., 0.1 M, 0.5 M, and 1 M. The list of prepared photocatalysts is presented in Table 1. The density of mannitol solutions was measured by the picometric method, and the viscosity was determined with the Ostwald viscometer. The obtained results are summarized in Table 5.

3.2. Characterization

The crystalline phase and the prepared photocatalysts' purity were characterized by X-ray powder diffraction on a D2 Phaser (Bruker, Billerica, MA, USA). Diffraction patterns were taken over the 2θ in the range 20–70°. X-ray photoelectron spectroscopy with ThermoFisher Scientific Escalab 250Xi (Waltham, MA, USA) was used to determine the chemical state of the elements and the composition of the surface of the BiOX. The morphology of the products was characterized by a field-emission scanning microscope (JEOL JSM-7610F FEG SEM), (JEOL Ltd., Akishima, Japan). UV–Vis diffuse reflectance spectra were recorded on a UV-2600 UV–VIS Spectrometer (Shimadzu Corp., Kyoto, Japan) using $BaSO_4$ as a reference. Photoluminescence spectra were measured on Perkin Elmer limited LS50B (Perkin Elmer, Waltham, MA, USA) using 315 nm as an excitation wavelength.

3.3. Photocatalytic Activity

The photocatalytic activity was evaluated by degradation of RhB dye, hexavalent chromium anion in the form of $Cr_2O_7^{2-}$ and cytostatic drug 5-FU. The concentration of RhB, 5-FU was 15 mgL^{-1} at pH 6.5, and Cr(VI) was 20 mgL^{-1} at pH 3. The concentration of photocatalysts during RhB and 5-FU degradation was 0.2 gL^{-1}, and Cr(VI) was photoreduced in the presence of 0.5g L^{-1} photocatalyst. The volume of the solution in the tests was 15 mL. Photocatalytic degradation experiments were conducted as follows: the photocatalyst was immersed in a selected micropollutant solution and magnetically stirred in the dark for 30 min to achieve the predetermined adsorption–desorption equilibrium; then, the solution was exposed to UV–Vis light irradiation of 150 W medium pressure mercury lamp for 120 min. At particularly time intervals, 1 mL of samples were collected, and the photocatalysts particles were removed immediately. The concentration of RhB was analyzed with a Perkin Elmer Lambda XLS+ spectrophotometer by monitoring the absorption peak at 553 nm. The Cr(VI) concentration was determined according to ISO PN-EN 18412. The 5-FU degradation was measured by HPLC analysis using Perkin Elmer Series 200 equipped with UV detector and Phenomenex C-18 column (150 mm × 4.6 nm, 2.6 μm) with parameters as follows: detection wavelength of 266 nm, the mobile phase was acetonitrile/water 2:98 (v/v), the retention time (t_R) was 4.9 min. The experiments were repeated three times, and the accuracy expressed as the relative standard deviation of three independent measurements did not exceed 3%. The sorption test was conducted in the dark for 120 min and revealed that after 30 min, the adsorption–desorption equilibrium was achieved for selected micropollutants and prepared BiOX.

4. Conclusions

In this study, a series of BiOX (X = Cl, Br, I) photocatalysts via a solvothermal process in mannitol solution were fabricated. It was found that the role of mannitol was strongly dependent on its concentration (in the range of 0–1 M). The mannitol solution's viscosity and density played a crucial role in the phase composition, morphology, and surface defect formation of the obtained photocatalysts. Mannitol simultaneously could act as a solvent, soft template, and structure-directing agent in the synthesis of BiOX (X = Cl, Br, I). The mannitol concentration affected particles' size; the higher its concentration, the smaller and more unified nanoparticles were formed. The presence of polyxydroksyl alcohol in the synthesis of BiOX had a high impact on the (110) and (102) facets growth, and the I_{110}/I_{102} ratios increased as the mannitol concentration increased.

Moreover, the photocatalysts synthesized in mannitol solutions characterized the prolonged life of pair charge carriers e^-/h^+ caused due to the appearance of surface defects ($Bi^{(3-x)+}$, Bi^{4+}, and Bi^{5+}). The defects with positive charge were responsible for the higher adsorption of negatively charged ions $Cr_2O_7^{2-}$, and RhB with aromatic rings. The increase in photocatalytic rate and performance toward the RhB, 5-FU, and Cr(VI) removal resulted from the higher adsorption ability and better separation of e^-/h^+ charge carriers. The photocatalytic degradation toward RhB, 5-FU, and Cr(VI) over the prepared BiOX was followed as a pseudo-first-order kinetic model.

Among all studied samples, BiOCl_0M and BiOBr_0.5M exhibited the best activity toward RhB. Based on the degradation rate of Rhodamine B, the optimal concentration of the mannitol solution used in the BiOI synthesis was 0.5M. The BiOX prepared in 0.5M mannitol solution, regardless of halogen, showed the highest efficiency of Cr(VI) removal under UV–Vis irradiation. Moreover, comparing the photocatalytic activities of the BiOCl, BiOBr, and BiOI series, it was found that the optimal mannitol concentration used in the preparation of the material were 1 M, 0.1 M, and 0.1 M for BiOCl, BiOBr, and BiOI, respectively. The possible mechanism of crystal growth has been proposed.

Mannitol is a simple and environmentally friendly compound. Our presented studies demonstrated that BiOX prepared in mannitol solution could be useful for efficiently removing a wide range of micropollutants.

Author Contributions: Conceptualization, A.B.-G., P.W., and E.M.S.; methodology, A.B.-G. and P.W.; validation, P.W. and A.B.-G.; formal analysis, P.W., A.B.-G., and E.M.S.; investigation, A.B.-G., P.W., K.S., A.M., J.R., and K.T.; resources, A.B.-G. and E.M.S.; data curation, A.B.-G.; writing—original draft preparation, A.B.-G., P.W., and E.M.S.; writing—review and editing, A.B.-G., P.W., and E.M.S.; visualization, P.W. and A.B.-G.; supervision, E.M.S.; project administration and funding acquisition, A.B.-G. and E.M.S. All authors have read and agreed to the published version of the manuscript.

Funding: This research was funded by the National Science Center (PL), grant number DEC-2017/01/X/ST5/01136, and the Ministry of Education and Science (PL), grant number DS 531-T020-D596-21.

Data Availability Statement: Data is contained within the article.

Conflicts of Interest: The authors declare no conflict of interest.

References

1. Ye, L.; Su, Y.; Jin, X.; Xie, H.; Zhang, C. Recent Advances in BiOX (X = Cl, Br and I) Photocatalysts: Synthesis, Modification, Facet Effects and Mechanisms. *Environ. Sci. Nano* **2014**, *1*, 90. [CrossRef]
2. Yang, Y.; Zhang, C.; Lai, C.; Zeng, G.; Huang, D.; Cheng, M.; Wang, J.; Chen, F.; Zhou, C.; Xiong, W. BiOX (X = Cl, Br, I) Photocatalytic Nanomaterials: Applications for Fuels and Environmental Management. *Adv. Colloid Interface Sci.* **2018**, *254*, 76–93. [CrossRef]
3. Liu, Y.; Hu, Z.; Yu, J. Fe Enhanced Visible-Light-Driven Nitrogen Fixation on BiOBr Nanosheets. *Chem. Mater.* **2020**, *32*, 1488–1494. [CrossRef]
4. Wu, S.; Wang, C.; Cui, Y.; Wang, T.; Huang, B.; Zhang, X.; Qin, X.; Brault, P. Synthesis and Photocatalytic Properties of BiOCl Nanowire Arrays. *Mater. Lett.* **2010**, *64*, 115–118. [CrossRef]
5. Mi, Y.; Li, H.; Zhang, Y.; Du, N.; Hou, W. Synthesis and Photocatalytic Activity of BiOBr Nanosheets with Tunable Crystal Facets and Sizes. *Catal. Sci. Technol.* **2018**, *8*, 2588–2597. [CrossRef]
6. Gao, M.; Zhang, D.; Pu, X.; Li, H.; Lv, D.; Zhang, B.; Shao, X. Facile Hydrothermal Synthesis of Bi/BiOBr Composites with Enhanced Visible-Light Photocatalytic Activities for the Degradation of Rhodamine B. *Sep. Purif. Technol.* **2015**, *154*, 211–216. [CrossRef]
7. Cheng, G.; Xiong, J.; Stadler, F.J. Facile Template-Free and Fast Refluxing Synthesis of 3D Desertrose-like BiOCl Nanoarchitectures with Superior Photocatalytic Activity. *New J. Chem.* **2013**, *37*, 3207. [CrossRef]
8. Wang, D.-H.; Gao, G.-Q.; Zhang, Y.-W.; Zhou, L.-S.; Xu, A.-W.; Chen, W. Nanosheet-Constructed Porous BiOCl with Dominant {001} Facets for Superior Photosensitized Degradation. *Nanoscale* **2012**, *4*, 7780. [CrossRef]
9. Zhang, D.; Li, J.; Wang, Q.; Wu, Q. High {001} Facets Dominated BiOBr Lamellas: Facile Hydrolysis Preparation and Selective Visible-Light Photocatalytic Activity. *J. Mater. Chem. A* **2013**, *1*, 8622. [CrossRef]
10. Hou, L.; Niu, Y.; Yang, F.; Ge, F.; Yuan, C. Facile Solvothermal Synthesis of Hollow BiOBr Submicrospheres with Enhanced Visible-Light-Responsive Photocatalytic Performance. *J. Anal. Methods Chem.* **2020**, 3058621. [CrossRef] [PubMed]
11. Hu, X.; Xu, Y.; Zhu, H.; Hua, F.; Zhu, S. Controllable Hydrothermal Synthesis of BiOCl Nanoplates with High Exposed {001} Facets. *Mater. Sci. Semicond. Process.* **2016**, *41*, 12–16. [CrossRef]
12. Jia, M.; Hu, X.; Wang, S.; Huang, Y.; Song, L. Photocatalytic Properties of Hierarchical BiOXs Obtained via an Ethanol-Assisted Solvothermal Process. *J. Environ. Sci.* **2015**, *35*, 172–180. [CrossRef] [PubMed]
13. Tian, F.; Xiong, J.; Zhao, H.; Liu, Y.; Xiao, S.; Chen, R. Mannitol-Assisted Solvothermal Synthesis of BiOCl Hierarchical Nanostructures and Their Mixed Organic Dye Adsorption Capacities. *CrystEngComm* **2014**, *16*, 4298–4305. [CrossRef]
14. Xiong, J.; Cheng, G.; Qin, F.; Wang, R.; Sun, H.; Chen, R. Tunable BiOCl Hierarchical Nanostructures for High-Efficient Photocatalysis under Visible Light Irradiation. *Chem. Eng. J.* **2013**, *220*, 228–236. [CrossRef]
15. Xiong, J.; Cheng, G.; Li, G.; Qin, F.; Chen, R. Well-Crystallized Square-like 2D BiOCl Nanoplates: Mannitol-Assisted Hydrothermal Synthesis and Improved Visible-Light-Driven Photocatalytic Performance. *RSC Adv.* **2011**, *1*, 1542. [CrossRef]
16. Bielicka–Giełdoń, A.; Wilczewska, P.; Malankowska, A.; Szczodrowski, K.; Ryl, J.; Zielińska-Jurek, A.; Siedlecka, E.M. Morphology, Surface Properties and Photocatalytic Activity of the Bismuth Oxyhalides Semiconductors Prepared by Ionic Liquid Assisted Solvothermal Method. *Sep. Purif. Technol.* **2019**, *217*, 164–173. [CrossRef]
17. Wilczewska, P.; Bielicka-Giełdoń, A.; Borzyszkowska, A.F.; Ryl, J.; Klimczuk, T.; Siedlecka, E.M. Photocatalytic Activity of Solvothermal Prepared BiOClBr with Imidazolium Ionic Liquids as a Halogen Sources in Cytostatic Drugs Removal. *J. Photochem. Photobiol. A Chem.* **2019**, *382*, 111932. [CrossRef]
18. Hao, H.-Y.; Xu, Y.-Y.; Liu, P.; Zhang, G.-Y. BiOCl Nanostructures with Different Morphologies: Tunable Synthesis and Visible-Light-Driven Photocatalytic Properties. *Chin. Chem. Lett.* **2015**, *26*, 133–136. [CrossRef]
19. Li, X.; Zhu, C.; Song, Y.; Du, D.; Lin, Y. Solvent Co-Mediated Synthesis of Ultrathin BiOCl Nanosheets with Highly Efficient Visible-Light Photocatalytic Activity. *RSC Adv.* **2017**, *7*, 10235–10241. [CrossRef]
20. Li, L.; Ai, L.; Zhang, C.; Jiang, J. Hierarchical {001}-Faceted BiOBr Microspheres as a Novel Biomimetic Catalyst: Dark Catalysis towards Colorimetric Biosensing and Pollutant Degradation. *Nanoscale* **2014**, *6*, 4627. [CrossRef]
21. Xing, H.; Ma, H.; Fu, Y.; Zhang, X.; Dong, X.; Zhang, X. Preparation of BiOBr by Solvothermal Routes with Different Solvents and Their Photocatalytic Activity. *J. Renew. Sustain. Energy* **2015**, *7*, 063120. [CrossRef]
22. Liu, Z.; Wu, B.; Xiang, D.; Zhu, Y. Effect of Solvents on Morphology and Photocatalytic Activity of BiOBr Synthesized by Solvothermal Method. *Mater. Res. Bull.* **2012**, *47*, 3753–3757. [CrossRef]
23. Li, R.; Ren, H.; Ma, W.; Hong, S.; Wu, L.; Huang, Y. Synthesis of BiOBr Microspheres with Ethanol as Self-Template and Solvent with Controllable Morphology and Photocatalytic Activity. *Catal. Commun.* **2018**, *106*, 1–5. [CrossRef]
24. Li, J.; Sun, S.; Qian, C.; He, L.; Chen, K.K.; Zhang, T.; Chen, Z.; Ye, M. The Role of Adsorption in Photocatalytic Degradation of Ibuprofen under Visible Light Irradiation by BiOBr Microspheres. *Chem. Eng. J.* **2016**, *297*, 139–147. [CrossRef]
25. Fang, Y.; Hua, T.; Feng, W.; Johnson, D.M.; Huang, Y. Mannitol Ligand-Assisted Assembly of BiOBr Photocatalyst in the Cationic Micelles of Cetylpyridinium Bromide. *Catal. Commun.* **2016**, *80*, 15–19. [CrossRef]

26. Hu, J.; Weng, S.; Zheng, Z.; Pei, Z.; Huang, M.; Liu, P. Solvents Mediated-Synthesis of BiOI Photocatalysts with Tunable Morphologies and Their Visible-Light Driven Photocatalytic Performances in Removing of Arsenic from Water. *J. Hazard. Mater.* **2014**, *264*, 293–302. [CrossRef] [PubMed]
27. Wang, X.; Li, F.; Li, D.; Liu, R.; Liu, S. Facile Synthesis of Flower-like BiOI Hierarchical Spheres at Room Temperature with High Visible-Light Photocatalytic Activity. *Mater. Sci. Eng. B* **2015**, *193*, 112–120. [CrossRef]
28. Gnayem, H.; Sasson, Y. Hierarchical Nanostructured 3D Flowerlike BiOCl$_x$Br$_{1-x}$ Semiconductors with Exceptional Visible Light Photocatalytic Activity. *ACS Catal.* **2013**, *3*, 186–191. [CrossRef]
29. Shi, X.; Chen, X.; Chen, X.; Zhou, S.; Lou, S.; Wang, Y.; Yuan, L. PVP Assisted Hydrothermal Synthesis of BiOBr Hierarchical Nanostructures and High Photocatalytic Capacity. *Chem. Eng. J.* **2013**, *222*, 120–127. [CrossRef]
30. Liu, J.; Hu, J.; Ruan, L.; Wu, Y. Facile and Environment Friendly Synthesis of Hierarchical BiOCl Flowery Microspheres with Remarkable Photocatalytic Properties. *Chin. Sci. Bull.* **2014**, *59*, 802–809. [CrossRef]
31. Zhao, Y.; Tan, X.; Yu, T.; Wang, S. SDS-Assisted Solvothermal Synthesis of BiOBr Microspheres with Highly Visible-Light Photocatalytic Activity. *Mater. Lett.* **2016**, *164*, 243–247. [CrossRef]
32. Wang, X.; Chen, H.; Li, H.; Mailhot, G.; Dong, W. Preparation and Formation Mechanism of BiOCl0.75I0.25 Nanospheres by Precipitation Method in Alcohol–Water Mixed Solvents. *J. Colloid Interface Sci.* **2016**, *478*, 1–10. [CrossRef]
33. Gao, X.; Zhang, X.; Wang, Y.; Peng, S.; Yue, B.; Fan, C. Rapid Synthesis of Hierarchical BiOCl Microspheres for Efficient Photocatalytic Degradation of Carbamazepine under Simulated Solar Irradiation. *Chem. Eng. J.* **2015**, *263*, 419–426. [CrossRef]
34. Yang, J.; Xie, T.; Zhu, Q.; Wang, J.; Xu, L.; Liu, C. Boosting the Photocatalytic Activity of BiOX under Solar Light via Selective Crystal Facet Growth. *J. Mater. Chem. C* **2020**, *8*, 2579–2588. [CrossRef]
35. Zhang, H.; Yang, Y.; Zhou, Z.; Zhao, Y.; Liu, L. Enhanced Photocatalytic Properties in BiOBr Nanosheets with Dominantly Exposed (102) Facets. *J. Phys. Chem. C* **2014**, *118*, 14662–14669. [CrossRef]
36. Dai, B.; Zhang, A.; Liu, Z.; Wang, T.; Li, C.; Zhang, C.; Li, H.; Liu, Z.; Zhang, X. Facile Synthesis of Metallic Bi Deposited BiOI Composites with the Aid of EDTA-2Na for Highly Efficient Hg0 Removal. *Catal. Commun.* **2019**, *121*, 53–56. [CrossRef]
37. Chang, C.; Zhu, L.; Fu, Y.; Chu, X. Highly Active Bi/BiOI Composite Synthesized by One-Step Reaction and Its Capacity to Degrade Bisphenol A under Simulated Solar Light Irradiation. *Chem. Eng. J.* **2013**, *233*, 305–314. [CrossRef]
38. Montoya-Zamora, J.M.; Martínez-de la Cruz, A.; López Cuéllar, E. Enhanced Photocatalytic Activity of BiOI Synthesized in Presence of EDTA. *J. Taiwan Inst. Chem. Eng.* **2017**, *75*, 307–316. [CrossRef]
39. Bárdos, E.; Márta, V.; Baia, L.; Todea, M.; Kovács, G.; Baán, K.; Garg, S.; Pap, Z.; Hernadi, K. Hydrothermal Crystallization of Bismuth Oxybromide (BiOBr) in the Presence of Different Shape Controlling Agents. *Appl. Surf. Sci.* **2020**, *518*, 146184. [CrossRef]
40. Hu, T.; Li, H.; Zhang, R.; Du, N.; Hou, W. Thickness-Determined Photocatalytic Performance of Bismuth Tungstate Nanosheets. *RSC Adv.* **2016**, *6*, 31744–31750. [CrossRef]
41. Zhang, G.; Cai, L.; Zhang, Y.; Wei, Y. Bi^{5+}, Bi$^{(3-x)+}$, and Oxygen Vacancy Induced BiOCl$_x$I$_{1-x}$ Solid Solution toward Promoting Visible-Light Driven Photocatalytic Activity. *Chem. Eur. J.* **2018**, *24*, 7434–7444. [CrossRef]
42. Heidari, S.Z.; Haghighi, M.; Shabani, M. Sunlight-activated BiOCl/BiOBr–Bi$_{24}$O$_{31}$Br$_{10}$ photocatalyst for the removal of pharmaceutical compounds. *J. Clean. Prod.* **2020**, *259*, 120679. [CrossRef]
43. Lin, Z.; Zhe, F.; Wang, Y.; Zhang, Q.; Zhao, X.; Hu, X.; Wu, Y.; He, Y. Preparation of interstitial carbon doped BiOI for enhanced performance in photocatalytic nitrogen fixation and methyl orange degradation. *J. Colloid Interface Sci.* **2019**, *539*, 563–574. [CrossRef]
44. Zhang, H.; Liu, L.; Zhou, Z. First-Principles Studies on Facet-Dependent Photocatalytic Properties of Bismuth Oxyhalides (BiOXs). *RSC Adv.* **2012**, *2*, 9224. [CrossRef]
45. Weng, S.; Fang, Z.; Wang, Z.; Zheng, Z.; Feng, W.; Liu, P. Construction of Teethlike Homojunction BiOCl (001) Nanosheets by Selective Etching and Its High Photocatalytic Activity. *ACS Appl. Mater. Interfaces* **2014**, *6*, 18423–18428. [CrossRef] [PubMed]
46. Gao, M.; Zhang, D.; Pu, X.; Li, H.; Li, J.; Shao, X.; Ding, K. BiOBr Photocatalysts with Tunable Exposing Proportion of {001} Facets: Combustion Synthesis, Characterization, and High Visible-Light Photocatalytic Properties. *Mater. Lett.* **2015**, *140*, 31–34. [CrossRef]
47. Xu, J.; Meng, W.; Zhang, Y.; Li, L.; Guo, C. Photocatalytic Degradation of Tetrabromobisphenol A by Mesoporous BiOBr: Efficacy, Products and Pathway. *Appl. Catal. B Environ.* **2011**, *107*, 355–362. [CrossRef]
48. Zhang, L.; Cao, X.-F.; Chen, X.-T.; Xue, Z.-L. BiOBr Hierarchical Microspheres: Microwave-Assisted Solvothermal Synthesis, Strong Adsorption and Excellent Photocatalytic Properties. *J. Colloid Interface Sci.* **2011**, *354*, 630–636. [CrossRef]
49. Arumugam, M.; Choi, M.Y. Recent Progress on Bismuth Oxyiodide (BiOI) Photocatalyst for Environmental Remediation. *J. Ind. Eng. Chem.* **2020**, *81*, 237–268. [CrossRef]
50. Xiao, X.; Xing, C.; He, G.; Zuo, X.; Nan, J.; Wang, L. Solvothermal Synthesis of Novel Hierarchical Bi$_4$O$_5$I$_2$ Nanoflakes with Highly Visible Light Photocatalytic Performance for the Degradation of 4-Tert-Butylphenol. *Appl. Catal. B Environ.* **2014**, *148*, 154–163. [CrossRef]
51. Fan, Z.; Zhao, Y.; Zhai, W.; Qiu, L.; Li, H.; Hoffmann, M.R. Facet-Dependent Performance of BiOBr for Photocatalytic Reduction of Cr(VI). *RSC Adv.* **2016**, *6*, 2028–2031. [CrossRef]

52. Hussain, M.B.; Khan, M.S.; Loussala, H.M.; Bashir, M.S. The Synthesis of a $BiOCl_xBr_{1-x}$ Nanostructure Photocatalyst with High Surface Area for the Enhanced Visible-Light Photocatalytic Reduction of Cr(VI). *RSC Adv.* **2020**, *10*, 4763–4771. [CrossRef]
53. Li, H.; Deng, F.; Zheng, Y.; Hua, L.; Qu, C.; Luo, X. Visible-Light-Driven Z-Scheme RGO/Bi_2S_3–BiOBr Heterojunctions with Tunable Exposed BiOBr (102) Facets for Efficient Synchronous Photocatalytic Degradation of 2-Nitrophenol and Cr(VI) Reduction. *Environ. Sci. Nano* **2019**, *6*, 3670–3683. [CrossRef]

Article

Study of the Direct CO₂ Carboxylation Reaction on Supported Metal Nanoparticles

Fabien Drault [1], Youssef Snoussi [1], Joëlle Thuriot-Roukos [1], Ivaldo Itabaiana, Jr. [1,2], Sébastien Paul [1] and Robert Wojcieszak [1,*]

[1] Univ. Lille, CNRS, Centrale Lille, Univ. Artois, UMR 8181-UCCS-Unité de Catalyse et Chimie du Solide, F-59000 Lille, France; fabien.drault@univ-lille.fr (F.D.); youssef.snoussi@univ-lille.fr (Y.S.); joelle.thuriot@univ-lille.fr (J.T.-R.); ivaldo.itabaiana@univ-lille.fr (I.I.J.); sebastien.paul@centralelille.fr (S.P.)
[2] Department of Biochemical Engineering, School of Chemistry, Federal University of Rio de Janeiro, Rio de Janeiro 21941-910, Brazil
* Correspondence: robert.wojcieszak@univ-lille.fr; Tel.: +33-(0)320-676-008

Abstract: 2,5-furandicarboxylic acid (2,5-FDCA) is a biomass derivate of high importance that is used as a building block in the synthesis of green polymers such as poly(ethylene furandicarboxylate) (PEF). PEF is presumed to be an ideal substitute for the predominant polymer in industry, the poly(ethylene terephthalate) (PET). Current routes for 2,5-FDCA synthesis require 5-hydroxymethylfurfural (HMF) as a reactant, which generates undesirable co-products due to the complicated oxidation step. Therefore, direct CO_2 carboxylation of furoic acid salts (FA, produced from furfural, derivate of inedible lignocellulosic biomass) to 2,5-FDCA is potentially a good alternative. Herein, we present the primary results obtained on the carboxylation reaction of potassium 2-furoate (K2F) to synthesize 2,5-FDCA, using heterogeneous catalysts. An experimental setup was firstly validated, and then several operation conditions were optimized, using heterogeneous catalysts instead of the semi-heterogeneous counterparts (molten salts). Ag/SiO₂ catalyst showed interesting results regarding the K2F conversion and space–time yield of 2,5-FDCA.

Keywords: carboxylation; metal nanoparticles; heterogeneous catalysis; FDCA; furoic acid

1. Introduction

Recently, the production of 2,5-furandicarboxylic acid (2,5-FDCA) from biomass has awakened interest [1–9]. 2,5-FDCA is one of the most important building blocks for the production of polymers, such as the poly(ethylene furandicarboxylate) (PEF), which can replace poly(ethylene terephthalate) (PET), derived from terephthalic acid (TA), a non-sustainable molecule. Two main routes have been studied in the literature for the 2,5-FDCA synthesis from C6 or C5 compounds transformation. Moreover, 5-hydroxymethylfurfural (HMF) oxidation to 2,5-FDCA has been widely studied, and although it has shown to have the best catalytic results, some problems regarding 2,5-FDCA selectivity are found due to the formation of unstable intermediate products [4–6,10–14]. Moreover, HMF is quite unstable and provokes serious problems during the oxidation process. In addition, HMF generally obtained from fructose need to be of very high purity. On the other hand, 2,5-FDCA synthesis from hemicellulose-derived chemicals is of great importance. Indeed, furfural could substitute HMF, as its industrial production from non-edible resources is a mature process. Production of furoic acid synthesis from furfural oxidation, using heterogeneous catalysts in a alkaline [15–17] or base free media [6,18], has been studied. Then, the C–H carboxylation of furoic acid with CO₂ can form 2,5-FDCA (Figure 1). This reaction has shown to be more selective than that from HMF [19–21].

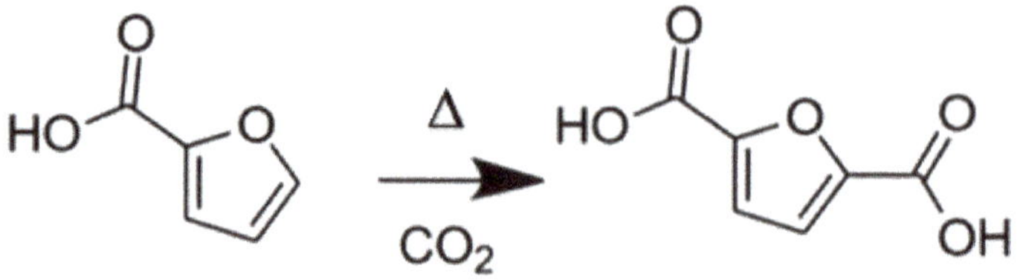

Figure 1. Reactional scheme of CO_2 carboxylation of furoic acid to 2,5-furandicarboxylic acid (2,5-FDCA).

However, the main problem in the 2,5-FDCA synthesis from furoic acid is the carboxylate group insertion on hydrocarbon C–H bonds [22,23]. As C1 feedstock, CO_2 presents thermodynamic and kinetic limitations [24]. Indeed, in the esterification of aromatic hydrocarbons with CO_2, a low equilibrium conversion at every temperature is obtained [24]. Consequently, several solutions have been studied to perform the direct C–H carboxylation, by using a base as a reagent, as previously developed by Kolbe and Schmitt [25–27], a Lewis acid [28], transition metal catalysts [29] and enzymes [19,30]. In terms of mechanism, those reagents could influence the mode of C–H cleavage that could be an electrophilic aromatic substitution, a C–H deprotonation by base or a C–H oxidation and subsequent CO_2 insertion. This reaction can take place both in basic or acidic conditions. In basic media, the use of a strong base deprotonates the C–H group with the most acidic proton, which is at position 5 in furoic acid (FA) [31], to form a strong nucleophilic carbon atom, being able to react with weakly electrophilic carbon dioxide. In acidic media, CO_2 is activated via coordination with a Lewis acid, leading to a reaction between the reactant and the activated CO_2 [32]. This reaction can occur at relatively low temperatures but requires high CO_2 pressure, and poor yield of the target product is reached due to the different parallel reactions that could occur. A general schematization of this process is illustrated in Figure 2.

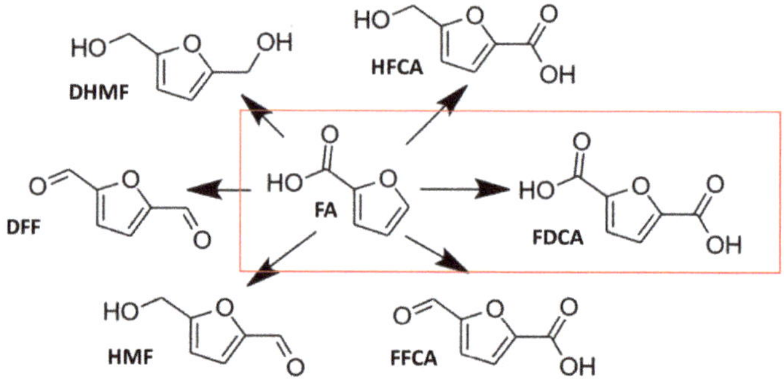

Figure 2. Possible carboxylation reaction routes from furoic acid (FA) to 2,5-furandicarboxylic acid (FDCA), 5-hydroxymethylfurfural (HMF), 2,5-dihydroxmethylfuran (DHMF), 5-hydromethyl-2-furancarboxylic acid (HFCA), 2,5-diformylfuran (DFF) and 5-formyl-2-furancarboxylic acid (FFCA).

In this context, the Henkel reaction of alkaline salts of aromatic acids to synthesize symmetrical diacids has been reported in the literature [33,34]. This reaction involves the thermal rearrangement or disproportionation of alkaline salts derived from aromatic acids to both the unsubstituted and the symmetrical aromatic diacids. This process is carried out under carbon dioxide or inert atmosphere, at high pressure, between 350 and 550 °C, producing potassium terephthalate and benzene from potassium benzoate in the presence of a metallic salt (e.g., cadmium, zinc . . .) [35–39]. Furthermore, the HCl acidification of potassium terephthalate produces TA, which is used to synthesize PET.

Alkaline salts of furoic acid disproportionate to produce 2,5-FDCA in a similar way to that observed in the Henkel reaction for TA synthesis [40–44]. However, formation of furan is also observed during the reaction. The latter could be hydrogenated to produce 1,4-butanediol [45]. Polycondensation of 1,4-butanediol and 2,5-FDCA can be performed in order to produce poly(1,4-butylene 2,5-furandicarboxylate) (PBF), which is a renewable alternative to PET [46].

For instance, Pan et al. [44] performed a reaction involving potassium furoate as reactant and $ZnCl_2$ as catalyst, under 38 Bar of CO_2, at 250 °C, for 3 h. They reported high selectivity of 86% to 2,5-FDCA, with a conversion of 61%. However, purity of 2,5-FDCA was not completely detailed.

Regarding the selectivity of the reaction, Thiyagarajan et al. [40,43] recently demonstrated the formation of asymmetrical diacids, as 2,4-FDCA in addition to the 2,5-FDCA. A series of catalytic tests using potassium 2-furoate as reactant in a Kugelrohr glass oven were performed, giving rise to FDCAs' formation. A yield of up to 91%, using CdI_2 as catalyst, at 260 °C, for 5.5 h, under a low flow of N_2 [40], was obtained. After esterification of the crude reaction mixture, they demonstrated the presence of both the 2,5-FDCA and the 2,4-FDCA asymmetrical diacids with a 70:30 molar ratio [40].

Several homogeneous or semi-heterogeneous catalysts (as CdI_2 and $ZnCl_2$ or Cs_2CO_3 and K_2CO_3, respectively) have shown their efficiency in the FDCA synthesis from furoic acid. However, the use of those types of catalysts complicates the purification/separation step of the desired product. The development of heterogeneous catalysts to overcome the homogeneous catalyst problematic could be a solution.

The main objective of this work is to study the possibility of using heterogeneous catalysts for the 2,5-FDCA synthesis from furoic acid derivatives. Experimental conditions were set based on those reported in the literature, for the Henkel reaction (devices, operation conditions and reactant/catalyst ratio). Additionally, we discuss herein which operation parameters could be optimized and which kind of heterogeneous catalyst could promote the 2,5-FDCA synthesis.

2. Results

2.1. Validation of the Reaction Setup

Regarding the 2,5-FDCA synthesis, one of the main objectives was using a Kugelrohr apparatus, under conditions similar to those reported in the literature [40]. Thiyagarajan et al. [40] performed the production of the 2,5-furan and 2,4-furandicarboxylic acid through Henkel reaction [33,36]. Typically, 10 g of potassium 2-furoate and 22 mol% of CdI_2 were mechanically mixed and loaded in a round flask, which was introduced in the Kugelrohr oven. The optimum operating conditions were 260 °C during 5.5 h, under a continuous flow of N_2 and a slow rotation of the reactor. A similar experiment using 10-times-less reactant and catalyst was performed in our laboratory. A comparison of our results from the NMR analysis with those obtained in the literature is shown in Table 1. Furthermore, in order to determine the chemical shift (in ppm) of FA and 2,5-FDCA peaks in DMSO using NMR, pure compounds were analyzed separately (Supplementary Materials Figures S1 and S2, respectively).

Table 1. Validation of the experimental setup for the Henkel reaction, using the Kugelrohr apparatus.

Experiment	X_{K2F} (%)	FDCAs Formation (%)	$S_{2,5-FDCA}$ (%)	$S_{2,4-FDCA}$ (%)
Literature [40]	92	91	70	30
This work	73	69	69	31

Conditions: 530 mg of CdI_2, 1 g of K2F, F_{N2} = 45 mL min^{-1}, 20 rpm, T = 260 °C, t = 5.5 h.

Results calculated from NMR analysis (Supplementary Materials Figure S3) showed a K2F conversion and a 2,5-FDCA formation slightly lower than that obtained by Thiyagarajan et al. [40]. However, the individual selectivity of furandicarboxylic acids remains of the same order. Regarding the results, it was concluded that the Kugelrohr glass oven is suitable for the Henkel reaction tests.

2.2. Dual Catalytic System (Ag/SiO$_2$ + CdI$_2$)

The Henkel reaction to produce 2,5-FDCA from K2F has shown good results using CdI$_2$ as catalyst [40]. However, the reaction temperature must reach at least 260 °C to achieve good conversions. At this temperature, CdI$_2$ starts to decompose (melting point is 387 °C), thus leading to a better interaction between the solid K2F and the semi-melted catalyst. In order to decrease the working temperature, one of the proposed solutions was to use a heterogeneous catalyst in addition to the CdI$_2$ catalyst, in CO$_2$ atmosphere. The temperature was decreased from 260 to 200 °C, to observe the evolution of the results compared to those in the literature [40]. Results are shown in Table 2.

Table 2. Influence of temperature for the 2,5-FDCA synthesis using Ag/SiO$_2$ and CdI$_2$ mixture on the K2F conversion and space–time yields (STY).

Entry	Catalyst	Temperature (°C)	Conversion (%)	STY$_{2,5\text{-FDCA}}$ (µmol kg^{-1} h^{-1})	STY$_{DFF}$ (µmol kg^{-1} h^{-1})
1	CdI$_2$	200	0	-	-
2	Ag/SiO$_2$/CdI$_2$	200	51	264	951
3	Ag/SiO$_2$/CdI$_2$	230	74	145	-
4	Ag/SiO$_2$/CdI$_2$	260	69	188	-
5	Ag/SiO$_2$	200	20	1203	-

Conditions: 17 mg of CdI$_2$, 35 mg of K2F, 50 mg of Ag/SiO$_2$, F$_{CO_2}$ = 45 mL min^{-1}, 20 rpm, T = 200–260 °C, t = 20 h.

Using a dual catalytic system seems to open the possibility to significantly decrease the reaction temperature, since the K2F conversion does not show a significant decrease (Table 2, Entries 2 and 4). However, 2,5-FDCA's yield is more affected by using lower temperature. Indeed, at 200 °C, the formation of the 2,5-diformyl furan (DFF) was observed. It was also the major product at this temperature. However, it confirms the beneficial effect of the Ag/SiO$_2$, since no activity was observed at 200 °C for CdI$_2$ alone (Table 2, Entry 1). In addition, Ag/SiO$_2$ alone already produces 2,5-FDCA, even if the K2F conversion decreases significantly (Table 2, Entry 5).

2.3. Effect of the Support

Taking into account good preliminary results obtained with the Ag/SiO$_2$ (Table 2), the screening of the different supports was performed. Silver and gold catalysts supported on different supports were tested, and the results are shown in the Figure 3.

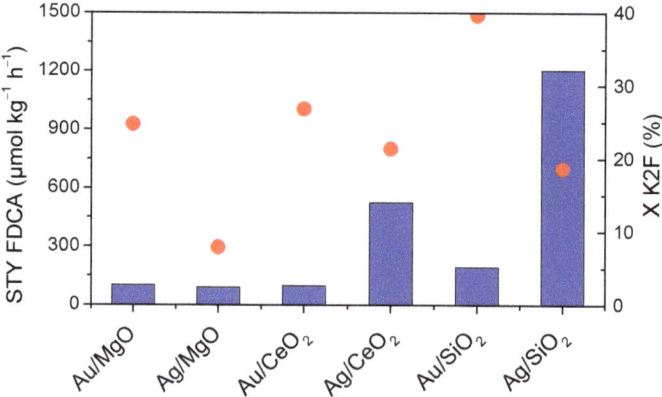

Figure 3. Effect of the support on gold and silver catalysts in the K2F carboxylation: K2F conversion (●) and the STY of 2,5-FDCA (■) (Conditions: Substrate/M = 9, F$_{CO_2}$ = 45 mL min^{-1}, 20 rpm, T = 200 °C, t = 20 h).

Regarding K2F conversion, the use of Au instead of Ag increased the activity, whatever the support. However, a lower 2,5-FDCA yield was obtained by using Au catalysts. Regarding the support, the use of MgO with Au or Ag leads to a low 2,5-FDCA production. On the contrary, Ag/CeO$_2$ and Ag/SiO$_2$ catalysts have shown similar activity (~20% of K2F conversion) and promote 2,5-FDCA production up to 524 and 1203 µmol kg^{-1} h^{-1}, respectively. In order to explain these results, it has to be considered that CeO$_2$ support has basic and redox properties, while MgO presents high basicity [47]. The redox properties of the CeO$_2$ could be the reason of an enhanced performance. Indeed, the oxygen vacancies of CeO$_2$ increase the adsorption capacity of CO$_2$ [47,48], which could promote the carboxylation reaction. However, the use of SiO$_2$ has shown a much better yield to 2,5-FDCA than CeO$_2$. That could be explained by the presence of acid sites in the support [49].

2.4. Effect of the Metal

Since the SiO$_2$ support has shown the best performances towards 2,5-FDCA synthesis, it has been selected for the screening of the metal phase in the heterogeneous catalyst. A series of M/SiO$_2$ catalysts were tested, in similar conditions, for comparison. K2F conversion and 2,5-FDCA space–time yield are shown in Figure 4.

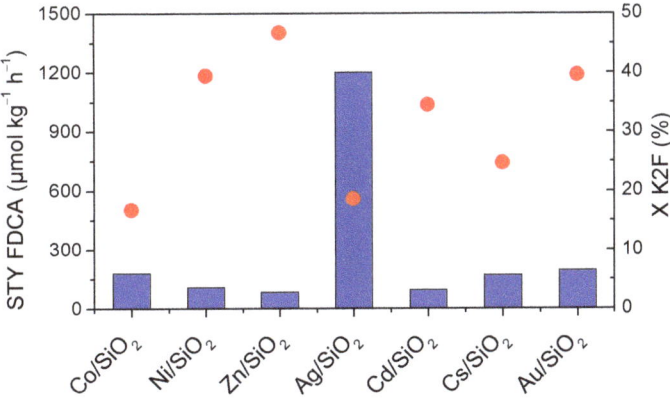

Figure 4. Effect of the metal, using SiO$_2$ supported catalysts on the conversion (●) of K2F and the STY of 2,5-FDCA (■) (Conditions: Substrate/M = 9, F$_{CO_2}$ = 45 mL min^{-1}, 20 rpm, T = 200 °C, t = 20 h).

High 2,5-FDCA production was observed only for Ag/SiO$_2$ catalyst with a conversion of 20% (Figure 4). Cs and Co supported on silica showed the lowest conversion values, between 17 and 25%. On the other hand, Zn and Cd supported on silica presented the highest conversion values, which are 47 and 35% respectively. Both metals have already shown promising results in the Henkel reaction, using CdI$_2$ and ZnCl$_2$.

Regarding the metal composition, only the catalyst containing Ag has shown interesting results in 2,5-FDCA synthesis, contrary to the other metals tested. An XRD diffractogram of the Ag/SiO$_2$ catalyst is shown in Supplementary Materials Figure S4. Diffraction peaks located at 34, 49 and 61° 2θ could correspond to Ag$_2$O (JCDS ICDD 00-042-9874) or Ag$_2$CO$_3$ (JCDS ICDD 04-017-5597), while the three other peaks at 38.1°, 44.2° and 64.4° 2θ are representative of the (111), (200) and (220) planes of metallic Ag (JCDS ICDD 00-001-1164), respectively. The mean crystallite sizes of Ag metallic and Ag$_2$O, calculated using Scherrer's equation, were of 25.3 and 9.5 nm, respectively.

One of the main differences between silver and the others metals comes from its particular electronic configuration ([Kr] 4d^{10} 5s^1 from group 11), typical from the so-called "coinage metals" group. Furthermore, it has been reported in the literature that Ag(I) salts promote the carboxylation of terminal alkynes [50–52]. The d^{10} electronic configuration of silver favors the activation of alkynes through its interaction with the C–C π-bond of alkyne.

Moreover, Lui et al. [53] have demonstrated that the use of heterogeneous Ag@MIL-101 catalysts promotes the capture and conversion of CO_2, overcoming the need of strong base or aggressive organometallic reagents to activate the hydrogen of the terminal alkyne. It was also shown that the adsorption of gaseous CO_2 on the Ag surface occurs differently compared to other metals such as Cu or Au. In case of oxidized Ag atoms, surface O atoms interact with gaseous CO_2 and form chemisorbed on the surface of the metal carbonic acid-like species. In these carbonic acid-like species, two oxygen atoms from CO_3^- are bonded to adjacent Ag bridging sites. The third oxygen atom forms a C double bond (C = O) perpendicular to the surface. This carbonic $O=CO_2^{\delta-}$ surface specie has a negative charge localized on the two oxygen atoms binding to the Ag surface [54].

The presence of Ag_2O could explain the 2,5-FDCA formation, which is produced only by using Ag catalysts. As previously reported, the use of Ag(I) promotes alkynes carboxylation. Consequently, in our case 2,5-FDCA formation could come from the interaction of Ag^+ and the reactant. Moreover, this hypothesis could explain the low 2,5-FDCA formation on Au catalysts, which only present metallic gold nanoparticles [55,56]. This parameter would need some deeper investigation to understand the role of silver in the reaction.

In conclusion, an effect of the support was observed from using supports with different acidity and redox properties. Supports with acidic sites and redox properties promote the production of 2,5-FDCA, while basic supports, like MgO, were demonstrated to have the lower catalytic performances. On the other hand, the screening of different metals supported on SiO_2 has shown that only Ag is leading to the formation of 2,5-FDCA, but with less conversion than Au, Zn, Cd and Ni. Therefore, for the optimization of the reaction conditions, the Ag/SiO_2 catalyst was used.

2.5. Effect of the Substrate/Metal Molar Ratio

In this section, the substrate/Ag molar ratio from 1 to 33 was studied. The catalytic results (K2F conversion and space–time yield (STY) of 2,5-FDCA) are presented in Figure 5.

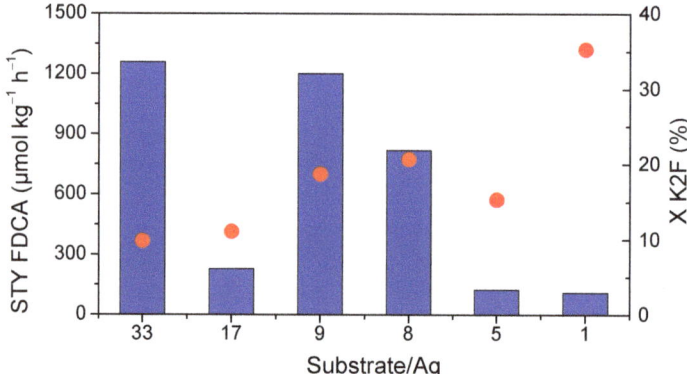

Figure 5. Effect of substrate/Ag molar ratio on the conversion of K2F (●) and the STY of 2,5-FDCA (■) (Conditions: Substrate/Ag = 1–33, F_{CO_2} = 45 mL min^{-1}, 20 rpm, T = 200 °C, t = 20 h).

As expected, the K2F conversion increases when the substrate/Ag molar ratio decreases. These results could be directly linked to a limitation of the available active sites on the Ag/SiO_2 catalyst. Regarding the 2,5-FDCA STY, a maximum of 1260 µmol kg^{-1} h^{-1} is obtained by using a substrate/Ag molar ratio equal to 26. The molar ratio of 9 seems to be an optimum value due to a 2,5-FDCA STY of 1203 µmol kg^{-1} h^{-1} near to the highest value, and to a K2F conversion of 20% instead of 9% (obtained for the molar ratio of 33). Surprisingly, the 1, 5 and 17 molar ratios have given a very low formation of 2,5-FDCA, which could come from a lack of physical interaction of the mixture, together with a kinetic limitation for the formation of 2,5-FDCA.

2.6. Effect of Reaction Temperature

The reaction temperature has been studied from 170 to 300 °C. The conversion and space–time yield are shown in Figure 6.

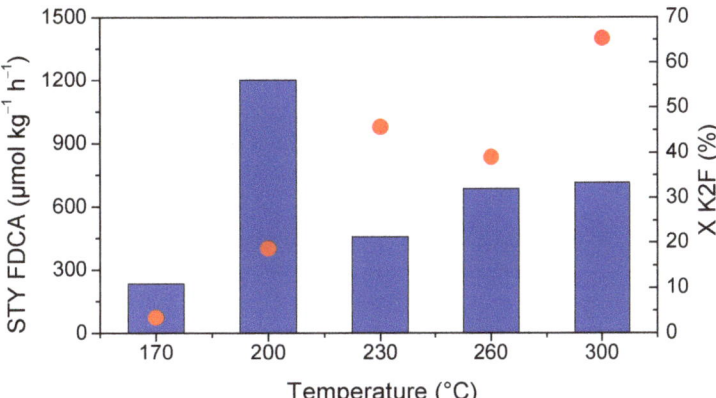

Figure 6. Influence of the reaction temperature on the catalytic performance of Ag/SiO$_2$ with the conversion of K2F (●) and the STY of 2,5-FDCA (■) (Conditions: Substrate/Ag = 9, F$_{CO_2}$ = 45 mL min^{-1}, 20 rpm, T = 170–300 °C, t = 20 h).

At 170 °C, a very low conversion of 6% was obtained. An increase of the temperature enhances conversion, with an optimum 2,5-FDCA space–time yield of 1203 µmol kg^{-1} h^{-1} at 200 °C. Furthermore, between 230 and 300 °C, an increase in K2F conversion to 2,5-FDCA reaching 65% was observed. However, only 715 µmol kg^{-1} h^{-1} of 2,5-FDCA yield at 300 °C was obtained.

3. Materials and Methods

3.1. Catalysts Preparation

A series of Au and Ag catalysts with a nominal metal content of 7 wt.% were prepared by wet impregnation, using water as solvent and using commercial SiO$_2$ (CARiACT Q-10), CeO$_2$ (Sigma-Aldrich, Saint Louis, MO, USA) and MgO (Sigma-Aldrich, Saint Louis, MO, USA) as support. Typically, for 1 g of catalyst, 10 mL of a 66 mmol L^{-1} of AgNO$_3$ solution (99%, Sigma-Aldrich, Saint Louis, MO, USA) was added to 0.93 g of the support. The mixture was kept under stirring (150 rpm), overnight, and the solvent was evaporated at 90 °C, using a vacuum, prior to a drying step at 80 °C, overnight. The obtained solids were afterwards calcined in air atmosphere for 4 h, at 300 °C.

M/SiO$_2$ catalyst series with M = Zn, Cs, Cd, Ni and Co were prepared by using the same method, by adjusting the concentration of the precursor solution in order to obtain a 7 wt.% metal content.

3.2. Catalytic Test

Potassium 2-furoate (K2F) was prepared by using furoic acid (97%, Sigma-Aldrich, Saint Louis, MO, USA) and KOH (85%, Sigma-Aldrich, Saint Louis, MO, USA) in a 1:1 molar ratio.

2,5-FDCA synthesis was performed in a Glass Oven B-585 Kugelrohr (Büchi), under CO$_2$ atmosphere. Typically, 1 mmol of K2F was mechanically mixed with 100 mg of Ag/SiO$_2$ and introduced into a round flask and placed inside the oven. Then, the solid mixture was slowly rotated at 20 rpm, at 200 °C, under a continuous flow of CO$_2$ (45 mL min^{-1}), during 20 h. Afterwards, the setup was cooled down for 1 h. The obtained black solid was dispersed in H$_2$O or MeOH, stirred for 1 h and filtered by using PTFE (2 µm), giving a pale-yellow filtrate and the remaining catalyst.

To analyze the reactions reagents and products, a gas chromatograph (GC, Agilent 7890B) apparatus equipped with a CP-Wax 52 CB GC column or a High-Performance Liquid Chromatography (HPLC, Waters 2410 RJ) apparatus equipped with a Shodex SUGAR SH-1011 column and UV detector, using 5 mM of H_2SO_4 (0.6 mL min^{-1}) as a mobile phase, was used. Conversion and STY were calculated by using the following formulas, Formulas (1) and (2), respectively.

$$X_{K2F} = (n_{0K2F} - n_{K2F})/n_{0K2F} \times 100, \qquad (1)$$

where X_{K2F} is the potassium furoate conversion (%), and n_{0K2F} and n_{K2F} are the initial and the non-reacted number of moles of potassium furoate, respectively (mol).

$$STY_i = n_i/(m_{catalyst} \times t_r), \qquad (2)$$

where STY_i is the space–time yield of a product i (μmol kg^{-1} h^{-1}), n_i is the number of moles of product i (μmol), $m_{catalyst}$ is the mass of catalyst (kg) and t_r is the reaction time (h).

4. Conclusions

A dual catalytic system containing a heterogeneous catalyst and CdI_2, a semi-homogeneous catalyst, was tested. Preliminary results confirmed the possibility to decrease the reaction temperature to 230 °C, obtaining an acceptable conversion (74%). However, a drastic decrease of the 2,5-FDCA STY was observed from 8489 to 154 μmol kg^{-1} h^{-1}, which could be explained by a high adsorption of the products on the heterogeneous catalyst. It is worth noting that an increase in the temperature of the reaction disfavored the adsorption and favored the catalytic conversion.

Several experiments have been performed to screen a support and a metal which could promote 2,5-FDCA production. SiO_2 support has shown the greatest promotion of 2,5-FDCA synthesis from K2F, contrary to CeO_2 and MgO. In addition, Ag/SiO_2 leads to the highest yields in 2,5-FDCA formation, while other monometallic (Ni, Co, Zn, Cd ...) catalysts showed much lower productivities.

Supplementary Materials: The following are available online, at https://www.mdpi.com/2073-4344/11/3/326/s1. Figure S1: Representative ^{1}H-NMR of furoic acid (FA). Figure S2: Representative 1H-NMR of isolated FDCA. Figure S3: Representative 1H-NMR of crude products obtained from the carboxylation of K2F to FDCA. Figure S4: XRD diffractogram of Ag/SiO_2.

Author Contributions: Conceptualization, F.D., Y.S. and R.W.; methodology, F.D.; formal analysis, F.D. and J.T.-R.; investigation, F.D. and Y.S.; resources, R.W.; data curation, F.D., J.T.-R., I.I.J. and R.W. writing—original draft preparation, F.D., and R.W.; writing—review and editing, F.D.; I.I.J., S.P. and R.W.; visualization, I.I.J. and S.P.; supervision, R.W.; project administration, R.W.; funding acquisition, R.W. All authors have read and agreed to the published version of the manuscript.

Funding: This research was funded by REGION HAUTS-de-FRANCE, grant number 3859523 FDCA STARTAIRR project and I-SITE ULNE grant V-Start'AIRR-18-001-Wojcieszak This study was supported by the French government through the Programme Investissement d'Avenir (I-SITE ULNE/ANR-16- IDEX-0004 ULNE) managed by the Agence Nationale de la Recherche.

Data Availability Statement: The data presented in this study are available on request from the corresponding author.

Acknowledgments: The REALCAT platform is benefiting from a state subsidy administrated by the French National Research Agency (ANR), within the frame of the "Future Investments" program (PIA), with the contractual reference "ANR-11-EQPX-0037". The European Union, through the ERDF funding administered by the Hauts-de-France Region, co-financed the platform. Centrale Lille, the CNRS and Lille University, as well as the Centrale Initiative Foundation, are thanked for their financial contribution to the acquisition and implementation of the equipment of the REALCAT platform. The Métropole Européen de Lille (MEL) for the "CatBioInnov" project is also acknowledged.

Conflicts of Interest: The authors declare no conflict of interest.

References

1. Corma, A.; Iborra, S.; Velty, A. Chemical Routes for the Transformation of Biomass into Chemicals. *Chem. Rev.* **2007**, *107*, 2411–2502. [CrossRef]
2. Zhou, H.; Xu, H.; Wang, X.; Liu, Y. Convergent Production of 2,5-Furandicarboxylic Acid from Biomass and CO_2. *Green Chem.* **2019**, *21*, 2923–2927. [CrossRef]
3. De Jong, E.; Dam, M.A.; Sipos, L.; Gruter, G.-J.M. Furandicarboxylic Acid (FDCA), A Versatile Building Block for a Very Interesting Class of Polyesters. In *Biobased Monomers, Polymers, and Materials*; Smith, P.B., Gross, R.A., Eds.; American Chemical Society: Washington, DC, USA, 2012; Volume 1105, pp. 1–13, ISBN 978-0-8412-2767-5.
4. Casanova, O.; Iborra, S.; Corma, A. Mécanisme: Biomass into Chemicals: One Pot-Base Free Oxidative Esterification of 5-Hydroxymethyl-2-Furfural into 2,5-Dimethylfuroate with Gold on Nanoparticulated Ceria. *J. Catal.* **2009**, *265*, 109–116. [CrossRef]
5. Teong, S.P.; Yi, G.; Zhang, Y. Hydroxymethylfurfural Production from Bioresources: Past, Present and Future. *Green Chem.* **2014**, *16*, 2015. [CrossRef]
6. Ferraz, C.P.; Silva, A.G.M.D.; Rodrigues, T.S.; Camargo, P.H.C.; Paul, S.; Wojcieszak, R. Furfural Oxidation on Gold Supported on MnO_2: Influence of the Support Structure on the Catalytic Performances. *Appl. Sci.* **2018**, *8*, 1246. [CrossRef]
7. Cavani, F.; Teles, J.H. Sustainability in Catalytic Oxidation: An Alternative Approach or a Structural Evolution? *ChemSusChem* **2009**, *2*, 508–534. [CrossRef]
8. Wojcieszak, R.; Itabaiana, I. Engineering the Future: Perspectives in the 2,5-Furandicarboxylic Acid Synthesis. *Catal. Today* **2019**, *354*, 211–217. [CrossRef]
9. Mabee, W.E.; Gregg, D.J.; Saddler, J.N. Assessing the Emerging Biorefinery Sector in Canada. *Appl. Biochem. Biotechnol.* **2005**, *123*, 765–778. [CrossRef]
10. Papageorgiou, G.Z.; Tsanaktsis, V.; Bikiaris, D.N. Synthesis of Poly(Ethylene Furandicarboxylate) Polyester Using Monomers Derived from Renewable Resources: Thermal Behavior Comparison with PET and PEN. *Phys. Chem. Chem. Phys.* **2014**, *16*, 7946–7958. [CrossRef]
11. Van Putten, R.-J.; van der Waal, J.C.; de Jong, E.; Rasrendra, C.B.; Heeres, H.J.; de Vries, J.G. Hydroxymethylfurfural, A Versatile Platform Chemical Made from Renewable Resources. *Chem. Rev.* **2013**, *113*, 1499–1597. [CrossRef]
12. Albonetti, S.; Lolli, A.; Morandi, V.; Migliori, A.; Lucarelli, C.; Cavani, F. Conversion of 5-Hydroxymethylfurfural to 2,5-Furandicarboxylic Acid over Au-Based Catalysts: Optimization of Active Phase and Metal–Support Interaction. *Appl. Catal. B Environ.* **2015**, *163*, 520–530. [CrossRef]
13. Davis, S.E.; Houk, L.R.; Tamargo, E.C.; Datye, A.K.; Davis, R.J. Oxidation of 5-Hydroxymethylfurfural over Supported Pt, Pd and Au Catalysts. *Catal. Today* **2011**, *160*, 55–60. [CrossRef]
14. Miao, Z.; Zhang, Y.; Pan, X.; Wu, T.; Zhang, B.; Li, J.; Yi, T.; Zhang, Z.; Yang, X. Superior Catalytic Performance of $Ce_{1-x}Bi_xO_{2-\delta}$ Solid Solution and $Au/Ce_{1-x}Bi_xO_{2-\delta}$ for 5-Hydroxymethylfurfural Conversion in Alkaline Aqueous Solution. *Catal. Sci. Technol.* **2015**, *5*, 1314–1322. [CrossRef]
15. Tian, Q.; Shi, D.; Sha, Y. CuO and Ag_2O/CuO Catalyzed Oxidation of Aldehydes to the Corresponding Carboxylic Acids by Molecular Oxygen. *Molecules* **2008**, *13*, 948–957. [CrossRef] [PubMed]
16. Hurd, C.D.; Garrett, J.W.; Osborne, E.N. Furan Reactions. IV. Furoic Acid from Furfural. *J. Am. Chem. Soc.* **1933**, *55*, 1082–1084. [CrossRef]
17. Taarning, E.; Nielsen, I.S.; Egeblad, K.; Madsen, R.; Christensen, C.H. Chemicals from Renewables: Aerobic Oxidation of Furfural and Hydroxymethylfurfural over Gold Catalysts. *ChemSusChem* **2008**, *1*, 75–78. [CrossRef]
18. Santarelli, F.; Wojcieszak, R.; Paul, S.; Dumeignil, F.; Cavani, F. Furoic Acid Preparation Method. Patent WO2017158106A1, 21 September 2017.
19. Payne, K.A.P.; Marshall, S.A. Enzymatic Carboxylation of 2-Furoic Acid Yields 2,5-Furandicarboxylic Acid (FDCA). *ACS Catal.* **2019**, *9*, 2854–2865. [CrossRef]
20. Shen, G.; Zhang, S.; Lei, Y.; Chen, Z.; Yin, G. Synthesis of 2,5-Furandicarboxylic Acid by Catalytic Carbonylation of Renewable Furfural Derived 5-Bromofuroic Acid. *Mol. Catal.* **2018**, *455*, 204–209. [CrossRef]
21. Zhang, S.; Lan, J.; Chen, Z.; Yin, G.; Li, G. Catalytic Synthesis of 2,5-Furandicarboxylic Acid from Furoic Acid: Transformation from C5 Platform to C6 Derivatives in Biomass Utilizations. *ACS Sustain. Chem. Eng.* **2017**, *5*, 9360–9369. [CrossRef]
22. Luo, J.; Larrosa, I. C−H Carboxylation of Aromatic Compounds through CO_2 Fixation. *ChemSusChem* **2017**, *10*, 3317–3332. [CrossRef]
23. Drault, F.; Snoussi, Y.; Paul, S.; Itabaiana, I.; Wojcieszak, R. Recent Advances in Carboxylation of Furoic Acid into 2,5-Furandicarboxylic Acid: Pathways towards Bio-Based Polymers. *ChemSusChem* **2020**, *13*, 5164–5172. [CrossRef] [PubMed]
24. Dabestani, R.; Britt, P.F.; Buchanan, A.C. Pyrolysis of Aromatic Carboxylic Acid Salts: Does Decarboxylation Play a Role in Cross-Linking Reactions? *Energy Fuels* **2005**, *19*, 365–373. [CrossRef]
25. Lindsey, A.S.; Jeskey, H. The Kolbe-Schmitt Reaction. *Chem. Rev.* **1957**, *57*, 583–620. [CrossRef]
26. Kolbe, H.; Lautemann, E. Constitution of Salicylic Acid and Its Bascity. *Liebigs Ann. Chem.* **1860**, 157–206. [CrossRef]
27. Schmitt, R. Beitrag Zur Kenntniss Der Kolbe'schen Salicylsäure Synthese. *J. Prakt. Chem.* **1885**, *1*, 397–411. [CrossRef]
28. Olah, G.A.; Török, B.; Joschek, J.P.; Bucsi, I.; Esteves, P.M.; Rasul, G.; Surya Prakash, G.K. Efficient Chemoselective Carboxylation of Aromatics to Arylcarboxylic Acids with a Superelectrophilically Activated Carbon Dioxide−Al_2Cl_6/Al System. *J. Am. Chem. Soc.* **2002**, *124*, 11379–11391. [CrossRef] [PubMed]

29. Dalton, D.M.; Rovis, T. C–H Carboxylation Takes Gold. *Nat. Chem.* **2010**, *2*, 710–711. [CrossRef] [PubMed]
30. Wuensch, C.; Glueck, S.M.; Gross, J.; Koszelewski, D.; Schober, M.; Faber, K. Regioselective Enzymatic Carboxylation of Phenols and Hydroxystyrene Derivatives. *Org. Lett.* **2012**, *14*, 1974–1977. [CrossRef]
31. Banerjee, A.; Dick, G.R.; Yoshino, T.; Kanan, M.W. Carbon Dioxide Utilization via Carbonate-Promoted C–H Carboxylation. *Nature* **2016**, *531*, 215–219. [CrossRef]
32. Tanaka, S.; Watanabe, K.; Tanaka, Y.; Hattori, T. EtAlCl$_2$/2,6-Disubstituted Pyridine-Mediated Carboxylation of Alkenes with Carbon Dioxide. *Org. Lett.* **2016**, *18*, 2576–2579. [CrossRef]
33. Raecke, B. Synthese von Di- und Tricarbonsäuren aromatischer Ringsysteme durch Verschiebung von Carboxyl-Gruppen. *Angew. Chem.* **1958**, *70*, 1–5. [CrossRef]
34. Raecke, B. A Process for the Production of Terephthalic Acid or Salts thereof or Derivatives thereof of Potassium Benzoate. Patent DE958920C, 28 February 1957.
35. Wang, Z. Henkel reaction. In *Comprehensive Organic Name Reactions and Reagents*; John Wiley & Sons: Hoboken, NJ, USA, 2010; pp. 1379–1382.
36. McNelis, E. Reactions of Aromatic Carboxylates. II. [1] The Henkel Reaction. *J. Org. Chem.* **1965**, *30*, 1209–1213. [CrossRef]
37. Patton, J.W.; Son, M.O. The Synthesis of Naphthalene-2,3-Dicarboxylic Acid by the Henkel Process. *J. Org. Chem.* **1965**, *30*, 2869–2870. [CrossRef]
38. Clayton, T.W.; Britt, P.F.; Buchanan, A.C. Decarboxylation of salts of aromatic carboxylic acids and their role in cross-linking reactions. *Prepr. Pap. Am. Chem. Soc. Div. Fuel Chem.* **2001**, *46*, 5.
39. Ogata, Y.; Tsuchida, M.; Muramoto, A. The Preparation of Terephthalic Acid from Phthalic or Benzoic Acid. *J. Am. Chem. Soc.* **1957**, *79*, 6005–6008. [CrossRef]
40. Thiyagarajan, S.; Pukin, A.; van Haveren, J.; Lutz, M.; van Es, D.S. Concurrent Formation of Furan-2,5- and Furan-2,4-Dicarboxylic Acid: Unexpected Aspects of the Henkel Reaction. *RSC Adv.* **2013**, *3*, 15678–15686. [CrossRef]
41. Dawes, G.J.S.; Scott, E.L.; Le Nôtre, J.; Sanders, J.P.M.; Bitter, J.H. Deoxygenation of Biobased Molecules by Decarboxylation and Decarbonylation—A Review on the Role of Heterogeneous, Homogeneous and Bio-Catalysis. *Green Chem.* **2015**, *17*, 3231–3250. [CrossRef]
42. Van Es, D.S. Rigid Biobased Building Blocks. *J. Renew. Mater.* **2013**, *1*, 61–72. [CrossRef]
43. Van Haveren, J.; Thiyagarajan, S.; Teruo Morita, A. Process for the Production of a Mixture of 2,4- Furandicarboxylic Acid and 2,5- Furandicarboxylic Acid (FDCA) via Disproportionation Reaction, Mixture of 2,4-FDCA and 2,5-FDCA Obtainable Thereby, 2,4-FDCA Obtainable Thereby and Use of 2,4-FDCA 2015. U.S. Patent US9284290B2, 15 March 2016.
44. Pan, T.; Deng, J.; Xu, Q.; Zuo, Y.; Guo, Q.-X.; Fu, Y. Catalytic Conversion of Furfural into a 2,5-Furandicarboxylic Acid-Based Polyester with Total Carbon Utilization. *ChemSusChem* **2013**, *6*, 47–50. [CrossRef]
45. Nakanishi, K.; Tanaka, A.; Hashimoto, K.; Kominami, H. Photocatalytic Hydrogenation of Furan to Tetrahydrofuran in Alcoholic Suspensions of Metal-Loaded Titanium(IV) Oxide without Addition of Hydrogen Gas. *Phys. Chem. Chem. Phys.* **2017**, *19*, 20206–20212. [CrossRef]
46. Ma, J.; Yu, X.; Xu, J.; Pang, Y. Synthesis and Crystallinity of Poly(Butylene 2,5-Furandicarboxylate). *Polymer* **2012**, *53*, 4145–4151. [CrossRef]
47. Martin, D.; Duprez, D. Mobility of Surface Species on Oxides. 1. Isotopic Exchange of $^{18}O_2$ with ^{16}O of SiO_2, Al_2O_3, ZrO_2, MgO, CeO_2, and CeO_2-Al_2O_3. Activation by Noble Metals. Correlation with Oxide Basicity. *J. Phys. Chem.* **1996**, *100*, 9429–9438. [CrossRef]
48. Yoshikawa, K.; Kaneeda, M.; Nakamura, H. Development of Novel CeO_2-Based CO_2 Adsorbent and Analysis on Its CO2 Adsorption and Desorption Mechanism. *Energy Proc.* **2017**, *114*, 2481–2487. [CrossRef]
49. Li, Y.; Zhang, R.; Du, L.; Zhang, Q.; Wang, W. Catalytic Mechanism of C–F Bond Cleavage: Insights from QM/MM Analysis of Fluoroacetate Dehalogenase. *Catal. Sci. Technol.* **2016**, *6*, 73–80. [CrossRef]
50. Fukue, Y.; Inoue, Y.; Oi, S. Direct Synthesis of Alkyl2-Alkynoates from Alk-l-Ynes, C02, and Bromoalkanes Catalysed by Copper(1) or Silver(1) Salt. *J. Chem. Soc. Chem. Commun.* **1994**, *18*, 2091. [CrossRef]
51. Sekine, K.; Yamada, T. Silver-Catalyzed Carboxylation. *Chem. Soc. Rev.* **2016**, *45*, 4524–4532. [CrossRef] [PubMed]
52. Manjolinho, F.; Arndt, M.; Gooßen, K.; Gooßen, L.J. Catalytic C–H Carboxylation of Terminal Alkynes with Carbon Dioxide. *ACS Catal.* **2012**, *2*, 2014–2021. [CrossRef]
53. Liu, X.-H.; Ma, J.-G.; Niu, Z.; Yang, G.-M.; Cheng, P. An Efficient Nanoscale Heterogeneous Catalyst for the Capture and Conversion of Carbon Dioxide at Ambient Pressure. *Angew. Chem.* **2015**, *127*, 1002–1005. [CrossRef]
54. Ye, Y.; Yang, H.; Qian, J.; Su, H.; Lee, K.-J.; Cheng, T.; Xiao, H.; Yano, J.; Goddard, W.A.; Crumlin, E.J. Dramatic Differences in Carbon Dioxide Adsorption and Initial Steps of Reduction between Silver and Copper. *Nat. Commun.* **2019**, *10*, 1875. [CrossRef]
55. Ayastuy, J.L.; Gurbani, A.; Gutiérrez-Ortiz, M.A. Effect of Calcination Temperature on Catalytic Properties of Au/Fe_2O_3 Catalysts in CO-PROX. *Int. J. Hydrog. Energy* **2016**, *41*, 19546–19555. [CrossRef]
56. Vigneron, F.; Caps, V. Evolution in the Chemical Making of Gold Oxidation Catalysts. *C. R. Chim.* **2016**, *19*, 192–198. [CrossRef]

Article

Influence of Water-Miscible Organic Solvent on the Activity and Stability of Silica-Coated Ru Catalysts in the Selective Hydrolytic Hydrogenation of Cellobiose into Sorbitol

Tommy Haynes [1], Sharon Hubert [1], Samuel Carlier [1], Vincent Dubois [2] and Sophie Hermans [1,*]

1. IMCN Institute, MOST Division, Université Catholique de Louvain, 1 Place Louis Pasteur, B-1348 Louvain-la-Neuve, Belgium; Tommy.Haynes@uclouvain.be (T.H.); Sharon.Hubert@uclouvain.be (S.H.); Samuel.Carlier@uclouvain.be (S.C.)
2. Department of Physical Chemistry and Catalysis, LABIRIS, 1 avenue Gryson, 1070 Brussels, Belgium; vidubois@spfb.brussels
* Correspondence: sophie.hermans@uclouvain.be; Tel.: +32-10-472810; Fax: (+32)-10-472330

Received: 29 November 2019; Accepted: 16 January 2020; Published: 23 January 2020

Abstract: Ruthenium nanoparticles supported on carbon black were coated by mesoporous protective silica layers (Ru/CB@SiO$_2$) with different textural properties (S_{BET}: 280–390 m^2/g, pore diameter: 3.4–5.0 nm) and were tested in the selective hydrogenation of glucose into sorbitol. The influence of key parameters such as the protective layer pore size and the solvent nature were investigated. X-ray photoelectron spectroscopy (XPS) analyses proved that the hydrothermal stability was highly improved in ethanolic solution with low water content (silica loss: 99% in water and 32% in ethanolic solution). In this work, the strong influence of the silica layer pore sizes on the selectivity of the reaction (shifting from 4% to 68% by increasing the pores sizes from 3.4 to 5 nm) was also highlighted. Finally, by adding acidic co-catalyst (CB–SO$_3$H), sorbitol was obtained directly through the hydrolytic hydrogenation of cellobiose (used as a model molecule of cellulose), demonstrating the high potential of the present methodology to produce active catalysts in biomass transformations.

Keywords: glucose; sorbitol; cellobiose; hydrothermal resistance; hydrolysis; hydrogenation; mesoporous silica; ruthenium; carbon

1. Introduction

For many years, the conversion of lignocellulosic biomass, an environmentally friendly and sustainable alternative resource to fossil fuels, into valuable chemicals and biofuels has drawn a lot of attention [1]. Therefore, cellulose, the main component of such biomass, has been widely studied and converted into various products such as hydrocarbons, oxygenated bio-oil, and sugar alcohols [2,3]. Enzymatic catalysis and fermentation allow breaking down cellulose and converting it into commodity chemicals [4]. Nevertheless, such biological processes suffer from limitations such as low efficiencies, limited scale of production, and narrow reaction conditions. Therefore, cellulose valorization transformations have been recently oriented toward solid heterogeneous catalysts. However, all these transformations are usually performed in aqueous media at higher temperature, implying the possible sintering of the catalytic active phase [5] and other deactivation processes.

Recently, our group has shown that the coverage of supported catalysts by a mesoporous silica layer could prevent the sintering of palladium nanoparticles dispersed on carbon black [6]. Nevertheless, it is well known that mesoporous silica materials such as MCM-41 or SBA-15 are poorly stable under hydrothermal conditions [7]. Although post-functionalization with hydrophobic moieties or the incorporation of aluminum atoms are well-known methods to improve the hydrothermal stability of these materials [8–12], some drawbacks are often silent up. For instance, aluminum doping in a silica matrix alters the host material structural order, leading to a more amorphous material [13,14]. In order to override the possible structural properties' modifications of siliceous mesoporous materials and increase the catalysts' hydrothermal stability, catalytic transformations of cellulose could alternatively be performed in less damaging water-miscible organic solvents. In a recent review, it has been reported that organic solvents have also a great impact on the catalysts' performances in biomass conversion [15]. Moreover, considerable efforts were made to find innovative solvents that were potentially less toxic, more biocompatible, and had improved reaction rates and product selectivities [16,17]. For instance, Mellmer et al. have shown that the rate of cellobiose hydrolysis is strongly improved by using γ-valerolactone as solvent compared to conversion in aqueous media [18]. More recently, it has been proven that by tuning the water content in dimethyl sulfoxide, as well as the temperature and the reaction time, it is possible to perform a non-catalytic conversion of cellobiose into 5-hydroxymethyl-2-furaldehyde [19]. In addition to the solvent nature in which the catalytic reaction is carried out, it has been also proven that the mesoporous silica material type, and indirectly some parameters such as the wall thickness, have a great influence on the material hydrothermal stability [7].

In this context, we propose to cover heterogeneous catalysts (ruthenium on carbon black) by protective mesoporous silica layers with different structural properties and to evaluate their performances in the one-pot conversion of cellobiose into sorbitol in water-miscible organic solvents (Figure 1). Ruthenium being the most effective metal for this reaction has been selected as the active metal in the present study [20,21]. Two surfactants have been selected (P123 and hexadecyltrimethylammonium bromide, or CTAB) for the mesoporous layers formation. These should lead to coating layers presenting mesopores of different sizes. In the first part, the positive impact of using an organic polar solvent (such as ethanol) and the influence of the mesoporous protective layer type on the catalyst's activity and stability in the selective hydrogenation of glucose into sorbitol will be highlighted. In a second part, preliminary tests in the hydrolytic hydrogenation of cellobiose into sorbitol will be carried out. Cellulose is an insoluble polymer with a complex and robust crystalline structure; therefore, simple and soluble model molecules of cellulose such as cellobiose are often used in biomass conversion studies [22–24]. Sorbitol, which is the hydrogenated form of glucose, is targeted in this work because it is a platform molecule that is used for the production of value-added chemicals [25–28]. This reaction is also a good model system to study both hydrolysis and hydrogenation reactions in one pot (Figure 1). In short, the goal of this paper is to successfully prepare a protected catalyst and then prove that the underlying active phase is still accessible and the protecting layer can remain intact in catalytic testing conditions.

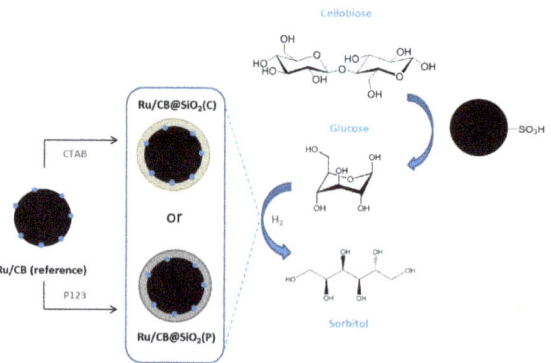

Figure 1. Overview of synthesized catalysts systems and catalytic applications tested.

2. Results and Discussion

A starting Ru/CB reference catalyst was prepared by homogeneous deposition–precipitation by using the urea method. This implies the slow precipitation of Ru hydroxide on the support followed by a reduction in N_2/H_2 atmosphere. Calcination was avoided because it is not compatible with the carbon support. Two different covered catalysts were subsequently prepared, using two different templating agents to create mesoporosity in the siliceous coating, namely Ru/CB@SiO$_2$(C) when using CTAB and Ru/CB@SiO$_2$(P) when using pluronic P123 as the templating agent. They were characterized by X-ray photoelectron spectroscopy (XPS), nitrogen physisorption, and SEM-EDX in order to identify the influence of their physicochemical characteristics on hydrothermal stability and activity in a biomass model reaction: the hydrolytic hydrogenation of cellobiose into sorbitol.

2.1. XPS

The surface atomic percentages in C1s, O1s, N1s, Si2p, and Ru3p measured by XPS for Ru/CB@SiO$_2$(C), Ru/CB@SiO$_2$(P) (covered catalysts) and Ru/CB (reference catalyst) are given in Table 1. As shown by these results, the covered materials exhibit a high level of oxygen and a lower carbon surface atomic percentage than the reference sample. Moreover, we can observe the emergence of a Si2p peak and a drastic fall of the Ru3p surface atomic percentage. All these observations confirm the silica layer deposition on a Ru/CB catalyst, regardless of the surfactant type (neutral or anionic) or the preparation conditions (acidic or basic). It can also be noted that the XPS spectra display the Ru3p$_{3/2}$ peak assigned to the metallic state (Figure S1 in the electronic supporting information). This result implies that the coverage of the catalysts surface by a silica layer does not affect the metal oxidation state. In the case of the covered sample using CTAB as a template (Ru/CB@SiO$_2$(C)), the higher level of nitrogen is imputed to residual surfactant in the smaller pores.

Table 1. X-ray photoelectron spectroscopy (XPS) analyses of reference and covered catalysts.

Sample	C1s	O1s	Si2p	N1s	Ru3p
Ru/CB	96.0	3.4	/	0.1	0.49
Ru/CB@SiO$_2$(C)	47.2	35.6	15.2	1.0	0.09
Ru/CB@SiO$_2$(P)	34.3	46.2	19.3	0.1	0.07

2.2. N_2 Physisorption

The covered samples have also been characterized by nitrogen physisorption (Figure 2) to evaluate the impact of surfactant type on the samples' textural properties. Ru/CB@SiO$_2$(C) exhibits a type IV (according to IUPAC) nitrogen adsorption–desorption isotherm with a H1 hysteris loop, which is typical of mesoporous materials with a narrow pore size distribution, as already proven in our recent paper with palladium nanoparticles as the active phase [6]. The isotherm obtained for Ru/CB@SiO$_2$(P) is also characteristic of mesoporous materials, but the H3 type hysteresis loop obtained implies a material possessing a framework with a wide pore size distribution. Moreover, the hysteresis branches' positions for this material shifted toward higher pressures. Since the capillary condensation pressure is a function of the pore diameter [29,30], the observed behavior shows that, as expected, covered material using pluronic surfactant (P123) displays larger pores than its CTAB counterpart. Indeed, the average pore diameter extracted from Barrett–Joyner–Halanda (BJH) curves is higher for Ru/CB@SiO$_2$(P) material (5 nm) than for Ru/CB@SiO$_2$(C) material (1.7 nm) (Figure S2). As expected, the BJH curve also shows a very large pore size distribution for the Ru/CB@SiO$_2$(P) catalyst. In the case of samples with narrow mesopores (such as Ru/CB@SiO$_2$(C)), it appears that the BJH method underestimates the sample pore size and that DFT (Density Functional Theory) analysis is more adapted to evaluate the pore size distribution [31]. Based on DFT analyses (Figure S2), Ru/CB@SiO$_2$(C) material contains essentially mesopores (3.4 nm) with some micropores. The total pores' volume consistently corresponds to the sum of values calculated by the Dubinin and the BJH methods in both cases (see Figure S2). In consequence, the specific surface area of the Ru/CB@SiO$_2$(P) sample is lower than that of the Ru/CB@SiO$_2$(C) material (280 m^2/g versus 390 m^2/g, respectively), which is also a direct influence of the pore diameter increase. In both cases, the Brunauer–Emmett–Teller (BET) surface area has been greatly increased compared to the uncovered catalyst (60 m^2/g).

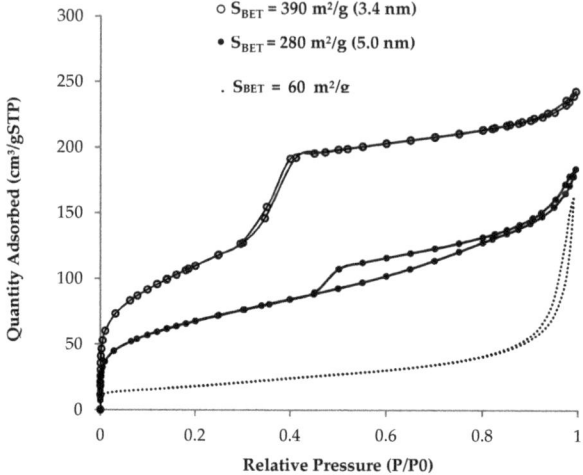

Figure 2. Nitrogen adsorption/desorption isotherms of (○) Ru/CB@SiO$_2$(C) and (•) Ru/CB@SiO$_2$(P) materials in comparison with the starting uncovered Ru/CB (dotted line) catalyst.

2.3. SEM-EDX

Then, the covered catalysts have been analyzed by SEM-EDX (Energy-Dispersive X-ray) and TEM, and compared to the Ru/CB reference sample (Figure 3). As it can be seen on SEM images, after coverage of the ruthenium-supported catalyst (Figure 3b,c), the samples' morphology has been only slightly modified, meaning that the silica layers are thin and homogeneously deposited around the Ru/CB catalysts. Moreover, EDX analyses (Figure S3) have revealed the presence of a distinguishable peak of silicon for covered samples, which is consistent with XPS analyses. The images also revealed that the silica layer is thicker in the case of the Ru/CB@SiO$_2$(C) catalyst. Based on TEM images, all the samples show narrow particle size distribution at ~1 nm, confirming that the silica coating does not affect the ruthenium particle size. TEM images of samples without Ru were also recorded (Figure S4), which confirmed the above observations.

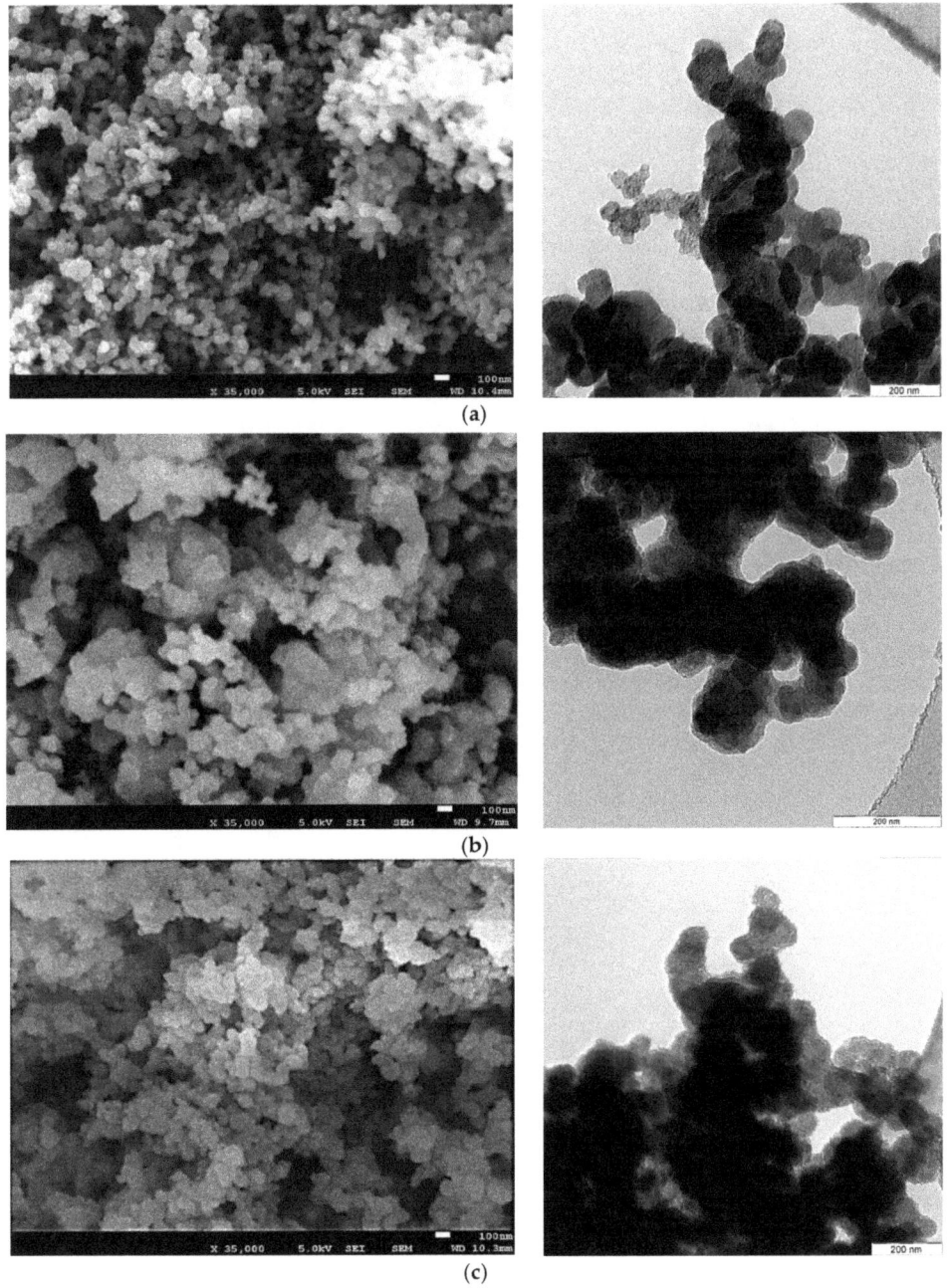

Figure 3. SEM/TEM images of (**a**) Ru/CB, (**b**) Ru/CB@SiO$_2$(C), and (**c**) Ru/CB@SiO$_2$(P).

2.4. Catalytic Tests

The catalytic performances of the synthesized catalysts were estimated in the hydrogenation of glucose into sorbitol. The glucose conversion and sorbitol selectivity are reported in Table 2. The reaction was first studied with a blank test in water (Entry 1). Although glucose is converted without catalyst (21%), the selectivity toward sorbitol is quite low (2%), leading essentially to fructose isomer. Since carbon black, which is used as support, could have a catalytic activity by itself, its catalytic performance was also evaluated (Entry 2): similar results to the blank were obtained. This test clearly shows that metallic nanoparticles are required to produce a high amount of sorbitol in this hydrogenation process. Indeed, when the reference (uncovered) Ru/CB catalyst is used (Entry 3), the selectivity toward sorbitol is higher (67%) together with a better activity (glucose conversion: 43%). This result makes sense, since it is well known that ruthenium is one of the best active phases for glucose hydrogenation (see Table S3 for a comparison of normalized activity values with catalysts from the literature) [32–35]. The covered catalyst Ru/CB@SiO$_2$(C) displays a conversion of 30% (Entry 4). According to elemental analyses (ICP) results (Table S1), this catalyst contains 1.8 wt % of ruthenium, while it was 2.3 wt % for the uncovered catalyst. Taking into account this dilution factor due to the addition of a silica layer (for the same engaged catalyst mass), we can calculate the activity as percentage conversion per mg of Ru engaged. The value obtained for Ru/CB@SiO$_2$(C) is very close to the reference catalyst (41%/mg and 47%/mg respectively, compare Entries 4 and 3). This means that the silica layer does not prevent the diffusion of glucose to reach the ruthenium nanoparticles that are the active phase for this transformation, even in the presence of some residual surfactant in the porous network. Furthermore, as shown by XPS measurements (Table S2), this residual surfactant has been almost completely removed during the catalytic test. Nevertheless, the selectivity toward sorbitol is surprisingly low (4%), indicating that the silica protective layer could have an impact on the reaction selectivity. Particles size and support effects can be discarded. Indeed, the ruthenium deposition and the carbonaceous support are the same as the reference catalyst, and the coverage methodology does not affect the particles' size, as already demonstrated in [6] and in TEM images. We believe that because the pores' sizes are below the limit for easy diffusion, the sugar molecules enter the pores with steric constraints and probably do not have a complete rotational degree of freedom (the concept of shape selectivity known for zeolites), with a strong effect on the selectivity. Indeed, based on the seminal paper by Beck and Schultz [36], the restrictions of glucose diffusion through the silica layers can be calculated, considering 3.4 nm and 5.0 nm as the pore diameters of our materials and 0.88 nm for the molecular diameter of glucose. This reveals that the mobility of glucose through the mesoporous channels of Ru/CB@SiO$_2$ catalysts is respectively 46% and 30% (for 3.4 nm and 5.0 nm pores) lower than its mobility in free solution.

Table 2. Conversion and selectivity for the hydrogenation of glucose after 2 h reaction at 150 °C.

Entry	Sample	Solvent	Glucose Conversion (%)	Activity [1] (%/mg)	Sorbitol Selectivity (%)
1	Blank	H$_2$O	21	/	2
2	CB	H$_2$O	16	/	7
3	Ru/CB	H$_2$O	43	47	67
4	Ru/CB@SiO$_2$(C)	H$_2$O	30	41	4
5 [2]	Ru/CB@SiO$_2$(C)-HT	H$_2$O	41	45	68
6	Blank	EtOH/H$_2$O	26	/	1
7	Ru/CB	EtOH/H$_2$O	92	100	74
8	Ru/CB@SiO$_2$(C)	EtOH/H$_2$O	69	94	4
9	Ru/CB@SiO$_2$(P)	EtOH/H$_2$O	43	82	58
10 [3]	Ru/CB@SiO$_2$(P)	EtOH/H$_2$O	50	81	66

[1] Normalized activity = glucose conversion per mg of ruthenium engaged. [2] Treated in hot water for 24 h. [3] Second run.

Moreover, after the catalytic test, XPS analyses have revealed that 99% of Si has been lost by the known desilication process in water (Table S2) [37]. As mentioned in the introduction, it is well established in the literature that silica materials are poorly stable in hydrothermal conditions [7]. Desilication is a slow phenomenon leading to a progressive loss of the silica layer in solution. So, during the whole catalytic test, the Ru/CB@SiO$_2$(C) catalyst is constantly covered by a silica layer that only decreases in thickness. Hence, the accessibility remains, and the selectivity is controlled by shape selectivity at all times, very differently from the uncovered catalyst. In order to confirm this hypothesis, we have treated the Ru/CB@SiO$_2$(C) in hot water for 24 h (to completely remove the silica layer). This "treated" catalyst has been then re-evaluated in the glucose hydrogenation reaction (Entry 5). Both its conversion and selectivity toward glucose are the same as that of the reference (uncovered) catalyst (Ru/CB), which corroborates our hypothesis.

Consequently, the next tests were carried out in an ethanol–water mixture (110:10, v/v) to minimize the silica losses. A new blank test has been performed in those conditions that shows relatively low activity and selectivity as in pure water (Entry 6). Astonishingly, in the case of the reference (uncovered) catalyst (Entry 7), the conversion has drastically increased (92%), while the selectivity is quite stable (74%). In the same way, the Ru/CB@SiO$_2$(C) covered sample shows higher conversion than in pure water (69%), but still a very low selectivity (4%) (Entry 8). Therefore, it appears that the solvent nature has an effective impact on the catalyst activity for this reaction, which was never reported in the literature, to the best of our knowledge. As expected, the stability of our catalysts has been also improved in an ethanol–water mixture with a much lower silicon loss (32% loss of the initial Si content according to XPS measurement compared to 99% loss in pure water, see Table S2). Moreover, the water-miscible organic solvent allows totally removing the residual surfactant. An additional TGA (Thermogravimetric Analysis) measurement of Ru/CB@SiO$_2$(C) after the catalytic test in ethanol/water media was carried out (see Figure S5). In order to clearly identify the temperature at which the CTAB is decomposed, a reference sample consisting of CTAB deposited on carbon black (CB) has been also analyzed. As can be noticed, CTAB is decomposed at ~250 °C under air. This mass loss does not appear in the case of Ru/CB@SiO$_2$(C) catalyst. These results clearly prove that any remaining CTAB surfactant is also totally removed during the catalytic test in organic media and does not prevent glucose from reaching the underlying ruthenium nanoparticles.

In the case of Ru/CB@SiO$_2$(P) material (Entry 9), the selectivity toward the desired product is much better (58%), which can be attributed to the larger pore diameter. As discussed above, the shape selectivity effect will be stronger for the smaller pores' layer. However, the observed difference in glucose conversion, in comparison with the reference catalyst (compare Entry 7 with 9), cannot be totally explained by the silica dilution factor as above (normalized activity calculated according to ruthenium mass engaged is 82%/mg in this case). Therefore, it would appear that some of the ruthenium nanoparticles are completely covered by the silica and consequently unreachable. In addition, the large pore distribution of this catalyst also implies the presence of micropores (see Figure S2), which could prevent the access to some ruthenium nanoparticles and explain the slight decrease in normalized activity. Nevertheless, the loss of Si after the catalytic test decreased even further (18% according to XPS measurement, see Table S2). This improvement in stability is not surprising. Indeed, Zhao et al. have already shown that mesoporous materials such as SBA-15, with larger pores, were more stable than MCM-41 under hydrothermal conditions [38]. More recently, Pollock et al. found that SBA-15 treated in liquid water at 155 °C lost its secondary pore network by a complete closure [39]. In our case, when Ru/CB@SiO$_2$(P) catalyst was reused for a second run (Entry 10), the conversion and selectivity are quite similar (if not better) than in the first run. The slight increase could be ascribed to relatively higher Ru loading due to slight Si loss (18% measured by XPS, as discussed above). In comparison, the uncovered Ru/CB catalyst, upon recycling, loses 16% activity (glucose conversion), but the selectivity is maintained. Moreover, the covered catalyst still exhibits a high specific surface area (227 m^2/g), the same isotherm curve shape, and similar pore size distribution compared with the fresh catalyst (Figure S6). These results imply that our material (more specifically the silica layer) is stable under these conditions. They also prove that the underlying ruthenium nanoparticles are still accessible and consequently that the pore network is still open.

The best-performing covered catalyst, namely Ru/CB@SiO$_2$(P), was finally tested in a model reaction of cellulose valorization: the catalytic transformation of cellobiose into sorbitol in one pot. Kinetic modeling has shown that hydrolysis is the rate-limiting step for the hydrolytic hydrogenation of cellobiose (or cellulose) to sorbitol (110–115 KJ/mol for the hydrolysis of cellobiose) [19,40]. Therefore, this step is a highly demanding reaction that usually requires strong acid sites irrespective of the reaction pathway to produce sorbitol, as shown in Figure 4.

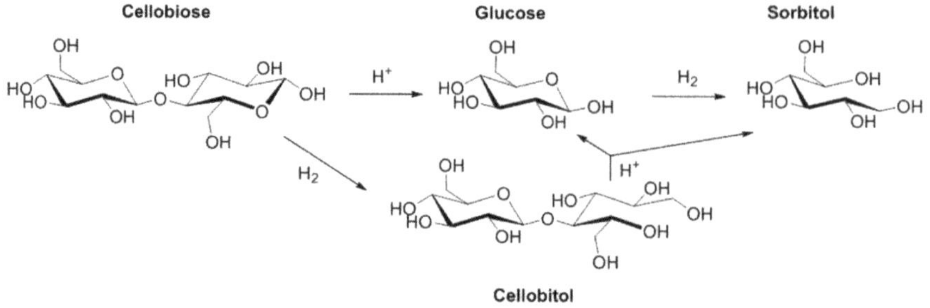

Figure 4. Reaction pathways for cellobiose conversion to sorbitol.

Nevertheless, as shown in Table 3, when the hydrolysis of cellobiose was performed without catalyst in pure water (Entry 1), glucose was obtained. Therefore, the hydrolytic hydrogenation of cellobiose in organic medium in the presence of our covered catalysts was performed with small fractions of water (Entry 2–4) to ensure the production of glucose without damaging the silica layer. Unfortunately, no glucose or sorbitol were obtained after 2 h of reaction with Ru/CB@SiO$_2$(P) catalyst (Entry 2), meaning that the small volume fraction of water (X_w = 8%) introduced is not enough to allow the hydrolysis reaction. By increasing this fraction to X_w = 33% (Entry 3) and X_w = 50% (Entry 4), the results remained unchanged. However, cellobiose is in the meantime mainly converted into cellobitol (up to 42%). Usually, it is known that the hydrolytic hydrogenation of cellobiose (or cellulose) to sorbitol occurs via the hydrolysis of glycosidic bonds followed by the hydrogenation of glucose into sorbitol. Nevertheless, as proven by Palkovits et al. [40], the cellobiose conversion could also proceed through an alternative pathway: the hydrogenation of a C–O bond on one of the glucose rings, leading to cellobitol and consecutive hydrolysis to sorbitol and glucose (Figure 4). This result proves again that our material, thanks to its large pores, allows the diffusion of larger biomolecules such as cellobiose through the silica layer to reach the underlying metallic nanoparticles that are the active phase for this transformation.

Table 3. Conversion and selectivity for the hydrogenation and hydrolysis of cellobiose at 150 °C for 2 h under 30 bars of pure hydrogen.

Entry	Sample	H$_2$O/EtOH (v/v)	Cellobiose Conversion (%)	Glucose Selectivity (%)	Ethyl Glucopyr. Selectivity (%)	Cellobitol Selectivity (%)	Sorbitol Selectivity (%)
1	blank	120/0	44	14	0	0	0
2	Ru/CB@SiO$_2$(P)	10/110	43	0	0	30	0
3	Ru/CB@SiO$_2$(P)	40/80	38	0	0	27	0
4	Ru/CB@SiO$_2$(P)	60/60	45	0	0	42	0
5	Ru/CB@SiO$_2$(P) + CB–SO$_3$H	10/110	69	13	19	6	2
6 [1]	Ru/CB@SiO$_2$(P) + CB–SO$_3$H	10/110	94	20	32	0	8

[1] Test carried out with 200 mg of Ru/CB@SiO$_2$(P).

Finally, an acidic material consisting of sulfonic acid moieties grafted on carbon black (CB-SO$_3$H) [41] was added (Entry 5) in the catalytic medium. Sulfonic acid functions are often used for cellobiose/cellulose hydrolysis reactions [42–45]. As can been seen, after 2 h, glucose is produced. This result confirms that acidic sites are essential to hydrolyse cellobiose in organic medium. Another compound has been also detected, which could be attributed to ethyl glucopyranoside. Experimental and theoretical studies have shown that this cellulose hydrolysis mechanism involves the glycosidic oxygen protonation (Figure 5) [46]. In this mechanism, a cyclic oxonium ion is formed and reacts with water to reestablish the anomeric center and regenerate H_3O^+ species. Based on this mechanism, a solvent such as ethanol could easily react with the second intermediate to produce ethyl glucopyranoside (Figure 5). More importantly, we have also observed in this test (Entry 5) the production of cellobitol and sorbitol. This means that the ruthenium nanoparticles of the covered catalyst are still active and accessible. Recently, Soisangwan et al. have shown that the conversion of cellobiose raised with the increasing ethanol concentration in subcritical water (liquid water between 100 and 374 °C under high pressure [usually from 1 to 6 MPa]) [47]. However, they also demonstrated that ethanol promotes the isomerization of disaccharides at the expense of a hydrolysis reaction [48]. In our case, isomerization is limited. Indeed, only a small amount of isomers such as fructose or cellobiulose are observed, demonstrating the good performances of our combined catalysts. Other (undetected) by-products such as 5-(hydroxymethyl)furfural (HMF) are formed as expected in this case. By increasing the amount of covered catalyst engaged (Entry 6), the conversion and the sorbitol selectivity are further improved. This also impacts positively the hydrolysis step, as expected. All these preliminary results in the hydrolytic hydrogenation of cellobiose show the high potential of the as-prepared protected catalyst in biomass valorization.

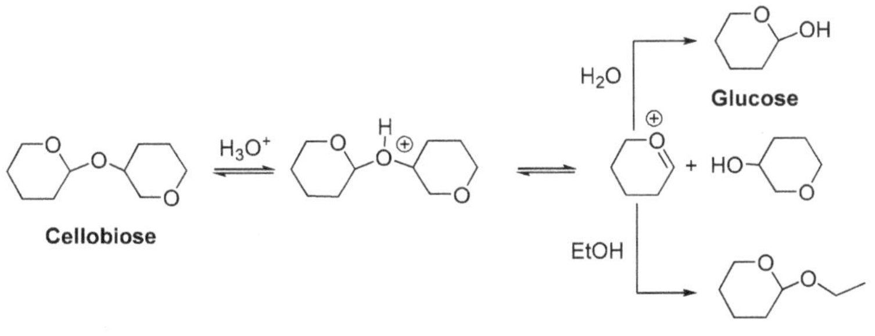

Figure 5. Mechanism of acid hydrolysis of glycosidic bonds in water or organic medium (hydroxyl groups have been voluntary omitted for more clarity).

3. Materials and Methods

3.1. Reagents

The carbon black support (CB) was received as 250G type from IMERYS GRAPHITE & CARBON. Hexadecyltrimethylammonium bromide (CTAB, 95%), (3-aminopropyl) trimethoxysilane (APTES, 99%), tetraethyl orthosilicate (TEOS, >98%), thionyl chloride ($SOCl_2$, >99%), Ruthenium(III) chloride hydrate, pluronic P123, sulfanilic acid (99%), and isopentyl nitrite (96%) were purchased from Sigma Aldrich and used as received. A commercial ultrasonic cleaner (VWR) was used for sonication.

3.2. Syntheses

3.2.1. Synthesis of Ru/CB Catalyst (Reference)

A non-covered Ru/CB catalyst used as a reference has been synthesized by an urea assisted deposition–precipitation procedure [49]. First, 600 mg of urea were dissolved in 300 mL of distilled water. Then, 1 g of CB was added, and the mixture was stirred for 15 min. Then, 73.5 mg of $RuCl_3$ (3.5 wt % Ru) were added. After stirring for 1 h, the mixture was heated to 120 °C for 1 h, cooled down to room temperature, and stirred overnight. Next, the solid was filtered out, washed with 250 mL of distilled water, and dried at 100 °C overnight. Finally, the catalyst was submitted to a thermal treatment under reducing atmosphere in a tubular oven STF 16/450 from CARBOLITE. The sample was placed into porcelain combustion boats and heated during 2 h at 600 °C (heating ramp and cooling ramp: 100 °C/h) under a stream of N_2/H_2 (95:5).

3.2.2. Synthesis of Ru/CB@SiO$_2$(C)

The synthesis of nanoparticles supported on carbon black (CB) covered by a thin mesoporous silica layer has been previously described for palladium catalysts [6]. In this study, it has been adapted for ruthenium as follows. First, 2 g of carbon black (CB) were introduced in a 250 mL round-bottom flask containing 100 mL of toluene. Then, 6 mL of SOCl$_2$ were added, and the mixture was heated for 5 h under reflux. Then, it was filtered out and extensively washed with toluene (500 mL). The resulting material (CB–Cl) was dried overnight under vacuum at 100 °C. Then, 1 g of CB–Cl was introduced in a 250 mL round-bottom flask containing 100 mL of dichloromethane. Then, 1 mL of APTES was added, and the mixture was stirred for 24 h at room temperature. Finally, the material (CB-APTES) was filtered out, washed with dichloromethane (250 mL) and methanol (250 mL), and dried overnight under vacuum at 100 °C. Then, ruthenium nanoparticles were deposited on CB-APTES (1 g) by a urea-assisted procedure, using the same synthesis conditions as described above, by engaging 73.5 mg of RuCl$_3$ (3.5 wt % Ru). After a thermal treatment for 2 h at 600 °C under reducing atmosphere (N$_2$/H$_2$, 95:5), 0.250 g of Ru/CB-APTES was introduced in a 100 mL round-bottom flask containing 25 mL of distilled water. Then, 10 mL of NaOH (0.1 M) were added, and the mixture was sonicated for 10 min. To this suspension, 0.571 g of CTAB was added, and the solution was heated at 60 °C. Then, 0.7 mL of TEOS was added dropwise within 30 min. This suspension was further stirred for 3 h 30 min, then charged into a propylene bottle, which was closed tightly and heated at 100 °C for 3 days. The product was filtered out, washed with ethanol (250 mL), and dried at 100 °C overnight. The CTAB template was removed by refluxing the solid material (Ru/CB@SiO$_2$) for 24 h in ethanol.

3.2.3. Synthesis of Ru/CB@SiO$_2$(P)

The synthesis of Ru/CB@SiO$_2$(P) was achieved following the procedure described above to prepare Ru/CB@SiO$_2$(C) with two modifications. First, during the sol–gel process with TEOS as a precursor, 35 mL of HCl 0.3 M was used instead of a NaOH (0.1 M)/water mixture. Second, 120 mg of pluronic P123 were added instead of 571 mg CTAB.

3.2.4. Synthesis of CB–SO$_3$H

The supports functionalization was carried out by a diazonium coupling method [41]. Typically, 1 g of carbon black was dispersed in 60 mL of distilled water. Then, 1.5 g of sulfanilic acid were added, and the suspension was stirred at 70 °C during 10 min. Afterwards, 1.2 mL of isopentyl nitrite were added at 30 °C, and the mixture was stirred during 16 h. Then, it was filtered out and washed with distilled water and ethanol. The resulting material, CB–SO$_3$H, was dried overnight at 100 °C.

3.3. Characterizations

The solid materials were characterized by X-ray photoelectron spectroscopy (XPS), scanning electron microscopy (SEM), transmission electron microscopy (TEM), elemental analyses (ICP), and N$_2$ physisorption.

XPS analyses were carried out at room temperature with an SSI-X-probe (SSX 100/206) photoelectron spectrometer from Surface Science Instruments (USA), equipped with a monochromatized microfocus Al X-ray source. Samples were stuck onto small sample holders with double-face adhesive tape and then placed on an insulating ceramic carousel (Macor®, Switzerland). Charge effects were avoided by placing a nickel grid above the samples and using a flood gun set at 8 eV. The binding energies were calculated with respect to the C-(C, H) component of the C1s peak fixed at 284.8 eV. Data treatment was performed using the CasaXPS program (Casa Software Ltd., UK). The peaks were decomposed into a sum of Gaussian/Lorentzian (85/15) after the subtraction of a Shirley-type baseline.

SEM images were obtained on a Field Emission Scanning Electron Microscope JEOL JSM–7600 F, equipped with an energy dispersive X-ray system. The powder samples were pressed onto double-face adhesive carbon tape adhered to an aluminum sample holder. Images were acquired at different acceleration voltages ranging between 3 and 15 keV, using an InLens detector. TEM images were obtained on an LEO 922 Omega Energy Filter Transmission Electron Microscope operating at 120 kV. The samples were suspended in hexane under ultrasonic treatment. A drop of the suspension was deposited on a holey carbon film supported on a copper grid (Holey Carbon Film 300 Mesh Cu, Electron Microscopy Sciences), which was dried overnight under vacuum at room temperature, before introduction in the microscope.

The elemental analyses (C, H, N, Si, and Ru) were carried out by MEDAC Ltd., UK by microgravimetry for C, H, N (direct measure) and by ICP after acid digestion for Si and Ru.

The pore texture of the covered catalysts was characterized by nitrogen adsorption–desorption isotherms. The measures were achieved at 77 K by using a Micromeritics ASAP 2020 analyzer. Before analysis, the samples (0.02–0.10 g) were degassed for 2 h at 200 °C with a heating rate of 10 °C/min under 0.133 Pa pressure. The analysis of the isotherms provided specific surface areas calculated with the Brunauer–Emmett–Teller (BET) equation, S_{BET}. The pore volume, V_p, of the samples and the pores' average diameter were calculated using the Barrett–Joyner–Halanda (BJH) and DFT models.

3.4. Catalytic Tests

The tests were carried out in a 250 mL stainless steel Parr autoclave. First, 1 g of cellobiose (or glucose) was added to 40 mg of catalyst in 120 mL of mQ (milliQ) water (or a mixture of ethanol/mQ water). Then, the autoclave was sealed, and the system was purged three times with nitrogen and once with pure hydrogen. Once the desired temperature has been reached (150 °C), 30 bars of hydrogen were introduced, and the mixture was stirred at 1700 rpm for 2 h. Then, the system was cooled down to room temperature, and the solution was filtrated. Then, the filtrate was diluted to 250 mL with mQ water and analyzed by HPLC.

HPLC analyses were performed with a Waters system equipped with a Waters 2414 refractive index (RI) detector (detector temperature = 30 °C). The column used is an Aminex HPX 87 C column, with mQ H_2O (18 MΩ.cm at 25 °C) as the eluent, a flux of 0.5 mL/min, a column temperature of 85 °C, and 25 µL of injected volume.

4. Conclusions

In the present work, catalysts covered by protective silica layers with very different morphologies and textural properties have been prepared. These catalysts have been fully characterized by XPS, SEM, ICP, and N_2 physisorption. The presence of the layer was confirmed by XPS and SEM/TEM. Two different templates were used, and both gave mesoporous silica but with different pore sizes: 3.4 nm when using CTAB and 5 nm when using pluronic P123. Moreover, the silica layer was thicker in the former case (Ru/CB@SiO_2(C) catalyst).

The covered catalysts were successfully tested in the hydrogenation of glucose into sorbitol. In comparison with uncovered catalyst, it was demonstrated that the diffusion of reactants/products through the mesoporous layer was possible to reach the underlying active phase. Moreover, the stability of protective silica layers was improved in organic medium at high temperature as confirmed by XPS. The solvent also had a strong influence on the catalysts performance: the conversion being doubled when moving from water to ethanol as solvent. Finally, the selectivity of the studied reaction was influenced by the silica layers' structure: the larger the pores, the higher the selectivity.

Preliminary tests in the hydrolytic hydrogenation of cellobiose were undertaken as well. These revealed that in the absence of acid sites, cellobiose is mainly converted into cellobitol (disaccharide hydrogenation product). This underlines the accessibility of ruthenium nanoparticles despite the protective silica layer and the need for strong acid sites to perform the hydrolysis in organic polar medium. On the contrary, when acidic material was added (heterogeneous acid catalyst consisting of sulfonic acid groups grafted on carbon black), sorbitol was formed, corresponding to the subsequent hydrogenation and hydrolysis steps from cellobiose. Therefore, the protected catalyst is still active in this combination one pot system. All these results prove that by using the present methodology, protected catalysts with good performances in model reactions for biomass valorization could be easily prepared, which opens the door to many studies with multifunctional catalytic systems.

Supplementary Materials: The following are available online at http://www.mdpi.com/2073-4344/10/2/149/s1, Figure S1: XPS Ru3p spectra for (**a**) Ru/CB, (**b**) Ru/CB@SiO$_2$(C), and (**c**) Ru/CB@SiO$_2$(P) catalysts. Figure S2: Pores size distribution obtained by DFT and BJH methods for (**a**) Ru/CB@SiO$_2$(C) and (**b**) Ru/CB@SiO$_2$(P) materials. Figure S3: SEM-EDX analyses of (**a**) Ru/CB, (**b**) Ru/CB@SiO$_2$(C), and (**c**) Ru/CB@SiO$_2$(P) catalysts. Figure S4: TEM images of (**a**) CB@SiO$_2$(C) and (**b**) CB@SiO$_2$(P) covered materials. Table S1: Elemental analysis by ICP of Ru/CB@SiO$_2$(C) and Ru/CB@SiO$_2$(P) catalysts (wt %). Table S2: XPS analyses (at %) of covered catalysts before and after catalytic tests in pure water (W) or ethanol–water mixture (W/E). Table S3: Comparison of normalized activity values with catalysts from literature. Figure S5: (**a**) Nitrogen adsorption–desorption isotherms at 77 K and pores size distributions from (**b**) BJH and (**c**) DFT for Ru/CB@SiO$_2$(P) catalyst after a catalytic test.

Author Contributions: Conceptualization and methodology—T.H., V.D. and S.H. (Sophie Hermans); experiment design, acquisition of data and analyses—T.H., S.H. (Sharon Hubert) and S.C.; writing—original draft preparation—T.H.; writing—review and editing—T.H., V.D. and S.H. (Sophie Hermans). All authors have read and agreed to the published version of the manuscript.

Funding: This research was funded by FRS-FNRS and UCLouvain.

Acknowledgments: We are grateful to Jean-François Statsijns for technical assistance and Matthieu Da Costa (Ghent University) for HPLC measurements. We also thank the IMERYS GRAPHITE & CARBON (Switzerland) firm for generous donations of carbon black.

Conflicts of Interest: The authors declare no conflict of interest.

References

1. Brethauer, S.; Studer, M.H. Biochemical conversion processes of lignocellulosic biomass to fuels and chemicals—A review. *Chimia (Aarau)* **2015**, *69*, 572–581. [CrossRef] [PubMed]
2. Zhou, C.; Xia, X.; Lin, C.; Tong, D.; Beltramini, J. Catalytic conversion of lignocellulosic biomass to fine chemicals and fuels. *Chem. Soc. Rev.* **2011**, *40*, 5588–5617. [CrossRef] [PubMed]
3. Climent, M.J.; Corma, A.; Iborra, S. Conversion of biomass platform molecules into fuel additives and liquid hydrocarbon fuels. *Green Chem.* **2014**, *16*, 516–547. [CrossRef]
4. Zhang, Y.-H.P.; Lynd, L.R. Toward an aggregated understanding of enzymatic hydrolysis of cellulose: Noncomplexed cellulase systems. *Biotechnol. Bioeng.* **2004**, *88*, 797–824. [CrossRef] [PubMed]
5. Wang, Y.; Rong, Z.; Wang, Y.; Wang, T.; Du, Q.; Wang, Y.; Qu, J. Graphene-based metal/acid bifunctional catalyst for the conversion of levulinic acid to γ-valerolactone. *ACS Sustain. Chem. Eng.* **2017**, *5*, 1538–1548. [CrossRef]
6. Haynes, T.; Ersen, O.; Dubois, V.; Desmecht, D.; Nakagawa, K.; Hermans, S. Protecting a Pd/CB catalyst by a mesoporous silica layer. *Appl. Catal. B Environ.* **2019**, *241*, 196–204. [CrossRef]
7. Xiong, H.; Pham, H.N.; Datye, A.K. Hydrothermally stable heterogeneous catalysts for conversion of biorenewables. *Green Chem.* **2014**, *16*, 4627–4643. [CrossRef]
8. Yang, H.; Zhang, G.; Hong, X.; Zhu, Y. Silylation of mesoporous silica MCM-41 with the mixture of Cl(CH2)3SiCl3 and CH3SiCl3: Combination of adjustable grafting density and improved hydrothermal stability. *Microporous Mesoporous Mater.* **2004**, *68*, 119–125. [CrossRef]
9. Castricum, H.L.; Mittelmeijer-Hazeleger, M.C.; Sah, A.; ten Elshof, J.E. Increasing the hydrothermal stability of mesoporous SiO2 with methylchlorosilanes—A "structural" study. *Microporous Mesoporous Mater.* **2006**, *88*, 63–71. [CrossRef]
10. Karpov, S.I.; Roessner, F.; Selemenev, V.F.; Belanova, N.A.; Krizhanovskaya, O.O. Structure, hydrophobicity, and hydrothermostability of MCM-41 organo-inorganic mesoporous silicates silylated with dimethoxydimethylsilane and dichloromethylphenylsilane. *Russ. J. Phys. Chem. A* **2013**, *87*, 1888–1894. [CrossRef]
11. Ribeiro Carrott, M.M.L.; Conceição, F.L.; Lopes, J.; Carrott, P.J.; Bernardes, C.; Rocha, J.; Ramôa Ribeiro, F. Comparative study of Al-MCM materials prepared at room temperature with different aluminium sources and by some hydrothermal methods. *Microporous Mesoporous Mater.* **2006**, *92*, 270–285. [CrossRef]
12. Russo, P.A.; Carrott, M.M.L.R.; Carrott, P.J.M. Effect of hydrothermal treatment on the structure, stability and acidity of Al containing MCM-41 and MCM-48 synthesised at room temperature. *Colloids Surf. A Physicochem. Eng. Asp.* **2007**, *310*, 9–19. [CrossRef]

13. Li, Y.; Yang, Q.; Yang, J.; Li, C. Mesoporous aluminosilicates synthesized with single molecular precursor (sec-BuO)$_2$AlOSi(OEt)$_3$ as aluminum source. *Microporous Mesoporous Mater.* **2006**, *91*, 85–91. [CrossRef]
14. Haynes, T.; D'hondt, T.; Morritt, A.L.; Khimyak, Y.Z.; Desmecht, D.; Dubois, V.; Hermans, S. Mesoporous aluminosilicate nanofibers with a low Si/Al ratio as acidic catalyst for hydrodeoxygenation of phenol. *ChemCatChem* **2019**, *11*, 4054–4063. [CrossRef]
15. Shuai, L.; Luterbacher, J. Organic solvent effects in biomass conversion reactions. *ChemSusChem* **2016**, *9*, 133–155. [CrossRef]
16. Gallo, J.M.R.; Alonso, D.M.; Mellmer, M.A.; Dumesic, J.A. Production and upgrading of 5-hydroxymethylfurfural using heterogeneous catalysts and biomass-derived solvents. *Green Chem.* **2013**, *15*, 85–90. [CrossRef]
17. Gu, Y.; Jérôme, F. Bio-based solvents: An emerging generation of fluids for the design of eco-efficient processes in catalysis and organic chemistry. *Chem. Soc. Rev.* **2013**, *42*, 9550–9570. [CrossRef]
18. Mellmer, M.A.; Martin Alonso, D.; Luterbacher, J.S.; Gallo, J.M.R.; Dumesic, J.A. Effects of γ-valerolactone in hydrolysis of lignocellulosic biomass to monosaccharides. *Green Chem.* **2014**, *16*, 4659–4662. [CrossRef]
19. Kimura, H.; Yoshida, K.; Uosaki, Y.; Nakahara, M. Effect of water content on conversion of D-cellobiose into 5-hydroxymethyl-2-furaldehyde in a dimethyl sulfoxide–water mixture. *J. Phys. Chem. A* **2013**, *117*, 10987–10996. [CrossRef]
20. Deng, W.; Tan, X.; Fang, W.; Zhang, Q.; Wang, Y. Conversion of cellulose into sorbitol over carbon nanotube-supported ruthenium catalyst. *Catal. Lett.* **2009**, *133*, 167–174. [CrossRef]
21. Adsuar-García, M.; Flores-Lasluisa, J.; Azar, F.; Román-Martínez, M. Carbon-black-supported Ru catalysts for the valorization of cellulose through hydrolytic hydrogenation. *Catalysts* **2018**, *8*, 572. [CrossRef]
22. Zhou, L.; Liu, Z.; Bai, Y.; Lu, T.; Yang, X.; Xu, J. Hydrolysis of cellobiose catalyzed by zeolites—The role of acidity and micropore structure. *J. Energy Chem.* **2016**, *25*, 141–145. [CrossRef]
23. Peña, L.; Ikenberry, M.; Ware, B.; Hohn, K.L.; Boyle, D.; Sun, X.S.; Wang, D. Cellobiose hydrolysis using acid-functionalized nanoparticles. *Biotechnol. Bioprocess Eng.* **2011**, *16*, 1214–1222. [CrossRef]
24. Bootsma, J.A.; Shanks, B.H. Cellobiose hydrolysis using organic–inorganic hybrid mesoporous silica catalysts. *Appl. Catal. A Gen.* **2007**, *327*, 44–51. [CrossRef]
25. Cortright, R.D.; Davda, R.R.; Dumesic, J.A. Hydrogen from catalytic reforming of biomass-derived hydrocarbons in liquid water. *Nature* **2002**, *418*, 964–967. [CrossRef]
26. Huber, G.W.; Chheda, J.N.; Barrett, C.J.; Dumesic, J.A. Production of liquid alkanes by aqueous-phase processing of biomass-derived carbohydrates. *Science* **2005**, *308*, 1446–1450. [CrossRef]
27. Kamm, B. Production of platform chemicals and synthesis gas from biomass. *Angew. Chem. Int. Ed.* **2007**, *46*, 5056–5058. [CrossRef]
28. Isikgor, F.H.; Becer, C.R. Lignocellulosic biomass: A sustainable platform for the production of bio-based chemicals and polymers. *Polym. Chem.* **2015**, *6*, 4497–4559. [CrossRef]
29. Kruk, M.; Jaroniec, M.; Ko, C.H.; Ryoo, R. Characterization of the porous structure of SBA-15. *Chem. Mater.* **2000**, *12*, 1961–1968. [CrossRef]
30. Kruk, M.; Jaroniec, M.; Sayari, A. Application of large pore MCM-41 molecular sieves to improve pore size analysis using nitrogen adsorption measurements. *Langmuir* **1997**, *13*, 6267–6273. [CrossRef]
31. Quantachrome Instruments. Pore Size Analysis by Gas Adsorption and the Density Functional Theory. 2018. AZoM, viewed 27 August 2019. Available online: https://www.azom.com/article.aspx?ArticleID=5189 (accessed on 27 August 2019).
32. Kusserow, B.; Schimpf, S.; Claus, P. Hydrogenation of glucose to sorbitol over nickel and ruthenium catalysts. *Adv. Synth. Catal.* **2003**, *345*, 289–299. [CrossRef]
33. Tronci, S.; Pittau, B. Conversion of glucose and sorbitol in the presence of Ru/C and Pt/C catalysts. *RSC Adv.* **2015**, *5*, 23086–23093. [CrossRef]
34. Zhang, J.; Lin, L.; Zhang, J.; Shi, J. Efficient conversion of D-glucose into D-sorbitol over MCM-41 supported Ru catalyst prepared by a formaldehyde reduction process. *Carbohydr. Res.* **2011**, *346*, 1327–1332. [CrossRef] [PubMed]
35. Romero, A.; Nieto-Márquez, A.; Alonso, E. Bimetallic Ru:Ni/MCM-48 catalysts for the effective hydrogenation of D-glucose into sorbitol. *Appl. Catal. A Gen.* **2017**, *529*, 49–59. [CrossRef]
36. Beck, R.E.; Schultz, J.S. Hindered diffusion in microporous membranes with known pore geometry. *Science* **1970**, *170*, 1302–1305. [CrossRef]

37. Ravenelle, R.M.; Schüβler, F.; D'Amico, A.; Danilina, N.; van Bokhoven, J.A.; Lercher, J.A.; Jones, C.W.; Sievers, C. Stability of zeolites in hot liquid water. *J. Phys. Chem. C* **2010**, *114*, 19582–19595. [CrossRef]
38. Zhao, D. Triblock copolymer syntheses of mesoporous silica with periodic 50 to 300 angstrom pores. *Science* **1998**, *279*, 548–552. [CrossRef]
39. Pollock, R.A.; Gor, G.Y.; Walsh, B.R.; Fry, J.; Ghampson, I.T.; Melnichenko, Y.B.; Kaiser, H.; DeSisto, W.J.; Wheeler, M.C.; Frederick, B.G. Role of liquid vs. vapor water in the hydrothermal degradation of SBA-15. *J. Phys. Chem. C* **2012**, *116*, 22802–22814. [CrossRef]
40. Negahdar, L.; Oltmanns, J.U.; Palkovits, S.; Palkovits, R. Kinetic investigation of the catalytic conversion of cellobiose to sorbitol. *Appl. Catal. B Environ.* **2014**, *147*, 677–683. [CrossRef]
41. Carlier, S.; Hermans, S. Highly efficient and recyclable catalysts for cellobiose hydrolysis: Systematic comparison of carbon nanomaterials functionalized with benzyl sulfonic acids. *Front. Chem.* **2020**. submitted.
42. Zhou, L.; Liu, Z.; Shi, M.; Du, S.; Su, Y.; Yang, X.; Xu, J. Sulfonated hierarchical H-USY zeolite for efficient hydrolysis of hemicellulose/cellulose. *Carbohydr. Polym.* **2013**, *98*, 146–151. [CrossRef] [PubMed]
43. Liu, Y.; Xiao, W.; Xia, S.; Ma, P. SO_3H-functionalized acidic ionic liquids as catalysts for the hydrolysis of cellulose. *Carbohydr. Polym.* **2013**, *92*, 218–222. [CrossRef] [PubMed]
44. Hu, L.; Li, Z.; Wu, Z.; Lin, L.; Zhou, S. Catalytic hydrolysis of microcrystalline and rice straw-derived cellulose over a chlorine-doped magnetic carbonaceous solid acid. *Ind. Crop. Prod.* **2016**, *84*, 408–417. [CrossRef]
45. Suganuma, S.; Nakajima, K.; Kitano, M.; Yamaguchi, D.; Kato, H.; Hayashi, S.; Hara, M. Hydrolysis of cellulose by amorphous carbon bearing SO_3H, COOH, and OH groups. *J. Am. Chem. Soc.* **2008**, *130*, 12787–12793. [CrossRef]
46. Rinaldi, R.; Schüth, F. Acid hydrolysis of cellulose as the entry point into biorefinery schemes. *ChemSusChem* **2009**, *2*, 1096–1107. [CrossRef]
47. Soisangwan, N.; Gao, D.-M.; Kobayashi, T.; Khuwijitjaru, P.; Adachi, S. Kinetic analysis for the isomerization of cellobiose to cellobiulose in subcritical aqueous ethanol. *Carbohydr. Res.* **2016**, *433*, 67–72. [CrossRef]
48. Gao, D.M.; Kobayashi, T.; Adachi, S. Production of keto-disaccharides from aldo-disaccharides in subcritical aqueous ethanol. *Biosci. Biotechnol. Biochem.* **2016**, *80*, 998–1005. [CrossRef]
49. Fang, B.; Chaudhari, N.K.; Kim, M.-S.; Kim, J.H.; Yu, J.-S. Homogeneous deposition of platinum nanoparticles on carbon black for proton exchange membrane fuel cell. *J. Am. Chem. Soc.* **2009**, *131*, 15330–15338. [CrossRef]

© 2020 by the authors. Licensee MDPI, Basel, Switzerland. This article is an open access article distributed under the terms and conditions of the Creative Commons Attribution (CC BY) license (http://creativecommons.org/licenses/by/4.0/).

Article

Influence of Pd and Pt Promotion in Gold Based Bimetallic Catalysts on Selectivity Modulation in Furfural Base-Free Oxidation

Hisham K. Al Rawas, Camila P. Ferraz, Joëlle Thuriot-Roukos, Svetlana Heyte, Sébastien Paul and Robert Wojcieszak *

Univ. Lille, CNRS, Centrale Lille, Univ. Artois, UMR 8181-UCCS-Unité de Catalyse et Chimie du Solide, F-59000 Lille, France; hisham.khalifehalrawas.etu@univ-lille.fr (H.K.A.R.); camila.palombo-ferraz@centralelille.fr (C.P.F.); joelle.thuriot@univ-lille.fr (J.T.-R.); svetlana.heyte@univ-lille.fr (S.H.); sebastien.paul@centralelille.fr (S.P.)
* Correspondence: Robert.wojcieszak@univ-lille.fr; Tel.: +33-(0)3-2067-6008

Abstract: Furfural (FF) has a high potential to become a major renewable platform molecule to produce biofuels and bio-based chemicals. The catalytic performances of Au_xPt_y and Au_xPd_y bimetallic nanoparticulate systems supported on TiO_2 were studied in a base-free aerobic oxidation of furfural to furoic acid (FA) and maleic acid (MA) in water. The characterization of the catalysts was performed using standard techniques. The optimum reaction conditions were also investigated, including the reaction time, the reaction temperature, the metal ratio, and the metal loading. The present work shows a synergistic effect existing between Au, Pd, and Pt in the alloy, where the performances of the catalysts were strongly dependent on the metal ratio. The highest selectivity (100%) to FA was obtained using Au_3-Pd_1 catalysts, with 88% using 0.5% Au_3Pt_1 with about 30% of FF conversion at 80 °C. Using Au-Pd-based catalysts, the maximum yield of MA (14%) and 5% of 2(5H)-furanone (FAO) were obtained by using a 2%Au_1-Pd_1/TiO_2 catalyst at 110 °C.

Keywords: bimetallic nanoparticles; gold catalysts; catalysis; oxidation; selectivity modulation

1. Introduction

The catalytic oxidation of bio-based molecules in general and of furanics in particular is a highly attractive process. In recent years, the production of biofuels via the hydrogenation of furfurals such as tetrahydrofuran and 2-methyltetrahydrofuran has been reported [1]. However, the oxidation of furfural can also lead to the formation of many interesting molecules such as furoic acid (FA), maleic acid (MA), and succinic acid (SA) [2]. Moreover, the oxidation of furfural to furoic acid is not easy because the overoxidation of products can also be obtained. The formation of SA is possible by passing 2(3H)-furanone as an intermediate, while MA can be produced by using 2(5H)-furanone as an intermediate. However, very often the rate of these competitive reactions is likely to be limited by the decarboxylation of furoic acid [3]. Furfural oxidation reactions can proceed by using different methods such as chemical oxidation, biochemical transformation, and homogeneous or heterogeneous catalytic conversions [4–6]. Although the presence of a base allows for higher reaction rates, higher feed concentrations, better product solubility, and lower adsorption of the products on the catalyst surface [7], many disadvantages arise when controlling for the selectivity to the desired product and in avoiding forming other byproducts. Moreover, there are also several difficulties in the separation process that can occur.

In previous research, the use of inorganic bases such as KOH or NaOH were commonly used for the oxidation reaction of furfural. It has already been proven by Besson et al. [8] and Wojcieszak et al. [9] that the use of a base in the oxidation of furfural results in its degradation and in the formation of humins, which present as a black precipitate. Moreover, the basic medium facilitates the C–C bond cleavage and leads to the formation of levulinic acid

and formic acid (compounds with a low molar weight). Thus, the design of heterogeneous catalytic systems capable of maintaining a high activity and selectivity without the use of a homogeneous base is still a huge challenge in the oxidation of furfural under noncontrolled pH [10,11]. It is well-known that, contrary to the conventional catalysts based on Pt and Pd, Au-based catalysts can offer a better resistance to water and O_2, and thus present more stability and selectivity in the oxidation of organic compounds in water. However, due to the absence of a base promoter and the formation of organic products or intermediates such as carboxylic acid, Au could be more easily deactivated. The introduction of a second metal to form the bimetallic nanoparticles has been successfully demonstrated in the oxidation of furfural under base-free conditions [12]. The alloying of metals such as Au and Pd, and Au and Pt in the catalysts used combines the advantages of different components at the atomic level, and as a result enhances the activity and the selectivity to the desired products. Moreover, it may help to slow or prevent catalyst deactivation, as well as to prevent the leaching of metals from the catalyst support in bimetallic Au-Pd-based catalysts [13]. The arrangement of the metal NPs influence the catalytic performance of the bimetallic catalytic systems. Thus, the design of the catalyst by controlling the structure of the bimetallic NPs is needed to obtain high catalytic performance [12]. The advantage of an addition of a second metal was clearly shown when working under an uncontrolled pH. To this end, Pt showed a better effect when compared to Pd, which means that Pt had a boosting effect wherein only gluconic acid was produced [14]. Furthermore, this high catalytic activity was attributed to the presence of a core–shell structure, as charges could be easily transferred between the core and the shell; thus oxygen was activated, leading to the attack of the desired functional group. Therefore, the catalytic activity changed according to the composition of the system. The synergy and the interaction between the two metals in the catalyst was mainly due to the geometric and the electronic effects. This synergistic effect was proven to be one of the major factors that strongly influences the catalytic performance, which has already been reported [14,15]. Moreover, the choice to use bimetallic Au-Pd and Au-Pt was made in order to avoid using a base that was mandatory when a monometallic gold catalyst was used as a catalyst in a water/oxygen system. In this article, preliminary studies applying a smart methodology based on the experimental design (Table S1) was used to better understand the role of the noble bimetallic catalyst on the oxidation of the furfural orientation. Sol immobilization method was applied for the synthesis. This method proved its capacity to produce metal particles with random alloy structures [16,17].

2. Results

2.1. X-Ray Diffraction (XRD)

The XRD analysis was used to determine the morphology of the prepared catalysts. The results obtained for the bimetallic Au-Pt/TiO_2 and Au-Pd/TiO_2 catalysts with different metal loading and metal molar ratios are presented in Figures 1 and 2.

The diffractograms presented in Figures 1 and 2 confirmed that the TiO_2 support used (P25 from Sigma) is a mixture of two known structures (anatase and rutile). The rutile phase is about 5% when taking into account the ratio between (110) the plane of the rutile and (101) the plane of the anatase. In addition, no modification of the support was observed during the sol-immobilization with the metals. As expected, no diffraction peaks from the metals (Au, Pt, and Pd) were observed. This indicates that the metal nanoparticles are well-dispersed on the surface of the support and that their particle sizes are very small (less than 3 nm as confirmed by TEM analysis). Any additional diffraction peaks from TiO_2 were observed before and after the synthesis, which confirms that no modification of the support occurred during the synthesis. The same observations can be made for the prepared Au-Pd- and Au-Pt-based catalysts with the 0.5, 1.25, and 2 wt.% of metal loading, respectively. All XRD patterns can be found on Figures S1–S6.

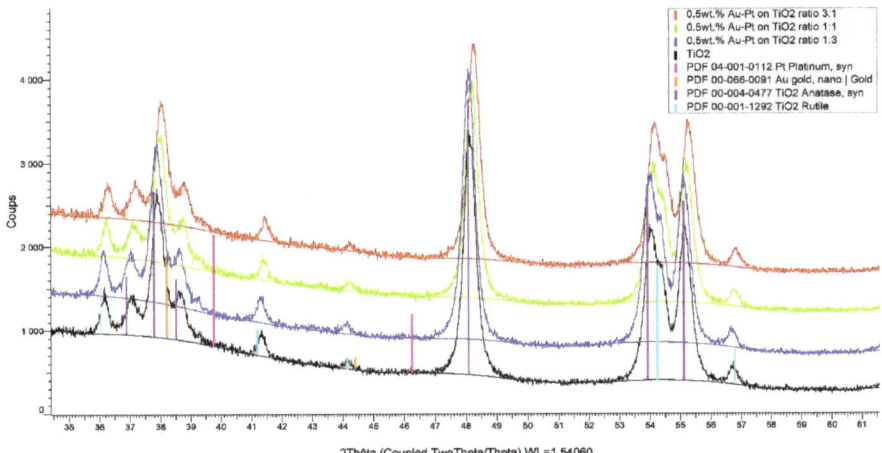

Figure 1. X-ray diffractograms of 0.5 wt.% of Au-Pt/TiO$_2$ catalysts with Au-Pt molar ratio of 3:1, 1:1, and 1:3.

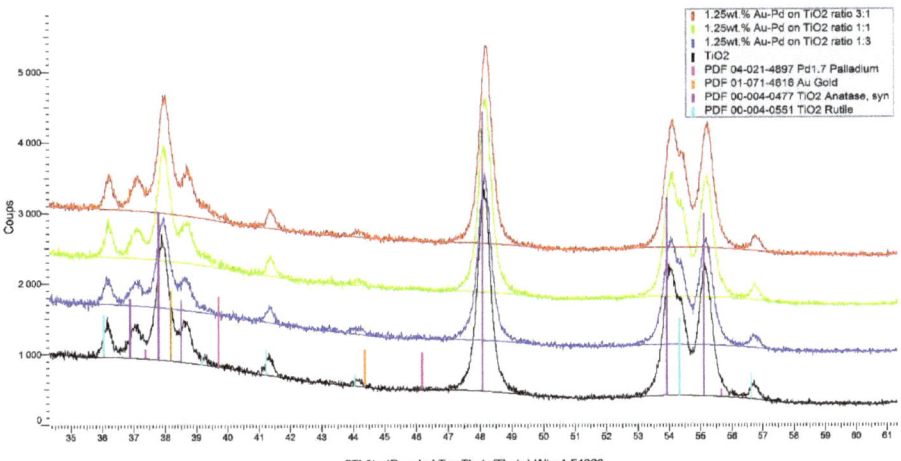

Figure 2. X-ray diffractograms of 1.25 wt.% of Au-Pd/TiO$_2$ catalysts with Au-Pd molar ratio of 3:1, 1:1, and 1:3.

2.2. ICP-OES

Initially, the ICP-OES analysis was performed to determine the real metal content in the catalysts that were prepared on the bench (theoretical value is 2 wt.%). The ICP results are shown in Table 1.

Table 1. ICP-OES analysis of the prepared catalysts.

Catalyst.	Au (wt.%)	Pd (wt.%)	Pt (wt.%)
2% Au/TiO$_2$	2.09	-	-
2% Pd/TiO$_2$	-	1.98	-
2% Pt/TiO$_2$	-	-	0.91
2% Au$_1$Pd$_1$/TiO$_2$	1.30	0.70	-
2% Au$_1$Pt$_1$/TiO$_2$	1.08	-	0.66

The ICP analysis for monometallic catalysts showed that the Au and Pd contents were close to the nominal values (2%). However, a very low Pt content was obtained for the monometallic Pt/TiO$_2$ catalyst. The same tendency was observed for the bimetallic 1:1 systems.

2.3. X-Ray Fluorescence (XRF)

The XRF analysis was performed in order to study the chemical composition of the prepared bimetallic Au-Pt/TiO$_2$ and Au-Pd/TiO$_2$ catalysts with different metal loading and molar ratios, and the results are presented in Tables S2 and S3.

As for the ICP, the XRF confirmed that the method used for the metal deposition on TiO$_2$ supports reaching the expected metal loading in the case of the monometallic catalysts. However, better results were obtained when the Au was used in excess (Pt:Au ratio of 1:3).

2.4. Transmission Electron Microscopy (TEM)

Three samples (2% Au/TiO$_2$; 2% Au$_1$Pd$_1$/TiO$_2$; 2% Au$_1$Pt$_1$/TiO$_2$) were characterized by the TEM to determine the average size of the metals on the support. Each sample was doped on the carbon side of a tiny copper grid and placed on the sample holder which was then inserted in the TEM machine. The sol-immobilization method has been shown to consistently produce catalysts with small nanoparticles with an exceptionally narrow particle size distribution. Images of the samples showed that the immobilized nanoparticles were well dispersed with an average metal particle size of 3 nm for all samples. In the case of the bimetallic catalysts, some larger aggregates were observed, as can be seen in Figure 3, However, the homogeneous distribution of the gold nanoparticles (between 1 and 4 nm) on the surface was also observed.

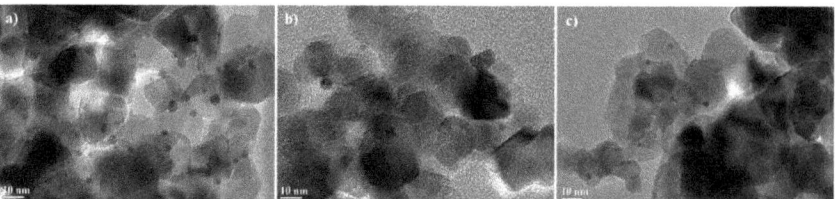

Figure 3. Transmission electron microscopy (TEM) of (**a**) 2%Au/TiO$_2$, (**b**) 2%Au$_1$Pd$_1$/TiO$_2$, (**c**) and 2%Au$_1$Pt$_1$/TiO$_2$.

2.5. Catalytic Tests: Base-Free Furfural Oxidation

The study of gold-based catalysts has grown considerably in recent years following the demonstration that their activity increases significantly when used in the form of nanoparticles. This was especially well demonstrated for the oxidation of carbohydrates [12]. However, the catalytic studies also demonstrated that the support played crucial roles in the aerobic oxidation of furfural, as it modifies the geometric or electronic state, offers better dispersions of active sites of the metal nanoclusters, and enhances the adsorption of the reactant and the reaction intermediates. Firstly, we tested three different monometallic catalysts supported on titanium(IV) oxide (TiO$_2$) in order to identify the effect of metal on the catalytic properties for the base-free FF oxidation. The results are given in Table 2.

Table 2. Catalytic results obtained for monometallic catalysts (T = 110 °C, P(O$_2$) = 15 bar, t = 2 h, 600 rpm, FF/Au = 50).

Catalyst	FF Conversion (%)	FA Selectivity (%)	Carbon Balance (%)
2% Au/TiO$_2$	44	92	96
2% Pd/TiO$_2$	5	29	98
2% Pt/TiO$_2$	11	9	89
TiO$_2$	3.7	-	97

The pH value of the solution decreases from 6 (initially) to 3 (after the reaction) due to the formation of furoic acid. As expected, only the gold-based catalysts showed high activity in this reaction and under the experimental conditions studied. Degradation and/or adsorption of the substrate was observed for pure TiO_2 oxide.

In addition, the effect of the reaction time on the oxidation of furfural from 2 to 14 h using the Au/TiO_2 catalyst was studied. Figure 4 shows that the maximum selectivity towards furoic acid (96%) was obtained after 4 h with a maximum carbon balance (98%). Although selectivity and carbon balance decreased slightly along the reaction, an increase in the yield of furoic acid of 22% was observed, reaching a maximum of 79% after 8 h. However, the selectivity and the carbon balance started decreasing after 8 h, reaching a minimum of 80% and 81%, respectively, after 14 h, which suggests the degradation of FA with time. Taking these results into account, the reaction time of 4 h was chosen for further studies.

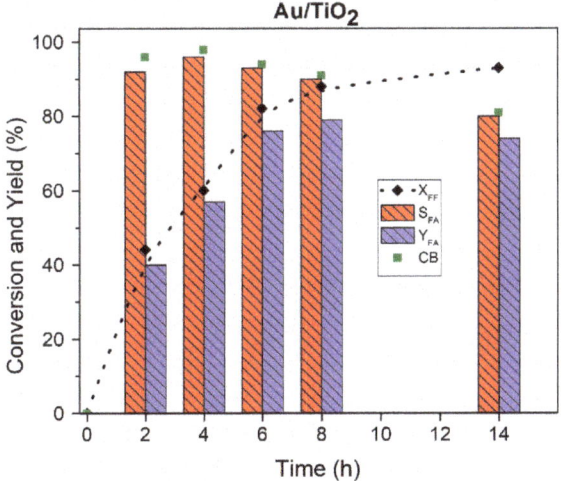

Figure 4. Effect of reaction time on the catalytic performance of 2 wt.% Au/TiO_2 catalyst ($P(O_2)$ = 15 bar, T = 110 °C, 600 rpm, FF/Au = 50).

The effect of the reaction temperature on furfural oxidation was studied using the Au/TiO_2 catalyst, and the results are presented in Table 3. At 130 °C, the catalyst displayed a higher conversion of FF and a higher yield of FA than at 110 °C after 2 and 4 h, respectively. At a higher temperature (130 °C) the degradation of furfural is more pronounced, as illustrated by the lower carbon balance values. However, a relatively high furoic acid yield of 72% could be obtained after 2 h of reaction at 130 °C.

Table 3. Effect of reaction temperature on the oxidation of furfural using 2% Au/TiO_2 catalyst. ($P(O_2)$ = 15 bar, 600 rpm, FF/Au = 50).

T (°C)	Time (h)	FF Conversion (%)	FA Selectivity (%)	Carbon Balance (%)
80	2	18	45	90
110	2	44	92	96
130	2	83	87	90
80	4	29	72	92
110	4	60	96	98
130	4	95	66	68

Taking into account the results presented above, and in order to minimize the FF degradation, further tests with bimetallic catalysts were performed at 80 °C and 4 h in a Screening Pressure Reactor. We studied the catalytic performance of the Au_xPt_y and Au_xPd_y bimetallic systems supported on TiO_2 with different molar ratios (1:3, 1:1, and 3:1) and metal loading (0.5, 1.25, and 2%) in the oxidation of furfural to FA and MA, and the results are presented in Figure 5. As expected, the oxidation of furfural using the bimetallic Au-Pd and Au-Pt catalysts at 80 °C gave better catalytic results than the monometallic gold, palladium, and platinum. Indeed, only 60% furfural conversion was observed for the Au/TiO_2 catalyst after 4 h of reaction at 110 °C, and 29% at 80 °C.

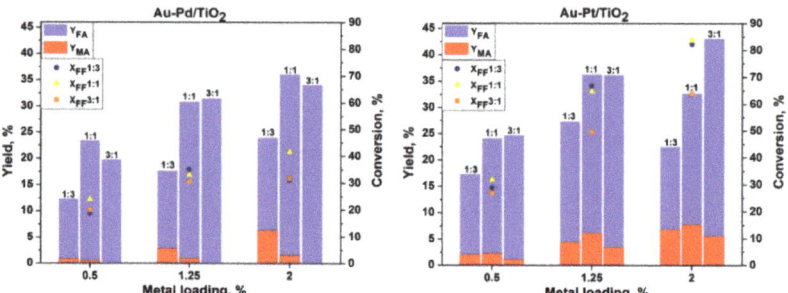

Figure 5. Catalytic results obtained for $Au-Pd/TiO_2$ and $Au-Pt/TiO_2$ catalysts in base-free oxidation of furfural (T = 80 °C, t = 4 h, 600 rpm, P(air) = 15 bar).

3. Discussion

To determine the specific surface area of the analyzed catalysts, nitrogen adsorption and desorption analyses on monometallic catalysts (2% Au/TiO_2, 2% Au_1Pd_1/TiO_2, and 2% Au_1Pt_1/TiO_2) were performed. The pore volume, pore size, and the values of the surface areas of the catalysts were calculated. The results are shown in Table 4.

Table 4. Textural properties of the catalysts.

Catalyst	Surface Area (m^2/g)	Pore Volume (cm^3/g)	Pore Size (nm)
2% Au/TiO_2	19.1	0.09	18.8
2% Au_1Pd_1/TiO_2	42.9	0.13	12.3
2% Au_1Pt_1/TiO_2	37.4	0.12	13.4
TiO_2 (P25)	55.1	0.25	16.2

The Brunauer–Emmett–Teller (BET) analysis showed that different porosities were obtained for the tested samples. The BET surface area of the Au_1Pd_1/TiO_2 and Au_1Pt_1/TiO_2 samples were twice as large as that of the Au/TiO_2 sample. The very low surface area of this catalyst is probably due to the pore blockage, as could be deduced from the significant decrease in the pore volume when compared to the TiO_2 support. The differences between the catalysts were also observed in the ICP and XRF analyses (Figure 6).

Observing the ICP values, which are quantitative data, some differences were observed between the theoretical and measured values that could be due to several factors. However, in considering the intrinsic error of the robotic system in the distribution of reagents and support (error estimated at 5%), values very close to the expected values for the catalysts were obtained. The exceptions were the catalysts rich in Pd and Pt at 0.5 and 1.0 wt.%, respectively, which always had a lower metal loading than was expected. The same observation was made already for the monometallic Pt/TiO_2 catalyst prepared on the bench (Table 1). The XRF analysis is a semiquantitative and nondestructive technique that can be very useful for the quantification of metals in catalysts, and especially in this case where the catalysts were synthesized in an automated way. Thus, the values were reasonably similar to the values obtained by the ICP for the Au-Pt catalysts but not for the

Au-Pd catalysts with loadings of 0.5 and 1.25 wt.%, respectively. The overestimation of the Pd concentration in the Au-Pd catalysts using the XRF in comparison with that of the ICP is due to the Rh tube that is used as the X-ray source, which has an energy close to that of Pd and which induces an interference of lines on the spectra while quantifying this element. No differences observed in the TEM analysis confirmed that the sol-immobilization method permits us to obtain small metal particle sizes and a good particle distribution. All these parameters significantly affected the catalytic properties of the catalysts.

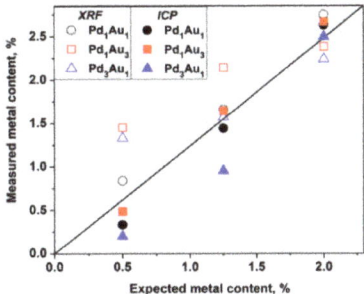
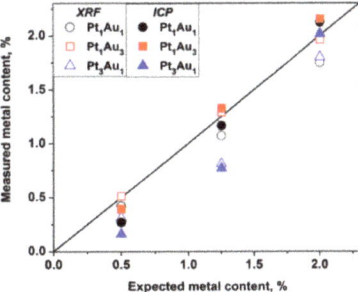

Figure 6. Real metal loading measured by XRF and ICP compared with the theoretical values (black line).

All the tested catalysts were active in the base-free oxidation of furfural at 80 °C, as shown in Figure 5. It could be seen that generally the increase in the catalyst metal loading increases the conversion of furfural. However, this increase depends on the composition of the catalyst. As expected, the higher catalytic activity was observed for catalysts with a lower Au content. The activity of the bimetallic catalysts with a high Au content is comparable to the results observed for the monometallic Au/TiO$_2$ catalyst (Figure 4). Much higher activity was observed for the Au-Pt catalysts when compared to the Au-Pd catalysts. This boosting effect was already observed for the Au-Pt catalysts in the glucose oxidation reaction, where a better effect of Pt was observed in comparison to Pd [14].

An increase in the metal loading from 0.5% to 2% using the Au$_3$Pt$_1$/TiO$_2$ catalyst leads to the increase in furfural conversion by 37%, reaching 64% with a 38% and 6% yield of FA and MA, respectively. The same behavior could be seen in terms of the FF conversion using Au$_3$Pd$_1$/TiO$_2$, which increases to reach only 32% but with 100% carbon balance and selectivity to FA, with neither MA nor 2-furanone (FAO) being formed. This is a remarkable result, as for the first time 100% selectivity to furoic acid with 100% carbon balance was observed in the base-free oxidation of furfural. Indeed, the increase in the catalytic activity in the oxidation reactions was well established using Au-Pd bimetallic catalysts [16–18]. The Au-Pd nanoparticles were found to be two times more active than Pd alone for the oxidation reaction of alcohols, where it was proposed that Au is an electronic promoter of Pd [19,20].

The effect of the catalyst's composition was also reported, whereby increasing the Pd content to an optimum ratio of Au:Pd in ca. 1:3 leads to the increase in the catalytic activity, while there is a progressive decrease in the activity with a further increase in the Pd content [21]. It may be concluded that there should be an optimum ratio between both metals that will maintain a high activity as well as a high selectivity. It is highly probable that with the changing concentrations of metals relative to the support material, the electronic and the geometric structures of the individual clusters change significantly, thereby affecting the bonding between the reactant and the catalyst, and as a result altering the catalytic performance. By considering this, it can be concluded that studying the composition and the structure of the Au-Pd and Au-Pt bimetallic catalysts is a key factor for designing more active catalysts and to find the relationships between their structures and the activities.

The catalytic activity changed significantly depending on the composition of the catalytic system. Table 5 shows that the selectivities to FA and MA depend on the ratio of metals in each catalyst. At the same FF conversion (about 30%), gold-rich catalysts promote the oxidation of FF to FA, reaching 100% of selectivity using 1.25% of the Au_3Pd_1-based catalyst and a maximum of 88% using 0.5% of Au_3Pt_1 catalyst (Table 5). These values decrease with the increase in the quantity of the second added metal, but at the same time the selectivity to MA increases to 7% and 8% using 1.25% of the Au_3Pd_1 and 0.5% of Au_3Pt_1 catalysts, respectively. This suggests that the selectivity of FA is strongly dependent on the ratio of Au, and that MA is dependent on the ratio of Pt or Pd. Therefore, Au has a beneficial effect on maintaining a high selectivity to furoic acid and limits the formation of byproducts. Moreover, the addition of platinum and palladium presents a boosting effect toward the ring-opening reactions and the formation of maleic acid. In addition, the formation of products such as FAO intermediate and two other unknown products were observed in small quantities, which could explain the decrease in the carbon balance. These observations clearly indicate a tandem pathway for the conversion of FF to MA via FA and FAO. Moreover, as has been proposed, maleic acid could be formed by the decarboxylation of furfural to furan, which then gives a 2-furanone intermediate (Figure 7) [22].

Table 5. Selectivity to FA and MA using 0.5% Au-Pt and 1.25% Au-Pd catalysts (T = 80 °C, t = 4 h, P(air) = 15 bar, 600 rpm).

Metal Loading (wt.%)	Catalyst	X_{FF} (%)	S_{FA} (%)	S_{MA} (%)	CB (%)
0.5	Au_1-Pt_3/TiO_2	29	53	7	89
	Au_1-Pt_1/TiO_2	32	68	7	92
	Au_3-Pt_1/TiO_2	27	88	4	98
1.25	Au_1-Pd_3/TiO_2	35	42	8	84
	Au_1-Pd_1/TiO_2	33	91	3	98
	Au_3-Pd_1/TiO_2	31	100	0	100

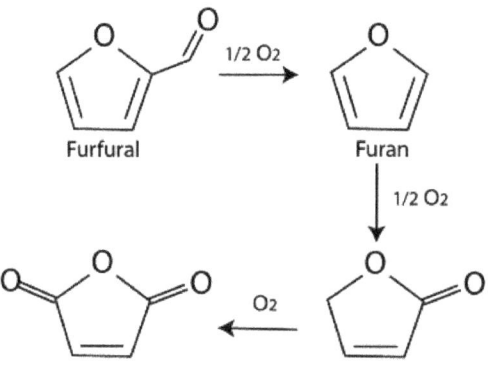

Figure 7. Reaction pathway for the formation of maleic anhydride from FF through FAO [22].

For comparing the effect of each metal, the two catalysts (2% Au_3Pt_1 and 2% Au_3Pd_1) were chosen. The results showed that the Pt-based catalyst was able to convert 64% of FF after 4 h with a 6% yield of MA, while only 32% of the conversion was obtained using the Au-Pd catalyst with no formation of MA. However, the selectivity to FA reached 100% with the Au-Pd catalyst instead of 60% for the Au-Pt sample. Both Pd and Pt in the bimetallic catalyst produced furoic acid as a major product. These results further demonstrate that 2% Au_3Pd_1/TiO_2 is a promising catalyst for the base-free aerobic oxidation of FF to FA in water. It could also be seen that with the Au-Pt catalysts the maximum yield of MA (8%) was achieved using the 2% Au_1-Pt_1/TiO_2 catalyst while only 1.6% was achieved using the 2% Au_1-Pd_1/TiO_2 catalyst. It seems that the reaction temperature strongly affects the catalytic properties of the Au-Pt and Au-Pd nanoparticles, whereby each catalyst behaved differently during the oxidation of FF. Therefore, when performing the same tests at 110 °C, a full conversion (100%) of furfural was observed. However, no products were detected using the Au-Pt based catalysts, with a very low carbon balance (in some cases reaching 0%). This indicates the overoxidation of the furfural and the formation of low molecular weight molecules or condensation products (via the ring opening reaction and the formation of maleic acid). Regarding the Au-Pd-based catalysts, a maximum yield of MA (14%) and (5%) of FAO was obtained using the 2% Au_1-Pd_1/TiO_2 catalyst at 110 °C, while the carbon balance was very low with a complete conversion of FF, also indicating the overoxidation and the degradation of FF, but at a lower extent when compared to the Au-Pt catalyst. It was previously reported that the overoxidation and the poisoning from byproducts could be responsible for the deactivation of Pd- or Pt-based catalysts when they are used in the liquid phase with oxygen as an oxidant [23]. By comparing the monometallic Au-, Pt-, and Pd-based catalysts to the bimetallic catalysts, the monometallic catalysts were by far less active than the bimetallic catalysts in terms of furfural conversion, and the formation of FA, MA, and FAO intermediates. From a functional point of view, it is important to state that better catalytic performances could be obtained with the highly active bimetallic catalysts than with the gold-based monometallic catalysts when working under certain conditions (such as 80 °C, air as oxidant) and with specific chemical compositions (such as Au-Pd and Au-Pt). A good correlation appeared between the performances of these catalysts, and in the chemical composition and the loading of both metals. The nature of the second metal (Pd or Pt) which mainly governs the catalytic activity seems to also have a strong impact on the oxidation in base-free conditions. Relatively high yields of MA observed for the Au-Pd catalysts indicate a radical mechanism of the reaction which favors the ring opening pathway.

The recyclability of the Au/TiO_2 catalyst was also studied. The recycling tests were performed using the same methodology as described in the Experimental Part. The catalyst was dried under air atmosphere at 80 °C overnight after each run. The results are given in Figure 8.

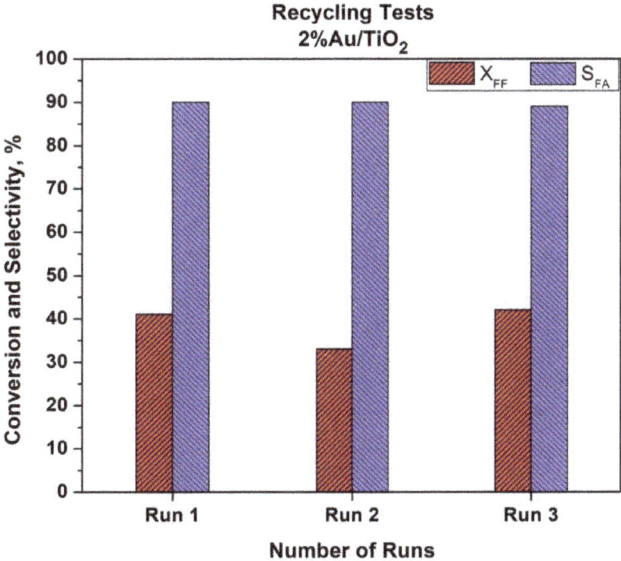

Figure 8. Recyclability tests using Au/TiO_2 catalyst (T = 110 °C, t = 2 h, P(O_2) = 15 bar, 600 rpm).

As can be seen from Figure 8, a slight decrease in the catalytic activity was observed. After the first run, a 20% drop in furfural conversion was observed. This drop was not confirmed after the third run when initial activity was achieved. This could be due to the problem with the catalyst recovery after the first test. The ICP analysis of the post reactant solution did not show a leaching of gold. We can also expect that a temperature of 110 °C is quite low to observe Au particle size growth. It is worth mentioning that selectivity to furoic acid remained both stable and high (90%) during the three recycling tests. More in-depth studies are in progress to study the recyclability and stability of the bimetallic catalysts in the batch and in the flow conditions.

4. Materials and Methods

Gold(III) chloride solution ($HAuCl_4$, 99.9%, Sigma–Aldrich, Saint Louis, MO, USA; 30%), potassium tetrachloropalladate(II) (K_2PdCl_4, 99.99%, Sigma–Aldrich, Saint Louis, MO, USA), chloroplatinic acid (H_2PtCl_6, 8 wt.% solution, Sigma–Aldrich, Saint Louis, MO, USA), sodium borohydride ($NaBH_4$, 98%, Sigma–Aldrich, Saint Louis, MO, USA), poly(vinyl alcohol) (PVA, MW 9000–10,000, 80% hydrolyzed, Sigma–Aldrich, Saint Louis, MO, USA), 2-Furaldehyde (FF, $C_5H_4O_2$, 99%, Sigma–Aldrich, Saint Louis, MO, USA), 2-furoic acid (FA, 98%, Sigma-Aldrich), and maleic acid (MA, >99%, Sigma-Aldrich). Supports: titanium(IV) oxide (TiO_2 P25, 99.5%, Sigma–Aldrich, Saint Louis, MO, USA). Initially the monometallic gold, palladium, and platinum catalysts supported on titanium(IV) oxide were synthesized on the bench using the sol-immobilization method for comparison with their equimolar bimetallic catalysts (Au_1Pd_1/TiO_2 and Au_1Pt_1/TiO_2), and by using 2 wt.% total metal loading.

The 2 wt.% Au, Pd, and Pt supported on TiO_2 catalysts were prepared by a sol-immobilization method using $NaBH_4$ as a reducing agent and polyvinyl alcohol (PVA) as a stabilizing agent to prevent the NPs from aggregation, as well as to control the size of the particles being formed (Figure 9). Briefly, 1.2 mL of a 2 wt.% aqueous solution of poly (vinyl alcohol) (PVA/Au (w/w) = 1.2) solution was dropped into a 200 mL metal precursor–water solution under stirring. In order to obtain the metallic nanoparticles, the freshly prepared solution of the reducing agent $NaBH_4$ (0.1 M, $NaBH_4$/metal (mol/mol) = 5) was added drop by drop to the PVA–metal solution and stirred for 30 min. TiO_2 support was then added and the pH was adjusted to 2 by the addition of H_2SO_4. Two hours later, the catalyst was filtered using a filter paper and a Büchner funnel, washed with hot water at 70 °C (3 times, 30 mL each) and ethanol (3 times, 30 mL each), and dried overnight at 80 °C.

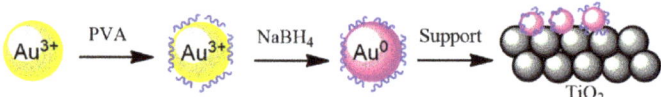

Figure 9. Method of sol-immobilization with PVA for Au/TiO_2.

Bimetallic Au-Pd and Au-Pt systems have been synthesized using the REALCAT platform by applying the Design of Experiments (DoE) methodology (Table S1) to study two different parameters (metal loading and metals ratios). The sol-immobilization method was applied (Figure 10) to deposit Au, Pd, and Pt NPs onto the surface of TiO_2, in which the total metal loading and metal ratio (Au:Pd and Au:Pt) were varied.

Three different metal loadings were studied: 0.5 wt.%, 1.25 wt.%, and 2 wt.%. The molar ratios between gold and the second metal were varied at 3:1, 1:1, and 1:3, respectively. A Response Surface Design (Central Composite) was created. It was designed for three factors to model curvature data and identify the factor settings that optimize the response. The Central Composite Design (CCD) was used with two factors at three levels of evaluation: low (−1), high (+1), and central (0) levels. The central level was also repeated three times to evaluate the error of the DoE model. The aforementioned levels are used in the CCD to easily represent the minimum, the central, and the maximum values of a factor influencing the catalyst activity. In the CCD, all the possible combinations of high and low levels are studied for the two factors. Hence, the total number of conditions for the CCD was 11 by each metal as calculated: $2^n + 3$ repetitions at 0 level (where n = 2, the number of factors in this study, and 3 is the number of levels). As the metal are non-numerical factors, no zero level can be applied for this parameter. Therefore, two CCDs were conducted: one for Pt (11 experiences) and the second one for Pd (11 experiences), equaling 22 experiences for both metals. When varying all the parameters, a matrix was built using the "Minitab" software. The responses studied were the FF conversion, the MA and FA yield, and the MA and FA selectivity.

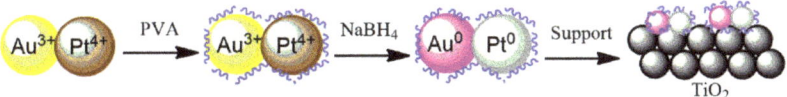

Figure 10. Method of sol-immobilization with PVA for Au-Pt/TiO$_2$ systems.

The prepared catalysts were characterized by using different techniques. X-Ray Diffraction (XRD) analysis was performed using a Bruker D8 Advance Powder X-ray diffractometer (Billerica, MA, USA) to determine the morphology of the prepared catalyst. X-Ray Fluorescence (XRF) (M4 Tornado) analysis was performed using an Energy Dispersive X-Ray Fluorescence (EDXRF) spectrometer provided from Bruker (Billerica, MA, USA) to study the chemical composition of the prepared catalysts. The elemental analysis for determining the real metal content in each catalyst was performed by Inductively Coupled Plasma-Optical Emission Spectroscopy 720-ES ICP-OES (Agilent, Santa Clara, CA, USA) with axial viewing and simultaneous CCD detection. The metal particle size in the supported catalysts was determined using Transmission Electron Microscopy (TEM) with a TEM/STEM FEI TECNAI F20 microscope (Hillsboro, Ore. USA) combined with an Energy Dispersive X-ray Spectrometer (EDS) (Hillsboro, OR, USA) at 200 kV. The surface area, pore volume, and distribution of the pore size were determined by nitrogen adsorption/desorption at 77.35 K using a TriStar II Plus and a 3Flex apparatus from Micromeritics (Norcross, GA, USA). The Brunauer–Emmett–Teller (BET) method was used for determining the specific surface area of the prepared materials.

The catalytic tests were performed in a Top Industry autoclave (batch reactor, Vaux le Penil, France). The reactant ((FF) = 24.7 mM) solutions were prepared by diluting a 43 µL of furfural in 21 mL of H$_2$O and stirring the solution to dissolve the furfural before adding it into the vessel. Initially, 1 mL of furfural aqueous solution was taken off for HPLC analysis (t$_0$) and the desired amount of catalyst was added in the autoclave. The reactor was purged three times with pure oxygen before reaching 15 bar. Then, the heating system was started, and the reaction was initiated after reaching 110 °C. The reaction was carried out at 15 bar, 600 rpm during 2, 4, 6, 8, or 14 h, respectively. At the end of the reaction, the catalyst was filtered off and 1 mL of the final solution was diluted for HPLC analysis in a Phenomenex column (ROA, organic acid H$^+$; 300 × 7.8 mm). Sulfuric acid (5 mmol/L^{-1}) was used as a mobile phase with a flow rate of 0.60 mL/min, and the products were detected on a UV-Vis detector at 253 nm. Catalytic furfural oxidation using the bimetallic catalysts were carried out on the REALCAT platform in a Screening Pressure Reactors system (SPR) from UnchainedLabs (UK). The required amounts of catalysts were placed in each reactor. Then, 2 mL of an aqueous solution of furfural ((FF) = 24.7 mM) was injected. The catalytic tests were performed under air pressure (15 bar) with stirring at 600 rpm for 4 h at 110 °C. After the reaction, the catalyst was filtered off and 1 mL of the final solution was diluted for HPLC analysis. The liquid products were analyzed by a High-Performance Liquid Chromatography (HPLC, Shimadzu, Japan) equipped with a UV detector SPD-20A operated at wavelengths of 210 nm and 253 nm, respectively, and a Bio-Rad Aminex HPX-87H column (7.8 × 300 mm) operated at 60 °C. Diluted H$_2$SO$_4$ (5 mM, 0.7 mL/min) was used as a mobile phase. Commercial standards (furfural, furoic acid, and maleic acid) were used for the calibration of the HPLC set up.

5. Conclusions

In this article, the catalytic performances of Au_xPt_y and Au_xPd_y bimetallic catalyst systems supported on TiO_2 were studied. The catalysts were prepared by applying a high throughput experimentation. Using the Design of Experiments (DoE), the nature of the second metal, the metal loading, and the molar ratio between Au and the second metal were studied. By comparing the monometallic Au-, Pt-, and Pd-based catalysts to the bimetallic counterparts, the synergetic effect of alloying was evidenced. The monometallic catalysts were by far less active than the bimetallic catalysts in terms of FF conversion, and in the formation of FA, MA, and FAO intermediates. The results obtained confirmed that the combination of metals have a positive effect on catalyst activity and selectivity. The sol-immobilization method using PVA was effective, as it leads to the formation of very small metal nanoparticles with an average particle size of 3 nm for all samples, as observed by TEM. Moreover, the ICP-OES analysis showed very close real metal loading values in comparison to those that were expected. A simple change in the metal-to-metal ratio and the metal loading significantly improved the catalytic properties, which offers the advantage of fine tuning the catalytic system. Increasing the metal loading leads to the increase in the FF conversion for all the catalysts studied. Both Pd and Pt alloyed to Au catalysts produced FA as the major product, and FAO and MA as minor products. More interestingly, Pd-Au systems were able to achieve much higher selectivity towards FA, where the highest selectivity (100%) to FA was obtained by using the Au_3-Pd_1 catalyst, and 88% using the 0.5% Au_3Pt_1 catalyst with about 30% of the FF conversion at 80 °C. Relatively high yields of MA (14%) were observed for Au-Pd catalysts, which indicate a radical mechanism of the reaction that favors the ring opening pathway. However, higher temperatures favor the degradation and the overoxidation of FF and leads to the formation of low molecular weight molecules or condensation products (via the ring opening reaction and the formation of maleic acid).

Supplementary Materials: The following are available online at https://www.mdpi.com/article/10.3390/catal11101226/s1, Figures S1–S6: XRD patterns, Table S1: DoE methodology, Tables S2 and S3: XRF and ICP results.

Author Contributions: Conceptualization, H.K.A.R. and C.P.F.; methodology, C.P.F.; formal analysis, J.T.-R.; investigation, H.K.A.R. and S.H.; data curation, H.K.A.R., S.H., J.T.-R., C.P.F., and R.W.; writing—original draft preparation, H.K.A.R. and R.W.; writing—review and editing, C.P.F. and J.T.-R.; S.H. and S.P.; visualization, R.W.; supervision, C.P.F. and R.W.; project administration, R.W.; funding acquisition, S.P. All authors have read and agreed to the published version of the manuscript.

Funding: This research received no external funding.

Institutional Review Board Statement: Not applicable.

Informed Consent Statement: Not applicable.

Data Availability Statement: Raw data are available on demand.

Acknowledgments: The REALCAT platform benefits from a state subsidy administered by the French National Research Agency (ANR) within the frame of the 'Future Investments' program (PIA) with the contractual reference ANR-11-EQPX-0037. The European Union, through the ERDF funding administered by the Hauts-de-France Region, has co-financed the platform. Centrale Lille, the CNRS, and Lille University, as well as the Centrale Initiatives Foundation, are thanked for their financial contributions to the acquisition and implementation of the equipment of the REALCAT platform. The Chevreul Institute (FR 2638) and the Ministère de l'Enseignement Supérieur, de la Recherche, et de l'Innovation are also acknowledged for supporting and partially funding this work.

Conflicts of Interest: The authors declare no conflict of interest.

References

1. Li, X.; Jia, P.; Wang, T. Furfural: A Promising Platform Compound for Sustainable Production of C4 and C5 Chemicals. *ACS Catal.* 2016, *6*, 7621–7640. [CrossRef]
2. Wojcieszak, R.; Santarelli, F.; Paul, S.; Dumeignil, F.; Cavani, F.; Gonçalves, R.V. Recent developments in maleic acid synthesis from bio-based chemicals. *Sustain. Chem. Process.* 2015, *3*, 9. [CrossRef]
3. Gong, L.; Agrawal, N.; Roman, A.; Holewinski, A.; Janik, M.J. Density functional theory study of furfural electrochemical oxidation on the pt (1 1 1) surface. *J. Catal.* 2019, *373*, 322–335. [CrossRef]
4. Zhang, Z.; Deng, K. Recent advances in the catalytic synthesis of 2,5-furandicarboxylic acid and its derivatives. *ACS Catal.* 2015, *5*, 6529–6544. [CrossRef]
5. Zhu, Y.; Shen, M.; Xia, Y.; Lu, M. Au/MnO2 nanostructured catalysts and their catalytic performance for the oxidation of 5-(hydroxymethyl)furfural. *Catal. Commun.* 2015, *C*, 37–43. [CrossRef]
6. Yang, B.; Dai, Z.; Ding, S.-Y.; Wyman, C.E. Enzymatic hydrolysis of cellulosic biomass. *Biofuels* 2011, *2*, 421–449. [CrossRef]
7. Comotti, M.; Pina, C.D.; Matarrese, R.; Rossi, M.; Siani, A. Oxidation of Alcohols and sugars using Au/C catalysts: Part 2. sugars. *Appl. Catal. A Gen.* 2005, *291*, 204–209. [CrossRef]
8. Besson, M.; Pinel, C.; Gallezot, P. Conversion of biomass into chemicals over metal catalysts. *Chem. Rev.* 2014, *114*, 1827–1870. [CrossRef]
9. Wojcieszak, R.; Cuccovia, I.M.; Silva, M.A.; Rossi, L.M. Selective oxidation of glucose to glucuronic acid by cesium-promoted gold nanoparticle catalyst. *J. Mol. Catal. A Chem.* 2016, *422*, 35–42. [CrossRef]
10. Biella, S.; Castiglioni, G.L.; Fumagalli, C.; Prati, L.; Rossi, M. Application of gold catalysts to selective liquid phase oxidation. *Catal. Today* 2002, *72*, 43–49. [CrossRef]
11. Delidovich, I.; Taran, O.; Matvienko, G.; Simonov, A.; Simakova, I.; Bobrovskaya, A.; Parmon, V. Selective oxidation of glucose over carbon-supported Pd and Pt catalysts. *Catal. Lett.* 2010, *140*, 14–21. [CrossRef]
12. Ferraz, C.P.; Costa, N.J.S.; Teixeira-Neto, E.; Teixeira-Neto, Â.A.; Liria, C.W.; Thuriot-Roukos, J.; Machini, M.T.; Froidevaux, R.; Dumeignil, F.; Rossi, L.M.; et al. 5-Hydroxymethylfurfural and furfural base-free oxidation over AuPd embedded bimetallic nanoparticles. *Catalysts* 2020, *10*, 75. [CrossRef]
13. Davis, S.E.; Ide, M.S.; Davis, R.J. Selective oxidation of alcohols and aldehydes over supported metal nanoparticles. *Green Chem.* 2012, *15*, 17–45. [CrossRef]
14. Comotti, M.; Pina, C.D.; Rossi, M. Mono- and bimetallic catalysts for glucose oxidation. *J. Mol. Catal. A Chem.* 2006, *251*, 89–92. [CrossRef]
15. Heidkamp, K.; Aytemir, M.; Vorlop, K.-D.; Prüße, U. Ceria Supported gold–platinum catalysts for the selective oxidation of alkyl ethoxylates. *Catal. Sci. Technol.* 2013, *3*, 2984–2992. [CrossRef]
16. Sankar, M.; Dimitratos, N.; Miedziak, P.J.; Wells, P.P.; Kiely, C.J.; Hutchings, G.J. Designing bimetallic catalysts for a green and sustainable future. *Chem. Soc. Rev.* 2012, *41*, 8099–8139. [CrossRef]
17. Tiruvalam, R.C.; Pritchard, J.C.; Dimitratos, N.; Lopez-Sanchez, J.A.; Edwards, J.K.; Carley, A.F.; Hutchings, G.J.; Kiely, C.J. Aberration corrected analytical electron microscopy studies of sol-immobilized Au + Pd, Au{Pd} and Pd{Au} catalysts used for benzyl alcohol oxidation and hydrogen peroxide production. *Faraday Discuss.* 2011, *152*, 63–86. [CrossRef] [PubMed]
18. Silva, T.A.G.; Teixeira-Neto, E.; López, N.; Rossi, L.M. Volcano-like behavior of Au-Pd core-shell nanoparticles in the selective oxidation of alcohols. *Sci. Rep.* 2014, *4*, 5766. [CrossRef] [PubMed]
19. Enache, D.I.; Edwards, J.K.; Landon, P.; Solsona-Espriu, B.; Carley, A.F.; Herzing, A.A.; Watanabe, M.; Kiely, C.J.; Knight, D.W.; Hutchings, G.J. Solvent-free oxidation of primary alcohols to aldehydes using Au-Pd/TiO2 catalysts. *Science* 2006, *311*, 362–365. [CrossRef] [PubMed]
20. Dimitratos, N.; Villa, A.; Wang, D. Pd and Pt catalysts modified by alloying with Au in the selective oxidation of alcohols. *J. Catal.* 2006, *244*, 113–121. [CrossRef]
21. Pritchard, J.; Kesavan, L.; Piccinini, M.; He, Q.; Tiruvalam, R.; Dimitratos, N.; Lopez-Sanchez, J.A.; Carley, A.F.; Edwards, J.K.; Kiely, C.J.; et al. Direct synthesis of hydrogen peroxide and benzyl alcohol oxidation using Au−Pd catalysts prepared by sol immobilization. *Langmuir* 2010, *26*, 16568–16577. [CrossRef] [PubMed]
22. Alonso-Fagúndez, N.; Granados, M.L.; Mariscal, R.; Ojeda, M. Selective conversion of furfural to maleic anhydride and furan with VOx/Al2O3 catalysts. *ChemSusChem* 2012, *5*, 1984–1990. [CrossRef] [PubMed]
23. Mallat, T.; Baiker, A. Oxidation of alcohols with molecular oxygen on platinum metal catalysts in aqueous solutions. *Catal. Today* 1994, *19*, 247–283. [CrossRef]

MDPI
St. Alban-Anlage 66
4052 Basel
Switzerland
Tel. +41 61 683 77 34
Fax +41 61 302 89 18
www.mdpi.com

Catalysts Editorial Office
E-mail: catalysts@mdpi.com
www.mdpi.com/journal/catalysts

www.ingramcontent.com/pod-product-compliance
Lightning Source LLC
LaVergne TN
LVHW070153120526
838202LV00013BA/1048